Probability Theory and Stochastic Modelling

Volume 100

The Probability Theory and Stochastic Modelling series is a merger and continuation of Springer's two well established series Stochastic Modelling and Applied Probability and Probability and Its Applications. It publishes research monographs that make a significant contribution to probability theory or an applications domain in which advanced probability methods are fundamental. Books in this series are expected to follow rigorous mathematical standards, while also displaying the expository quality necessary to make them useful and accessible to advanced students as well as researchers. The series covers all aspects of modern probability theory including:

- Gaussian processes
- Markov processes
- Random Fields, point processes and random sets
- Random matrices
- Statistical mechanics and random media
- Stochastic analysis

as well as applications that include (but are not restricted to):

- Branching processes and other models of population growth
- Communications and processing networks
- Computational methods in probability and stochastic processes, including simulation
- Genetics and other stochastic models in biology and the life sciences
- Information theory, signal processing, and image synthesis
- Mathematical economics and finance
- Statistical methods (e.g. empirical processes, MCMC)
- Statistics for stochastic processes
- Stochastic control
- Stochastic models in operations research and stochastic optimization
- Stochastic models in the physical sciences

More information about this series at http://www.palgrave.com/gp/series/13205

Christiane Cocozza-Thivent

Markov Renewal and Piecewise Deterministic Processes

Springer

Christiane Cocozza-Thivent
Lyon, France

ISSN 2199-3130 ISSN 2199-3149 (electronic)
Probability Theory and Stochastic Modelling
ISBN 978-3-030-70446-9 ISBN 978-3-030-70447-6 (eBook)
https://doi.org/10.1007/978-3-030-70447-6

Mathematics Subject Classification: 60K15, 60J25, 60J75, 60K20, 60K30, 90B25, 90B05

This Springer imprint is published by the registered company Springer Nature Switzerland AG
The registered company address is: Gewerbestrasse 11, 6330 Cham, Switzerland

In memory of Claude

Preface

This book is about Piecewise Deterministic Markov Processes (PDMP) and their close connection to Markov renewal processes. A PDMP makes it possible to model a deterministic mechanism whose behavior is modified by jumps that occur at random times. For example, in queues jumps are due to the arrivals and departures of customers. In storage models, jumps occur at the time of deliveries. In insurance risk models, jumps in cash flow are created by the repayment of claims. In the above examples the jumps are caused by events independent of the deterministic evolution between them, but this is not always the case. The probability distributions of the jump times and their locations may depend on the deterministic evolution. In reliabilities studies, jumps come from breakdowns and repairs of the equipment or of its components and are due to aging and wear. In earthquake modeling, stress is created by tectonic forces and jumps are the result of earthquakes generated by these forces.

Our goal and focus are different from those of Davis [51], they complement each other. In [51], which has become a reference book, martingales have an essential role. M. H. A. Davis entirely describes the extended generator of the PDMPs he is studying, which we call classic PDMPs, and applications are about optimal control. The present book is the result of a thought on models, the desire to understand the role of their assumptions and the will to practice on the simplest ones. Markov renewal processes and more generally marked point processes play a central role. They allow to define more general PDMPs, to give equations of the Chapmann-Kolmogorov type that lead to numerical schemes, to provide indicators on the performance of systems including cost evaluation, to get stability results in a very general framework based on Jacod's works on semi-Markov chains.

The deterministic behavior between jumps of a classic PDMP is governed by an ordinary differential equation (ODE). The introduction of a hybrid space, which is the product of a countable state by open subsets of euclidean spaces, makes it possible to have different ODEs according to the regions of the state space in which the process is located. But it makes the notations more cumbersome and may artificially complicate the study. Following Jacobsen [66], we only assume that the deterministic evolution is given by a deterministic Markov flow, that is a continuous function which provides the state at a given time according to the state at the initial time, the state at a time later than t depending only on the state at time t. Using ODE terminology, we could say that characteristics may merge, making it impossible to reverse time.

The probability distribution of the elapsed time between two successive jumps depends on the state at the first of the two jumps. We refer to them as interarrival distributions. In a classic PDMP, interarrival distributions are of a special type. They mainly have densities. In this book we define PDMPs with general interarrival distributions. This generalization enables to take a step back and better understand the origin of some properties. Of course, it also allows to handle more models.

Our approach is linked to the origins of this work. The initial idea was to become familiar with PDMP by studying models that are easy to grasp thanks to a very simple deterministic evolution. The Completed Semi-Markov Process (CSMP) seemed to be a good testing ground. This is a semi-Markov process to which is added a complementary variable to get a Markov process. The semi-Markov process is constant between jumps. The complementary variable, equal to the time elapsed since the last jump, evolves linearly between jumps.

It turns out that the simplicity of the deterministic evolution of a CSMP is not its only advantage, far from it. Indeed, any classic PDMP is a deterministic function of a CSMP, this function is the flow of the ODE vector field. This suggests to define a PDMP as a function of a CSMP, this function being a deterministic Markov flow. A coherence condition that connects the flow and the CSMP renewal kernel makes the resulting process a Markov process. Thus results on CSMPs give effortlessly results on PDMPs. The coherence condition, which is intuitively natural, holds for classic PDMPs and allows the study of PDMP in a more general framework. By specifying gradually the interarrival distributions, we highlight the contribution of the assumptions.

When we assume that the interarrival distributions are of the same type as that of a classic PDMP, we are less restrictive. This gives much more flexibility in modeling. For example, this makes it possible in many models to avoid introducing a hybrid space.

One of our objectives is to provide numerical schemes to compute quantities of interest for the applications and to prove the convergence of these schemes. For this we need to get Chapmann-Kolmogorov equations for a sufficiently large class of functions. But there are too few functions in the extended generator domain, even for classic PDMPs because of Dirac measures in the interarrival distributions due to jumps at the state space boundary. Therefore we develop more general equations. They have an additional term that is an integral with respect to the intensity of a marked point process. This intensity is of interest in performance evaluations and can be approximated by numerical schemes.

Intensities of marked point processes have a major role to get Kolmogorov equations and also in stability studies that we carry out in a general framework by adapting results of Jacod [67] to CSMP. These intensities are not random as in Brémaud [20]. They are measures that only give a mean value, but they are defined for general marked point processes, without any assumptions. Here the intensity of a marked point process provides the average number of marks belonging to a given set for a given period of time.

A chapter is devoted to numerical schemes which are finite volume schemes derived from the Chapmann-Kolmogorov type equations previously developed. They approximate measures and not functions as usually. The pros and cons of explicit and implicit schemes are discussed using examples.

The book ends with an opening to a class of processes that we call Switching Processes. They are got by replacing the deterministic evolution between jump times by Markov processes. We hope it will inspire researchers for future work.

This book is aimed at researchers, graduate students and engineers who would like to be initiated to a model that is interesting both from a mathematical point of view and for its rich potential in applications.

I am grateful to my colleagues Michel, Sophie and Robert for getting me started on PDMPs, with a special thought for Robert who introduced us to the numerical analysis with patience and passion.

Many thanks to Romain for his guidance in the English language, which was very useful to me.

I would like to thank the referees for the advice they have given me on how to improve this book and in particular their encouragement to give many examples.

I would like to express my very deep gratitude to my husband Claude for his encouragements, his support and his patience during the lengthy gestation of this work, whose completion he did not get to see.

Lyon, France Christiane Cocozza-Thivent

Contents

Notations

1_A	Indicator function, $1_A(x)$ (resp. 1_A) equals 1 if $x \in A$ (resp. A is true) and 0 if not
$:=$	By definition
a.e.	Almost everywhere
a.s.	Almost surely
A_t	Elapsed time since the last jump time, p. 26
$B(E)$	Set of bounded Borel functions defined on E, p. 129
$\mathcal{B}(E)$	Borel σ-field of a topological space E
β	Regular conditional probability distribution of Y_{n+1} given $Y_n = x, T_{n+1} - T_n = \mathrm{v}$, p. 42
$C^{1,2}$	p. 200
cad-lag	Right continuous with left limits
Card(A)	Cardinal of the set A
A^c	Complement of A, if $A \subset E$ then $A^c = E \backslash A$
$\mathcal{X}\psi$	p. 104
$*$	Convolution, Section 2.2 p. 17, p. 226
CSMP	Completed semi-Markov process
δ_a	Dirac measure at a, $\int f(x)\delta_a(dx) = f(a)$
$\Delta f(s)$	$= f(s) - f(s_-) = f(s) - \lim_{u \to s, u < s} f(u)$
dx	Lebesgue measure
dF_x	Regular probability distribution of $T_{n+1} - T_n$ given $Y_n = x$, p. 42
$\partial_i \psi$	p. 107
$\partial_\phi f$	p. 109
$\partial_{t,\phi}$	p. 109
$\check{E}$	p. 42
$\widehat{E}$	p. 42
$\equiv, \equiv_{loc}$	p. 123
$\mathbb{E}_x$	p. 21, p. 30
$\mathbb{E}_{x,u}$	p. 30

$\|\| \cdot \|\|$	Norm of R^d
$\|\|f\|\|_\infty$	$= \sup_x \|f(x)\|$
$\mathcal{F}_t$	σ-field generated by the Markov renewal process observed up to time t, p. 26
$\bar{F}_x(t)$	$dF_x(]t, +\infty])$, p. 21
iid:	Independent and identically distributed
$\lim_{s\uparrow t}$	$= \lim_{\substack{s\to t,\\ s<t}}$
Γ_i, Γ_i^*	p. 53, p. 54
iff	If and only if
$L\varphi$	p. 117
$\mathcal{L}f$	p. 121
$\min(a,b), a \wedge b$	Minimum of a and b
$\max(a,b)$	Maximum of a and b
a_+	$= \max(a,0)$
$\mathbb{N}$	Set of non-negative integers
$\mathbb{N}^*$	Set of positive integers
N	Semi-Markov kernel of the Markov renewal process
N_t	$= \sum_{n\geq 1} 1_{\{T_n \leq t\}}$, p. 22
$N_0^{x,u}$	p. 22
ODE	Ordinary differential equation
$\mathbb{P}_x$	p. 21, p. 30
$\mathbb{P}_{x,u}$	p. 30
$\mathbb{R}_+$	$= [0, +\infty[$
$\mathbb{R}_+^*$	$=]0, +\infty[$
$\bar{\mathbb{R}}_+$	$= [0, +\infty]$
$\mathbb{R}_-$	$=] - \infty, 0]$
$\mathbb{R}_-^*$	$=] - \infty, 0[$
$\bar{\mathbb{R}}_-$	$= [-\infty, 0]$
r_1	p. 85
R	$= \sum_{n\geq 0} N^{*n}$, p. 33
$\underline{\sigma}$	p. 96
σ	p. 98
σ_i	p. 98
X_{t-}	$X_{t-} = \lim_{\substack{s_n\to t\\ s_n<t}} X_{s_n}$
SP	Switching Process
$t^*(x)$	p. 42
$\tau_s g$	p. 201
r.v.	Random variable
W_t	p. 25
(Y,T), $(Y_n, T_n)_{n\geq 1}$	Markov renewal process
Z_t	p. 26

Chapter 1
Introduction

Over the few last years, increasing attention has been paid to piecewise deterministic Markov processes (PDMPs) because of their strong potential to describe a variety of situations. They make it possible to represent phenomena with a deterministic behavior that is modified by random jumps.

Even if they are not always identified as such, PDMPs are used in a variety of fields. These include queueing systems (Browne and Sigman [22], Davis [50], Last [77]), storage models (Boxma et al. [17], Çinlar and Pinsky [28], Davis [50], Harrison and Resnick [63, 64]), earthquake modeling (Borovkov and Vere-Jones [16], Brémaud and Foss [21], Last [77], Ogata and Vere-Jones [84], Vere-Jones [94, 95], Zheng [98]), analysis of insurance risk (Albrecher and Thonhauser [2], Dassios and Embrechts [49], Embrechts and Schmidli [56], Rolski et al. [88], Schäl [92]), predictive reliability (Cocozza-Thivent and Eymard [33], Cocozza-Thivent and Kalashnikov [34], Cocozza-Thivent and Roussignol [35, 36], De Saporta et al. [52, 53], Eymard and Mercier [58], Lair et al. [76], Last [77], Last and Szekli [78], Lorton [82]), biology (Boxma et al. [18], Buckwar and Riedler [23], Crudu et al. [46, 47], Fontbona et al. [60], Goreac and Rotenstein [61], Koureta et al. [74], Rudnicki and Tyran-Kaminńska [90]), communication networks (Chafaï et al. [25], Graham and Robert [62]). Other references can be found in Schäl [91].

In analysis of insurance risks, jumps occur when the company has to pay a claim. In stock models, jumps are caused by deliveries. In earthquake modeling, jumps arise when tectonic forces are released and an earthquake occurs. In predictive reliability, jump times are times of failure or completion of maintenance (corrective, i.e., repair or preventive).

This general class of nondiffusion stochastic models was identified by M.H.A. Davis in 1984 [50] and called piecewise-deterministic processes (PDP) or piecewise-deterministic Markov processes (PDMP). Around the same time, the study of such models on $\mathbb{R}$, called jump processes with drift, was also the dissertation topics of

C. Cocozza-Thivent, *Markov Renewal and Piecewise Deterministic Processes*,
Probability Theory and Stochastic Modelling 100,
https://doi.org/10.1007/978-3-030-70447-6_1

R. Wobst [97]. Davis's book, *Markov Models and Optimization*, [51] published in 1993, has been a major contributor in bringing attention to PDMPs and is a reference book. In recent years, interest in these models has resurfaced because of their strong modeling potential.

A path of a piecewise deterministic process $(X_t)_{t\geq 0}$ on E is defined from its initial state X_0, a flow ϕ, and jump times $(T_n)_{n\geq 1}$. The flow ϕ provides the deterministic evolution between jumps: when $x \in E$ is the initial state or the state right after a jump, $\phi(x,t) \in E$ is the state after a period t without a jump. Hence $X_t = \phi(X_0, t)$ for $t < T_1$, $X_t = \phi(X_{T_n}, t - T_n)$ for $T_n \leq t < T_{n+1}$. We assume that $t \to \phi(x,t)$ is continuous for all x. Define $Y_n = X_{T_n}$. When Y_n contains all the information on the past up to time T_n to get the probability distribution of the future, $(Y_n, T_n)_{n\geq 1}$ is called a Markov renewal process. The fact that Y_n summarizes the past is expressed as follows: the conditional probability distribution of $(Y_{n+1}, T_{n+1} - T_n)$ given $(Y_p, T_p)_{p\leq n}$ depends only on Y_n. The conditional probability distribution of $(Y_{n+1}, T_{n+1} - T_n)$ given $Y_n = x$ $(n \geq 1)$ is denoted by $N(x, dy, dv)$, where N is the kernel of the Markov renewal process. The probability distributions of $T_{n+1} - T_n$ given $Y_n = x$ $(x \in E, n \geq 1)$ are the interarrival probability distributions.

Example 1.1 A stock, for example of raw material used to manufacture a product, decreases at a rate c. It is supplied by deliveries that take place at times T_n, where $(T_n)_{n\geq 0}$ is a zero-delayed renewal process, that is, $T_0 = 0$ and $(T_{n+1} - T_n)_{n\geq 0}$ are i.i.d. random variables with probability distribution dF (interarrival distribution). The delivered quantities are i.i.d. random variables with probability distribution μ; they are independent of $(T_n)_{n\geq 1}$. Then the stock level $(X_t)_{t\geq 0}$ is a piecewise deterministic process. We have $E = \mathbb{R}_+$, $\phi(x,t) = \max(x - ct, 0)$. Define $Y_n = X_{T_n}$ and D as a random variable with probability distribution μ; then $(Y_n, T_n)_{n\geq 1}$ is a Markov renewal process, and its kernel is $N(x, dy, dv) = \beta(x, v; dy)\, dF(v)$, where $\beta(x, v; dy)$ is the probability distribution of $\phi(x, v) + D$.

When $c = 1$, X_t is also the residual work at time t (or virtual waiting time, i.e., the waiting time of a customer arriving at time t) of a GI/G/1 queue. When $c = 0$ and the arrivals are a Poisson process, i.e., dF is an exponential distribution, it is the well-known compound Poisson process.

Example 1.2 (*Dassios and Embrechts* [49], *Davis* [51] *(21.11)*) The cash flow of an insurance company increases at speed c as long as there is no claim to pay or debt to be reimbursed. Claims occur at times T_n, which are modeled by a renewal process with interarrival distribution dF, the amounts are i.i.d. random variables with the same probability distribution μ. When the company does not have enough cash to pay, it borrows at a rate b. Then the cash flow $(X_t)_{t\geq 0}$ is a piecewise deterministic process. When $x \geq 0$, we have $\phi(x,t) = x + ct$ for all t. When $x < 0$, $\frac{\partial}{\partial t}\phi(x,t) = c + b\,\phi(x,t)$ as long as $\phi(x,t) < 0$; hence assuming $-c/b < x < 0$, we have $\phi(x,t) = (x + c/b)e^{bt} - c/b$ if $t < t_0 = \frac{1}{b}\log(c/(c+bx))$, $\phi(x,t) = c(t - t_0)$ if $t \geq t_0$. When $x \leq -c/b$, we have $\phi(x,t) = (x + c/b)e^{bt} - c/b$ for all t. Define $Y_n = X_{T_n}$ and D as a random variable with probability distribution μ. Then $(Y_n, T_n)_{n\geq 1}$ is a Markov renewal process with kernel $N(x, dy, dv) = \beta(x, v; dy)\, dF(v)$, where $\beta(x, v; dy)$ is the probability distribution of $\phi(x, v) - D$.

In the above examples, the stock level, respectively the cash flow, has no influence on deliveries (times and quantities), respectively on the claims (times and amounts). In contrast, in stress models, the jump times depend on the stress level. This dependence appears in the hazard rates that characterize the probability distributions of the time between two consecutive jumps.

It is equivalent to give the probability density function f of a positive random variable T or its hazard rate h. In short, $\mathbb{P}(T \in (t, t + \Delta t) \mid T > t) = h(t)\Delta t + o(\Delta t)$. We have $\mathbb{P}(T > t) = e^{-\int_0^t h(s)\,ds}$, $f(t) = h(t)\,e^{-\int_0^t h(s)\,ds}$ (Appendix A, Sect. A.3).

Example 1.3 Earthquakes are due to tectonic forces. We assume that the evolution of these forces is a deterministic process described by ϕ, where $\phi(x, t)$ stands for the tectonic forces when the time elapsed since the previous earthquake is t and the tectonic forces just after this earthquake were x. The T_n are the times when an earthquake occurs, and Y_n is the tectonic forces just after the earthquake that occurs at time T_n. The tectonic forces at time t are $\phi(Y_n, t - T_n)$ when $T_n \leq t < T_{n+1}$. Assume that the hazard rate at which earthquakes occur depends on the tectonic forces in the following way: the probability that an earthquake occurs in the time interval $(t, t + \Delta t)$ given that the last earthquake was at time 0 and the tectonic forces are τ_t at time t is $\lambda(\tau_t)\,\Delta t + o(\Delta t)$. Assume also that the probability distribution of the tectonic forces right after the earthquake given that these forces are x right before is $Q(x; dy)$. Then $(Y_n, T_n)_{n\geq 1}$ is a Markov renewal process; its kernel is $N(x, dy, dv) = \lambda(\phi(x, v))\,e^{-\int_0^v \lambda(\phi(x,s))\,ds}\,Q(\phi(x, v); dy)\,dv$.

When right before time 0 there is a delivery in Example 1.1 or a claim payment in Example 1.2 or an earthquake in Example 1.3, the probability distribution of (Y_1, T_1) is $N(x, dy, dv)$ for some x. The Markov renewal process is said to be zero-delayed. We then define $Y_0 = x$, $T_0 = 0$, $Z_t = Y_n$ when $T_n \leq t < T_{n+1}$ $(n \geq 0)$. By definition, $(Z_t)_{t\geq 0}$ is a zero-delayed semi-Markov process. Semi-Markov processes generalize jump Markov processes: the probability distribution of the length of stay in a given state and the next state does not depend on the past. The length of stay in a state is not supposed to have an exponential distribution. This length of stay and the next state are not supposed to be independent while they are in a jump Markov process, that is, $N(x, dy, dv)$ is not assumed to be a product measure.

Example 1.4 1. The simplest nontrivial example of a semi-Markov process is given by the alternating renewal process. In predictive reliability, it is the basic model for equipment that has failures and repairs. The equipment is up (state e_1) or down (state e_0). Failure times are $(T_{2n+1})_{n\geq 0}$; completion of repair times are $(T_{2n})_{n\geq 0}$. The equipment is new at time 0. At completion of repair, the equipment is as good as new. Hence $(T_n)_{n\geq 0}$ is a zero-delayed alternating renewal process, i.e., $T_0 = 0$, $T_{n+1} - T_n$ $(n \geq 0)$ are independent, $T_{2n+1} - T_n$ $(n \geq 0)$ have the same probability distribution dF_1, which is the probability distribution of failure-free operating periods, $T_{2n} - T_{2n-1}$ $(n \geq 1)$ have the same probability distribution dF_0, which is the probability distribution of repair periods. Define $Y_{2n} = e_1$ $(n \geq 0)$, $Y_{2n+1} = e_0$ $(n \geq 0)$, then $(Y_n, T_n)_n$ is a zero-delayed Markov renewal process. The associated semi-

Markov process gives the state of the equipment at any time. We have $E = \{e_1, e_0\}$, $N(e_1, dy, dv) = \delta_{e_0}(dy)\, dF_1(v)$, $N(e_0, dy, dv) = \delta_{e_1}(dy)\, dF_0(v)$.

2. Assume now, moreover, that preventive maintenance occurs when the equipment reaches age a. If the failure comes exactly at the end of an operating time equal to a, it is assumed that it is the preventive maintenance that is carried out, because everything is ready, since it has been programmed. Clearly, the length of stay in state e_1 and the next state are not independent: if the length of stay in e_1 is strictly less than a, the next state is e_0, while if the length of stay is a, the next state is the preventive maintenance state when the preventive maintenance is not supposed to be instantaneous, and e_1 when it is supposed to be instantaneous.

When the preventive maintenance is not instantaneous, denote by e_2 the preventive maintenance state and by dF_2 the probability distribution of the preventive maintenance periods. Then we have $N(e_1, dy, dv) = 1_{\{v<a\}} dF_1(v)\, \delta_{e_0}(dy) + dF_1([a, +\infty])\, \delta_{e_2}(dy)\, \delta_a(dv)$, $N(e_2, dy, dv) = dF_2(v)\, \delta_{e_1}(dy)$. When the preventive maintenance is instantaneous, then $N(e_1, dy, dv) = 1_{\{v<a\}} dF_1(v)\, \delta_{e_0}(dy) + dF_1([a, +\infty])\, \delta_{e_1}(dy) \delta_a(dv)$.

3. Assume again that there is only corrective maintenance (i.e., repairs and no preventive maintenance) and in addition that dF_1 and dF_0 have densities. The hazard rate of the probability density function (pdf) of the operating periods (respectively of the repair periods) is called the failure rate (respectively the repair rate). Assume that the failure and repair rates depend on the temperature equipment. This temperature increases (respectively decreases) when the equipment is in operation (respectively in repair). When the temperature is τ at the beginning of an operating period (respectively a repair period), it is $\phi_1(\tau, t)$ (respectively $\phi_0(\tau, t)$) after an operating duration t (respectively after a repair duration t). The state of the equipment can be modeled by the first component of the semi-Markov process associated with Markov renewal process $(Y_n, T_n)_n$, where marks Y_n take values in $\{e_1, e_0\} \times \mathbb{R}$, the first component Y_n^1 is the state of the equipment at time T_n, and the second one Y_n^2 is the temperature of the equipment at time T_n. Its kernel is

$$N((e_i, \tau), dy_1, dy_2, dv) = \\ 1_{\mathbb{R}_+}(v)\, \lambda_i(\phi_i(\tau, v))\, e^{-\int_0^v \lambda_i(\phi_i(\tau,s))\, ds}\, \delta_{e_{1-i}}(dy_1)\, \delta_{\phi_i(\tau,v)}(dy_2)\, dv,$$

where λ_1 is the failure rate and λ_0 is the repair rate. The semi-Markov process is $(Z_t^1, Z_t^2)_{t\geq 0}$, Z_t^1 is the equipment state, and Z_t^2 is the temperature at the time of the last state change.

4. Assume now that the equipment is composed of two pieces. Each one is as in item 1 above. Define $\eta_t^i \in \{e_1, e_0\}$ as the state of piece i at time t, and $\eta_t = (\eta_t^1, \eta_t^2)$, η_t is piecewise constant but is not a semi-Markov process, unless each piece has constant failure and repair rates. Denote by $(T_n)_n$ its jump times and $Y_n = \eta_{T_n}$, $(Y_n, T_n)_n$ is not a Markov renewal process, because Y_n does not contain all the information about the past. For example, assume that T_n is a failure time of component 1, and component 2 is up but not new at this time; its age is missing, so one cannot know the probability distribution of its next failure. The time D_n^i elapsed since the last change of state of

piece i must be added. Define Y'_n as $Y'_n = (Y_n, D^1_n, D^2_n)$. Then $(Y'_n, T_n)_n$ is a Markov renewal process. Its associated semi-Markov process is (η_t, Z^1_t, Z^2_t), where Z^i_t is defined as follows: if $\eta^i_t = e_1$ (respectively $\eta^i_t = e_0$), then Z^i_t is the age (respectively the time elapsed since the beginning of the repair in progress) of piece i at the last time the equipment changed state. We have $Z^1_t = 0$ or $Z^2_t = 0$.

If some probability distribution of the length of stay in a state is not exponential, then a semi-Markov process is not a Markov process, because only exponential distributions are without memory. To get a Markov process, a supplementary variable must be added. Most often, the choice is the waiting time W_t defined as $W_t = T_{n+1} - t$ when $T_n \le t < T_{n+1}$. But considering the path description of a piecewise deterministic process, we choose the time A_t elapsed since the last jump, namely $A_t = t - T_n$ when $T_n \le t < T_{n+1}$.

However, the choice of A_t as the supplementary variable raises a difficulty. How is one to define A_t for $t < T_1$ if time 0 cannot be considered a jump time, that is, when right before time 0 there is not a delivery in Example 1.1, a claim payment in Example 1.2, an earthquake in Example 1.3, a failure or a completion of maintenance in Example 1.4 (cases 1 and 2)? These events must be assumed to have taken place at time $-u$ for some $u \ge 0$. The probability distribution of (Y_1, T_1) is then $N_0^{x,u}(dy, dv)$ for some $x \in E$, $N_0^{x,u}(dy, dv)$ being defined as follows: $N_0^{x,u}(dy, dv)$ is the conditional probability distribution of $(Y, T - u)$ given $T > u$, where the probability distribution of (Y, T) is $N(x, dy, dv)$. It is defined for u such that $\mathbb{P}(T > u) > 0$, i.e., $u < t^*(x) = \sup\{t > 0 : \mathbb{P}(T_{n+1} - T_n > t \mid Y_n = x) > 0\}$ $(n \ge 1)$.

Example 1.5 When the hazard rate of T is $h(t)$, the hazard rate of $T - u$ given $T > u$ is $h(t + u)$. For example, when the failure rate of an equipment is λ and at the initial time the equipment age is u, then its failure rate h is $h(t) = \lambda(u + t)$.

More generally, assume $N(x, dy, dv) = \ell(x, v)\, e^{-\int_0^v \ell(x,s)\, ds}\, \beta(x, v; dy)\, dv$. This means that interarrival distributions have densities and $\beta(x, v; dy)$ is the probability distribution of Y_{n+1} given $Y_n = x$, $T_{n+1} - T_n = v$ $(n \ge 1)$. We get $N_0^{x,u}(dy, dv) = \ell(x, u + v)\, e^{-\int_0^v \ell(x,u+s)\, ds}\, \beta(x, u + v; dy)\, dv$. In Example 1.3, when the last earthquake before time 0 was at time $-u$ and tectonic forces right after it were x, the probability that the next earthquake occurs in $(t, t + \Delta t)$ is $\lambda(\phi(x, u + t))\, \Delta t + o(\Delta t)$, and the probability that the tectonic forces right after it belong to the set A is $Q(\phi(x, u + v); A)$.

Actually, the initial distribution of every Markov renewal process with kernel N we consider in this book is $N_0^{x,u}$ or a mixture of such distributions. This restriction is not a handicap for the study of PDMPs. Measures $N_0^{x,u}$ have a key role in stability studies. When the initial distribution is $N_0^{x,u}$, we define $Z_t = x$, $A_t = u + t$ for $0 \le t < T_1$. Then the fundamental theorem on the change of the observation time (Theorem 2.1, Sect. 2.4) states that the post-t Markov renewal process $(Y^t_n, T^t_n)_{n \ge 1}$, i.e., the Markov renewal process observed from time t, given its past up to time t, is a Markov renewal process with kernel N, and the probability distribution of (Y^t_1, T^t_1) (initial distribution) is $N_0^{Z_t, A_t}$.

Process $(Z_t, A_t)_{t\geq 0}$ is a Markov process that we call a completed semi-Markov process (CSMP). We define a piecewise deterministic process as $(\phi(Z_t, A_t))_{t\geq 0}$, where ϕ is a deterministic flow, and $(Z_t, A_t)_{t\geq 0}$ a CSMP. It generalizes the usual path description of a PDMP, at least when $A_0 = 0$ or $A_0 > 0$ and $t > T_1$.

We now return to the earthquake model (Example 1.3). When the last earthquake before time 0 occurred at time $-u$ and the tectonic forces right after it were x, the tectonic forces at time t are $X_t = \phi(Z_t, A_t) = \phi(x, u+t)$ for $t < T_1$. However, in the path description they are said to be $\phi(X_0, t) = \phi(\phi(x, u), t)$. In fact, it is the same if the deterministic evolution is Markov, that is, $\phi(x, u+t) = \phi(\phi(x, u), t)$ (1) for all (x, u). We call it a deterministic Markov flow.

In Davis [51], ϕ is given by an ODE, namely $\frac{\partial}{\partial t}\phi(x,t) = \mathbf{v}(\phi(x,t)), \phi(x,0) = x$, where $\mathbf{v}$ is Lipschitz continuous. Such a ϕ satisfies (1). Assuming only that (1) holds makes modeling easier.

Example 1.6 The flow ϕ of Example 1.1 is a deterministic Markov flow, but it is not given by an ODE because of the discontinuity in the speed. We have $\frac{\partial}{\partial t}\phi(x,t) = \mathbf{v}(\phi(x,t))$ with $\mathbf{v}(x) = c$ for $x > 0$ and $\mathbf{v}(0) = 0$. Also in Example 1.2, ϕ is a deterministic Markov flow but $\phi(x,t)$ is not the solution of an ODE on $\mathbb{R}$: when $-c/b < x < 0$, there is a discontinuity in the speed at point 0 of the state space. To get a PDMP as studied in Davis [51], a hybrid space must be introduced. In Example 1.1 (respectively Example 1.2), we can basically take, for example, $E = \{a\}\times]-\infty, 0[\cup\{0\}$ (respectively $E = \{a\}\times]-\infty, 0[\cup\{b\}\times[0, +\infty[$), basically because $\{0\}$ and $[0, +\infty[$ are not open subsets of $\mathbb{R}$ as stated in [51], but considering the speed directions, we feel that it doesn't matter here. Hence jumps must be added when the PDMP reaches $\{a\}\times\{0\}$, the active boundary in Davis's terminology. Therefore, some of the interarrival probability distributions are those of the minimum of a random variable with probability distribution dF and a constant that is the time to reach this boundary following the flow.

In Jacobsen [66], the interarrival probability distributions have densities (no boundary), and the deterministic flow is only assumed to satisfy (1). Condition (1) is a very natural condition for a piecewise deterministic process to be a Markov process.

Under assumption (1), we prove that $X_t = \phi(Z_t, A_t)$ is a Markov process when condition $N_0^{x,u}(dy, dv) = N(\phi(x,u), dy, dv)$ holds for all $x \in E, u < t^*(x)$ (recall that $t^*(x) = \sup\{t > 0 : N(x, E\times]t, +\infty]) > 0\}$). This last condition, which is a compatibility condition between the flow and the probability distributions of the jumps, is quite natural, also considering the theorem about the change of the observation time.

Therefore, we define a PDMP as a process that can be written as $(\phi(Z_t, A_t))_{t\geq 0}$, where ϕ is a deterministic Markov flow; more precisely, for all $x \in E, t\in[0, t^*(x)] \rightarrow \phi(x,t)$ is continuous and

$$\forall(x, s, t) \in E \times \mathbb{R}_+^2,\ s + t \leq t^*(x),\ \phi(x, s+t) = \phi(\phi(x, s), t), \tag{1.1}$$

$(Z_t, A_t)_{t\geq 0}$ is a CSMP with kernel N, and the condition

$$\forall x \in E,\ \forall u < t^*(x),\ N_0^{x,u}(dy, dv) = N(\phi(x,u), dy, dv) \tag{1.2}$$

holds.

Example 1.7 Assume that the interarrival distributions have densities. Noting their hazard rates ℓ and $\beta(x, v; dy)$, the probability distribution of Y_{n+1} given $Y_n = x$, $T_{n+1} - T_n = v$ $(n \geq 1)$, we have $N(x, dy, dv) = 1_{\mathbb{R}_+}(v)\, \ell(x, v)\, e^{-\int_0^v \ell(x,s)\, ds}$ $\beta(x, v; dy)\, dv$. When (1.1) holds, condition (1.2) is $\ell(x, v) = \lambda(\phi(x, v))$, $\beta(x, v; dy) = Q(\phi(x, v); dy)$ for some λ and Q.

Hence if the tectonic forces of Example 1.3 evolve according to a deterministic Markov flow between earthquakes, then their evolution including effects of earthquakes is a PDMP.

Example 1.8 In Example 1.1 (respectively Example 1.2), ϕ is a deterministic Markov flow but (1.2) does not hold, unless dF is an exponential distribution. This is because the stock level (respectively the cash-flow) at a given time does not provide information on the probability distribution of the time until the next delivery (respectively the next claim). A supplementary variable must be added to get a Markov process. Considering the context, the best choice is the time elapsed since the last delivery. More details are given in Example 3.4, Sect. 3.1.

In classic PDMPs, i.e., PDMPs studied in Davis [51], each interarrival distribution is the probability distribution of the minimum of a random variable with a probability density function and a constant; the density is written using its hazard rate. More precisely,

$$N(x, dy, dv) = \Big(1_{\{v<\alpha(x)\}}\, \lambda(\phi(x, v))\, e^{-\int_0^v \lambda(\phi(x,s))\, ds}\, dv \tag{1.3}$$
$$+ 1_{\{\alpha(x)<+\infty\}}\, e^{-\int_0^{\alpha(x)} \lambda(\phi(x,s))\, ds}\, \delta_{\alpha(x)}(dv) \Big)\, Q(\phi(x, v); dy),$$

$\alpha(x) = \inf\{t > 0 : \phi(x, t) \in B\}$, where B is the topological boundary of the state space. On-site jumps are forbidden, that is, $Q(x; \{x\}) = 0$ for all x.

In Examples 1.1 and 1.2, when dF has a probability density function, N is as in (1.3), but it does not contain Dirac measures $\delta_{\alpha(x)}$ when we do not impose on ϕ that it be the flow of an ODE but only satisfy (1.1). If we want ϕ to be the flow of an ODE, a hybrid space must be defined, and Dirac measures appear (see Example 1.6).

It is better not to have Dirac measures in N, but we can't always avoid them. For example, they are unavoidable in Example 1.4 when preventive maintenance is carried out and in the following example.

Example 1.9 A fluid is stored in a tank and is used at some rate. When the tank level reaches a given minimum level, a pump is switched on to fill it. When the tank is full, the pump is switched off. This pump is subjected to failures, and its failure rate depends on its operating time. It is therefore necessary to make a distinction between periods when the pump is in operation and periods when it is at rest. The times when the pump switches from one of these states to the other are the times when the tank level reaches the minimum or maximum level, which brings Dirac measures in N. The model is specified in Example 3.14, Sect. 3.4.

We refer to a parametrized PDMP when N is similar to (1.3) but with fewer restrictions. In short, N is given as

$$N(x, dy, dv) = 1_{\{v<\alpha(x)\}}\, \lambda(\phi(x, v))\, e^{-\int_0^v \lambda(\phi(x,s))\, ds}\, Q(\phi(x, v); dy)\, dv \quad (1.4)$$
$$+1_{\{\alpha(x)<+\infty\}}\, e^{-\int_0^{\alpha(x)} \lambda(\phi(x,s))\, ds}\, q(\phi(x, v); dy)\, \delta_{\alpha(x)}(dv),$$

$\alpha(x) \in]0, +\infty]$ for all x. Condition (1.2) is then $\alpha(\phi(x, u)) = \alpha(x) - u$ for all x and $u \in [0, \alpha(x)[$. In the examples, $\alpha(x)$ is the time it takes for a particle starting from x to reach a subset of E by following the flow, but no link with the topological boundary of E is imposed on this subset, and it can be different depending on the region of the space in which x is located. However, condition $\alpha(x) > 0$ is imperative.

Example 1.10 A particle is moving on $[-1, +\infty[$ with speed 1. It starts at point -1. When it reaches 0, with probability p it jumps to a point in $]-1, 0[$ chosen with the uniform probability distribution. With probability $1 - p$ it continues with the same speed. The particle's position is a PDMP taking values in $[-1, +\infty[$. We have $\phi(x, t) = x + t$, N is as in (1.4) with $\lambda = 0$ (hence Q does not need to be defined), $\alpha(x) = \inf\{t > 0 : \phi(x, t) = 0\}$, $q(0, dy) = p\, 1_{]0,1[}(dy) + (1 - p)\, \delta_0(dy)$. Note that $\alpha(x) = |x|$ for $x \in [-1, 0[$, $\alpha(x) = +\infty$ for $x \geq 0$.

Introducing Q and q makes it possible to distinguish the effects of so-called purely random jumps, i.e., those appearing with rate λ, and jumps caused by Dirac measures. Moreover, jumps on site are possible. This allows one to add more jumps if needed for more information without introducing a hybrid space.

Example 1.11 In Example 1.1, we can add jumps when the stock becomes empty to get information about them, for example, the mean number of times the stock becomes empty in a given time interval (see Example 1.12 below). Denote by $(T_n^+)_{n\geq 1}$ the new increasing sequence we get (see Fig. 3.1, Sect. 3.1) and define mark Y_n as the stock level at time T_n^+. For the reason outlined in Example 1.8, $(Y_n, T_n^+)_{n\geq 1}$ is not a Markov renewal process when dF is not an exponential distribution. Define S_t as the stock level at time t, D_t as the elapsed time since the last delivery, $Y_n^+ = (Y_n, D_{T_n^+}) = (S_{T_n^+}, D_{T_n^+})$. Then $(Y_n^+, T_n^+)_{n\geq 1}$ is a Markov renewal process, and $(S_t, D_t)_{t\geq 0}$ is a PDMP (see the end of Example 3.4, Sect. 3.1). Note that $D_{T_n^+} = 0$ iff T_n^+ is some T_p and $S_{T_n^+} = 0$ iff T_n^+ is not some T_p.

For most PDMPs, on-site jumps hide some jump times when one is looking at its path. It is the same for semi-Markov processes but not for CSMPs, because A_t is reset to zero at each jump.

The topic of Davis [51] is in its title: *Markov Models and Optimization*. The objective is optimal control and martingales play an essential role. Martingales are deduced from the extended generator, and very particular attention is paid to its domain, which is fully characterized. Actually, the martingale approach of nondiffusion stochastic processes was initiated by P. Brémaud in his thesis [19] with application to communication theory. A little later, this approach was generalized by R. Boel, P. Varaiya, and E. Wong and applied also to control theory [14, 15].

In [71], J. Jacod and A.V. Skorokhod developed an extremely general framework in which no topological structure is assumed on the state space. Their definition of processes, which they call jump Markov processes, is not constructive but is based on the filtration of the process. They give the structure of the martingales, semimartingales, additive functionals, etc., in reliance on general process theory. Most results are given for quasi-Hunt jump Markov processes, for which the interarrival distributions cannot contain Dirac measures. Therefore, their model does not apply to classic PDMPs as soon as the constants mentioned above exist.

The purpose here is to have a more general approach to PDMPs than in Davis [51], to facilitate modeling and to take advantage of the underlying structure, i.e., Markov renewal. We also want to highlight the effects of the assumptions on interarrival distributions. As applications, we show how one can access performance indicators including cost evaluation, and then compute them using numerical schemes.

CSMPs are general PDMPs as defined above. Their deterministic evolution between jumps is especially simple, since their flow is $\phi((x_1, x_2), t) = (x_1, x_2 + t)$. Their study makes it possible to focus on the jump process without being disturbed by the deterministic behavior between jump times. Then we deduce results on PDMPs that are deterministic functions of CSMPs (recall that $X_t = \phi(Z_t, A_t)$). When very occasionally a result for PDMPs is not immediately inferred from a similar result for CSMPs, the proof for CSMPs guides the proof for PDMPs.

As much as possible we have sought to give testable assumptions and realistic criteria to check them.

Markov renewal processes as well as directly related continuous-time Markov processes, including CSMPs, are presented in Chap. 2. We give Markov renewal equations satisfied by CSMP marginal distributions and mention semiregenerative processes. We conclude the chapter with results on the asymptotic behavior of the solutions of Markov renewal equations. These results, the demonstration of which was a challenge around the 1980s, are very elaborate generalizations of Blackwell's renewal theorem. The asymptotic behavior of semiregenerative processes, and as a result of CSMPs, is deduced.

PDMPs are defined in Chap. 3. The definition of parametrized PDMPs is a little more elaborate than the one given above. It may seem a bit cumbersome at first glance. The reason is the possible presence of "boundaries" within the state space as in Example 1.10. It allows one to cover a few more examples than what is possible with the definition given above, and above all to allow discontinuities (such as that of α in Example 1.10) in later chapters, without complicating the state space. Then we briefly explain how to simulate PDMPs. Finally, examples are given to help one understand the formalism and the interest of the definitions.

Chapter 4 has its origin in a practical question: how can one estimate the probability that a multicomponent piece of equipment will be running over a whole time interval? Two methods are presented. The first is based on the evaluation of the marginal probability distributions of the process killed or stopped at the first time the equipment fails. The first section describes these processes. The second method, much less well known, consists in modifying the initial process by removing some jumps and calculating a functional of this new process. It is called a decomposition of the process. It is particularly interesting in the case of rare events.

Chapter 5 is crucial for the following. It is devoted to intensities of marked point processes related to CSMPs and PDMPs. They are not random intensities as in Brémaud [20]. They are defined even when the interarrival distributions have no density. Here the intensity gives the mean number of jumps in a given time interval whose marks belong to a given set. In addition to their own interest, they allow one to obtain and understand generalized Kolmogorov equations in the next chapter.

Example 1.12 Let us return to Example 1.11. Define ρ_+ as the intensity of the marked point process $(Y_n, T_n^+)_{n\geq 1}$, then $\rho_+(\{0\} \times [0, t])$ is the mean number of times the stock has become empty during the time interval $[0, t]$.

Now let us look at the first two cases of Example 1.4. Denoting by ρ the intensity of the Markov renewal process $(Y_n, T_n)_{n\geq 1}$, $\rho(\{e_0\} \times [0, t])$ is the mean number of failures up to time t, $\rho(\{e_1\} \times [0, t])$ is the mean number of completions of maintenance (repairs and preventive maintenance when there are some) up to time t. In case of noninstantaneous preventive maintenance, $\rho(\{e_2\} \times [0, t])$ is the mean number of preventive maintenances begun up to time t. Completions of repairs and preventive maintenance can be separated by considering the marked point process $(Y_n, T_{n+1} - T_n, T_{n+1})_{n\geq 0}$. Denoting by $\bar{\rho}$ its intensity, the mean number of failures up to time t is $\bar{\rho}(\{e_1\} \times [0, a[\times [0, t])$, and the mean number of preventive maintenances up to time t is $\bar{\rho}(\{e_1\} \times \{a\} \times [0, t])$.

We are mainly interested in the intensities of the semi-Markov process $(Y_n, T_n)_{n\geq 1}$, of $(Y_n, T_{n+1} - T_n, T_{n+1})_{n\geq 0}$, and of $(X_{T_{n+1}-}, T_{n+1})_{n\geq 0}$, where $(X_t)_{t\geq 0}$ is the PDMP associated to ϕ and $(Y_n, T_n)_n$. When $(T_n)_n$ is a renewal process, the cumulative function of its intensity is the renewal function; for a semi-Markov process it provides the potential function. Define ϱ as the intensity of the marked point process $(X_{T_{n+1}-}, T_{n+1})_{n\geq 0}$.

Example 1.13 In the stock example, modeled as in Example 1.8, the T_n are the times of a delivery, the PDMP is $X_t = (S_t, D_t)$, where S_t is the stock level, D_t the time elapsed since the last delivery, and the mean number of times the stock level is at most a at the time of a delivery is $\varrho([0, a] \times \mathbb{R}_+ \times]0, t])$.

For parametrized PDMPs, we make a distinction between the jumps that are due to the density part of the interarrival distributions and those that are due to the Dirac measures. When N is as in (1.4), we write $\varrho = \varrho_c + \sigma$, where ϱ_c (respectively σ) is the intensity of $(X_{T_{n+1}-}, T_{n+1})_{n\geq 0}$ restricted to T_{n+1} caused by purely random jumps, i.e., due to the density parts of interarrival distributions (respectively caused by Dirac measures). We have $\varrho_c(dx, ds) = \lambda(x)\,\pi_s(dx)\,ds$, where π_s is the probability distribution of X_s. The measure σ is concentrated on $\Gamma = \{\phi(x, \alpha(x)) : x \in E, \alpha(x) < +\infty\}$; it is (abusively) called the boundary measure. Measures ϱ_c and σ have a crucial role in the generalized Kolmogorov equations. They can be evaluated by numerical schemes. In the stock model of Example 1.11, σ is concentrated on $\{0\} \times \mathbb{R}_+ \times \mathbb{R}_+$, and $\sigma(\{0\} \times \mathbb{R}_+ \times [0, t])$ is the mean number of times the stock has become empty up to time t. This is how it is computed in Chap. 9.

Generalized Chapman–Kolmogorov equations are given in Chap. 6. We don't get them from martingales, as is usually done. We begin with CSMPs, more precisely, with $(Z_t, A_t, t)_{t\geq 0}$. The term that corresponds to the jumps is an integral with respect to $\bar{\rho}$, the intensity of $(Y_n, T_{n+1} - T_n, T_{n+1})_{n\geq 0}$. Then for a general PDMP, we write $g(X_t, t) = g(\phi(Z_t, A_t), t)$, and the integral with respect to $\bar{\rho}$ becomes an integral with respect to ϱ, which is the image of $\bar{\rho}$ under the map $(y, v, w) \to (\phi(y, v), w)$. For parametrized PDMPs, particular attention is paid to test functions g. To prove convergence of numerical schemes, we must show that Kolmogorov equations characterize PDMP marginal probability distributions π_t $(t \geq 0)$ and boundary measure σ. For this we have to write these equations for a large class of functions g, which must be allowed to have discontinuities on $\Gamma = \{\phi(x, \alpha(x)) : x \in E, \alpha(x) < +\infty\}$. To better understand this, let us return to Example 1.10. We have $E = [-1, +\infty[$, $\Gamma = \{0\}$. Take $g(x, t) = f(x)$; to get a Kolmogorov equation with test function f, the basic assumption is that $t \to f(\phi(x, t))$ is absolutely continuous. Actually, to get the characterization, it must be possible for f to have a discontinuity at the point 0. More precisely, it must be possible for f to be right continuous and to have a left-hand limit at the point 0. Why not left continuous with right-hand limit? Because of the flow direction. It is as if the state space had been split into $E_1 = \{1\} \times [-1, 0[$ and $E_2 = \{2\} \times [0, +\infty[$.
Let us return to general parametrized PDMPs. Generalized Kolmogorov equations consist of the sum of three terms. The first is for the deterministic evolution between jumps: since $t \in [0, t^*(x)[\to g(\phi(x, t), t)$ is, in short, supposed to be absolutely continuous, it is an integral with respect to Lebesgue measure. The second is an integral with respect to $\varrho_c(dx, ds) = \pi_s(dx)\, ds$, hence containing Lebesgue measure. The third is an integral with respect to σ.
Chapter 6 ends with a section about properties of the semigroup of parametrized CSMPs and PDMPs when N has no Dirac measures. This section has no impact on what follows.

Likewise for Chap. 7 on martingales, which is not used in the sequel. It presents classical results following the chronology specific to this book: we start with Markov renewal processes, then we move on to parametrized CSMPs and finally to parametrized PDMPs. Its last section is related to infinitesimal and extended generators.

Chapter 8 is about stability of CSMPs and PDMPs, and intensities of marked point processes related to them. It does not require any assumptions about interarrival distributions, since it is not based on the characterization of invariant measures using the extended generator of the process. Within the spirit of this book, we start with the stability of CSMPs, adapting a part of Jacod [67]. Markov renewal process intensity is a significant tool. In steady state, we get simple formulas for the intensities we are interested in. We immediately infer PDMP stationary measure and PDMP Harris recurrence from the CSMP results. As examples, we give results for renewal and alternating renewal processes, we find Asmussen's results on the virtual waiting time of queues and, in the stock model vocabulary, the steady state mean number of times the factory must be stopped for lack of supply during a given time interval. This chapter is a good illustration of the benefit of the approach adopted in this book. In the last section we state a convergence theorem of PDMP marginal distributions, as an application of results for semiregenerative processes given in Chap. 2.

Chapter 9 concerns numerical schemes. They are finite-volume schemes derived through Kolmogorov equations. For the two first schemes, it is assumed that the interarrival distributions have densities and the flow is given by an ODE. The first is explicit, while the second is implicit. These schemes are for approximation of measures, namely the PDMP marginal probability distributions $(\pi_t)_{t\geq 0}$. They are suggested by the integrodifferential equations satisfied by the densities of the π_t when π_0 and every $Q(x; dy)$ have probability density functions. The third scheme is an implicit scheme for general parametrized PDMPs. It gives not only an approximation of the π_t, but also of the boundary measure σ, and as a result of the intensity ϱ of $(X_{T_{n+1}-}, T_{n+1})_{n\geq 0}$. The method for proving the convergence of the schemes is described in the first section. Assumptions and references for the proof of the first two are given in the sections that describe them. For the third, the proof is detailed in Appendix C. These schemes only require $\phi(x,t)$ for small t; they do not require the resolution of ODEs when the flow is given by such equations. On the other hand, numerical schemes are particularly well adapted to sensitivity studies. Indeed, most of the computation time is devoted to determining the coefficients of the linear systems of equations to be solved. In changing a parameter of the model, only some coefficients have to be computed again. The second part of the chapter is devoted to examples. In all of them, π_t includes a Dirac measure. The pros and cons of explicit and implicit schemes are illustrated through Example 1.10. Example 1.11 provides an opportunity to get in addition the boundary measure σ.

For example, in stress models, wear on equipment can be better modeled by a stochastic process than by a deterministic process. Therefore, one may wonder about the contribution of the methodology developed throughout this book when the deterministic evolution between jumps is replaced by a Markov process. A beginning answer is given in Chap. 10. We give a path construction of what we call switching processes, and we present some properties of these processes. Their decomposition in the sense of Sect. 4.2 is mentioned. A condition similar to (1.2) is given for having a Markov process. Semiregenerative properties are highlighted. They allow one to state an asymptotic theorem and also to make the connection with processes that have started attracting some researchers who had earlier specialized in diffusion processes: when the evolution between jumps is a semimartingale, the Kolmogorov equations we get are those expected.

Appendix A contains a summary of notions that are used in this book, including conditional probability distributions, Markov chains, hazard rates with which we write probability density functions, and absolute continuity of functions, which is a basic property for the test functions of the Kolmogorov equations.

Appendix B shows how the case of interarrival distributions that are mixtures of a distribution with a probability density function and at most a countable number of

Dirac measures could be reduced to that of interarrival distributions that are as in (1.4).

In the highly technical Appendix C, the convergence of the third numerical scheme introduced in Chap. 9 is proved. It contains the result on the uniqueness of the solutions $((\pi_t)_{t\geq 0}, \sigma)$ of the generalized Kolmogorov equations for parametrized PDMPs, which is interesting in itself.

Throughout this book, random variables are defined on a probability space $(\Omega, \mathcal{F}, \mathbb{P})$. A point of Ω is denoted by ω. Most often, the random variables are specified by uppercase Latin or lowercase Greek letters such as $X, Y, Z, T, A, W, \xi, \ldots$.

Nontemporal variables usually take values in $(E, \mathcal{E})$, where E is a Polish space, i.e., a separable completely metrizable topological space, and $\mathcal{E}$ is its Borel σ-algebra. Since temporal variables T_n can be equal to infinity, a point $\Delta \notin E$ is added for the corresponding mark. The space $E \cup \{\Delta\}$ is endowed with the σ-algebra generated by $\mathcal{E}$ and Δ.

Chapter 2
Markov Renewal Processes and Related Processes

This chapter is an introduction to Markov renewal theory. It contains the fundamental theorem on the change of the observation time and the definition of a completed semi-Markov process (CSMP), on which all future developments will be based.

In Sect. 2.1 we clarify some definitions and recall very briefly the definition of a Markov chain with a general state space. Main definitions and basic results are given in Sects. 2.2 and 2.3. From Sect. 2.4, the basic assumption $\lim_n T_n = +\infty$ is assumed to hold throughout the book. The fundamental theorem on the change of the observation point is given in this section. The basic process for this book, the CSMP (completed semi-Markov process), is introduced in Sect. 2.5, as well as Markov renewal equations. In Sect. 2.6, other continuous-time processes related to a Markov renewal process are defined. In Sect. 2.7, we show how Markov renewal equations appear for semiregenerative processes and consequently what can be said about the long-time behavior of such processes. These last two sections can be skipped on a first reading.

Sections 2.3 to 2.5 are really essential. A Markov renewal process $(Y_n, T_n)_n$ on a space E is a marked point process with marks $Y_n \in E$ and jump times $T_n \in \mathbb{R}_+$. It is defined through its initial distribution (the probability distribution of the first jump time and its location) and its Markov renewal kernel N. A semi-Markov process is a continuous-time representation of a Markov renewal process; to get a Markov process, an additional random variable must be added. In the PDMP context, a good choice is the elapsed time A_t since the last jump, but it is not defined if t is less than the first jump. It wouldn't be a problem if it were enough to consider zero-delayed Markov renewal processes, that is, with a jump at time $t = 0$, but that would be too restrictive, in particular due to stationarity concerns. The theorem on the change of the observation time suggests that one consider also Markov renewal processes with a jump time before time 0, namely Markov renewal processes with an initial distribution as $N_0^{x,u}$. Assuming that the initial distribution is $N_0^{x,u}$ means that the last jump before time 0 is at time $-u$ and its mark is x; thus in such a situation, A_t may

C. Cocozza-Thivent, *Markov Renewal and Piecewise Deterministic Processes*, Probability Theory and Stochastic Modelling 100,
https://doi.org/10.1007/978-3-030-70447-6_2

be defined for all $t \geq 0$. This is why the initial distribution is often assumed to be $N_0^{x,u}$ for some $(x, u) \in E \times \mathbb{R}_+$ or a mixture of them. This restriction turns out not to be disruptive for PDMPs.

2.1 Kernels and General Markov Chains

Throughout this section, $(G, \mathcal{G})$, $(G_1, \mathcal{G}_1)$ and $(G_2, \mathcal{G}_2)$ are measurable spaces.

Definition 2.1 A kernel (respectively a probability kernel) from G_1 to G_2 is a map $M : G_1 \times \mathcal{G}_2 \to \mathbb{R}_+ \cup \{+\infty\}$ such that

- for all $x_1 \in G_1$, the map $A \in \mathcal{G}_2 \to M(x_1, A)$ is a σ-finite measure (respectively a probability distribution);
- for all $A \in \mathcal{G}_2$, the map $x_1 \to M(x_1, A)$ is measurable.

A kernel (respectively a probability kernel) on G is a kernel (respectively a probability kernel) from G to G.

If M is a kernel from G_1 to G_2 and $f : G_2 \to \mathbb{R}_+$ a measurable function, then $x \to \int_{G_2} f(y)\, M(x, dy)$ is measurable.

Notation Given two kernels M_1 and M_2 on G, the kernel $M_1 M_2$ on G is defined as

$$\int_G f(y)\, M_1 M_2(x, dy) = \int_G \left(\int_G f(z)\, M_2(y, dz) \right) M_1(x, dy) \qquad (2.1)$$

for every measurable function $f : G \to \mathbb{R}_+$.
We write $M_1 M_2(x, dz) = \int_{\{y \in G\}} M_1(x, dy)\, M_2(y, dz)$ and (2.1) more intuitively as $\int_G f(y)\, M_1 M_2(x, dy) = \int_{G^2} f(z)\, M_1(x, dy)\, M_2(y, dz)$.

Given a kernel M on G and a nonnegative measure μ on $(G, \mathcal{G})$, the measure μM on $(G, \mathcal{G})$ is defined as

$$\int_G f(x)\, \mu M(dx) = \int_G \left(\int_G f(y)\, M(x, dy) \right) \mu(dx) \qquad (2.2)$$

for every measurable function $f : G \to \mathbb{R}_+$. We write $\mu M(dy) = \int_{\{x \in G\}} \mu(dx)$ $M(x, dy)$ and (2.2) more intuitively as $\int_G f(x)\, \mu M(dx) = \int_{G^2} f(y)\, \mu(dx)$ $M(x, dy)$.

Definition 2.2 A sequence $Z = (Z_n)_{n \geq 0}$ of random variables (r.v.) with values in G is said to be a (time-homogeneous) Markov chain with state space G if for all $n \geq 0$, the conditional probability distribution of Z_{n+1} given $(Z_0, Z_1, \ldots, Z_n)$ depends only on Z_n.

The probability kernel M on G is said to be the probability kernel (or transition function) of the Markov chain Z if for all $x \in E$, $M(x, \cdot\,)$ is a conditional probability distribution of Z_1 given $Z_0 = x$, i.e.,

$$\mathbb{E}(f(Z_1) \mid Z_0) = \int_G f(z)\, M(Z_0, dz)$$

for every measurable function $f : G \to \mathbb{R}_+$.

The probability distribution of Z_0 is called the initial distribution of the Markov chain.

Proposition 2.1 *The following assertions are equivalent:*

1. *the sequence $(Z_n)_{n \geq 0}$ is a Markov chain with state space G, probability kernel M, and initial distribution μ;*
2. *for every $n \geq 0$ and measurable function $f : G^{n+1} \to \mathbb{R}_+$,*

$$\mathbb{E}(f(Z_0, Z_1, \ldots Z_n)) =$$
$$\int_{G^{n+1}} f(z_0, z_1, z_2, \ldots, z_n)\, \mu(dz_0)\, M(z_0, dz_1) M(z_1, dz_2) \ldots M(z_{n-1}, dz_n). \quad (2.3)$$

Proposition 2.2 *The sequence $(Z_n)_{n \geq 0}$ is a Markov chain iff for all $n \geq 0$, $p \geq 1$, and measurable function $f : G^{n+p+1} \to \mathbb{R}_+$, we have*

$$\mathbb{E}(f(Z_0, \ldots, Z_n, Z_{n+1}, \ldots, Z_{n+p}) \mid Z_0, \ldots, Z_n) = g(Z_0, \ldots, Z_n)$$

with $g(z_0, \ldots, z_n) = \mathbb{E}(f(z_0, \ldots, z_n, Z_1, \ldots, Z_p) \mid Z_0 = z_n)$.

Proposition 2.3 *Given a Markov chain $(Z_n)_{n \geq 0}$ with state space G and probability kernel M, the relations*

$$\mathbb{E}(f(Z_n) \mid Z_{n-1}) = \int_G f(z)\, M(Z_{n-1}, dz),$$

$$\mathbb{E}(f(Z_n) \mid Z_0 = z_0) = \int_G f(z)\, M^n(z_0, dz) = \mathbb{E}(f(Z_{p+n}) \mid Z_p = z_0)$$

hold for every $n \geq 1$, $p \geq 0$, and measurable function $f : G \to \mathbb{R}_+$.

2.2 Renewal Kernels and Convolution

Definition 2.3 A renewal kernel N on E is a kernel from E to $E \times \mathbb{R}_+$ such that $N(x, E \times \{0\}) = 0$ for all $x \in E$.

A semi-Markov kernel N on E is a renewal kernel such that $N(x, E \times \mathbb{R}_+) \leq 1$ for all $x \in E$. It is proper if $N(x, E \times \mathbb{R}_+) = 1$ for all $x \in E$.

The proper renewal kernel I is defined as

$$I(x, dy, du) = \delta_{x,0}(dy, du).$$

$I(x, E\times\{0\}) = \delta_{x,0}(E\times\{0\}) = 1!$
So Not satisfy Def 2.3 ?!

Recall that the convolution of two nonnegative measures μ_1 and μ_2 on $\mathbb{R}_+$ is defined as

$$\int_{\mathbb{R}_+} g\, \mu_1 * \mu_2 = \int_{\mathbb{R}_+^2} g(u+v)\, \mu_1(du)\, \mu_2(dv)$$

for every Borel function $g : \mathbb{R}_+ \to \mathbb{R}_+$.

Given two renewal kernels N_1 and N_2 on E, $N_1 * N_2$ is defined as

$$N_1 * N_2(x, dy, \cdot) = \int_E N_1(x, dy_1, \cdot) * N_2(y_1, dy, \cdot),$$

that is,

$$\int_{E\times\mathbb{R}_+} \varphi(y, v) N_1 * N_2(x, dy, dv) = \int_{(E\times\mathbb{R}_+)^2} \varphi(y_2, v_1 + v_2) N_1(x, dy_1, dv_1) N_2(y_1, dy_2, dv_2)$$

for any Borel function $\varphi : E \times \mathbb{R}_+ \to \mathbb{R}_+$.

We get

$$I * N = N * I = N.$$

Define

$$N^{*0} = I, \quad N^{*(n+1)} = N^{*n} * N \ (n \geq 0).$$

Given a nonnegative measure μ on $E \times \mathbb{R}_+$, a renewal kernel N on E, and a Borel function $\varphi : E \times \mathbb{R}_+ \to \mathbb{R}_+$, define

$$\mu * \varphi : \ \mu * \varphi(t) = \int_{E\times\mathbb{R}_+} 1_{[0,t]}(u)\, \varphi(x, t-u)\, \mu(dx, du),$$

$$N * \varphi : \ N * \varphi(x, t) = \int_{E\times\mathbb{R}_+} 1_{[0,t]}(v)\, \varphi(y, t-v)\, N(x, dy, dv),$$

$$\mu * N : \int_{E\times\mathbb{R}_+} \varphi(x, u)\, \mu * N(dx, du) = \int_{(E\times\mathbb{R}_+)^2} \varphi(x_2, v_1 + v_2)\, \mu(dx_1, dv_1)\, N(x_1, dx_2, dv_2).$$

We get

$$\mu * I = \mu,$$

$$(N_1 * N_2) * \varphi = N_1 * (N_2 * \varphi), \quad \mu * (N * \varphi) = (\mu * N) * \varphi.$$

From now on, $(E, \mathcal{E})$ **is a measurable Polish space**, that is, E is a separable completely metrizable topological space and $\mathcal{E}$ is its Borel σ-algebra. Given $\Delta \notin E$ one endows $E \cup \Delta$ with the σ-algebra generated by $\mathcal{E}$ and Δ.

2.3 Markov Renewal Processes

Definition 2.4 Let N be a semi-Markov kernel on E. The process $(Y, T) = (Y_n, T_n)_{n\geq 1}$ is said to be a (time homogeneous) Markov renewal process on E with kernel N if it is a marked point process (see Sect. A.4) with values in E such that for all $n \geq 1$, the conditional probability distribution of $(Y_{n+1}, T_{n+1} - T_n)$ given $(Y_1, T_1, \dots, Y_n, T_n)$ is the conditional probability distribution of $(Y_{n+1}, T_{n+1} - T_n)$ given Y_n and is provided by the kernel N, that is,

$$\begin{aligned}&\mathbb{E}(\varphi(Y_{n+1}, T_{n+1} - T_n)\, 1_{\{T_{n+1}<+\infty\}} \mid Y_1, T_1, \dots, Y_n, T_n) \\ &= 1_E(Y_n) \int_{E\times\mathbb{R}_+} \varphi(y, v)\, N(Y_n, dy, dv) \end{aligned} \tag{2.4}$$

for every Borel function $\varphi : E \times \mathbb{R}_+ \to \mathbb{R}_+$.

The restriction to $E \times \mathbb{R}_+$ of the probability distribution of (Y_1, T_1) is called the initial distribution of the Markov renewal process (Y, T).

The measures $\int_{\{y\in E\}} N(x, dy, dv) = \mathbb{P}(T_{n+1} - T_n \in (v, v + dv) \mid Y_n = x)$ are the interarrival distributions of the Markov renewal process.

Therefore, the initial distribution of a Markov renewal process (Y, T) on E is a nonnegative measure μ_0 on $E \times \mathbb{R}_+$ such that $\mu_0(E \times \mathbb{R}_+) \leq 1$ and the probability distribution of (Y_1, T_1) is $\mu_0 + (1 - \mu_0(E \times \mathbb{R}_+))\, \delta_{\Delta,+\infty}$.

Usually for Markov renewal processes the index is running in $\mathbb{N}$ as for renewal processes and Markov chains. Here it is running in $\mathbb{N}^*$ in accordance with the notation for Markov jump processes and piecewise deterministic Markov processes.

Example 2.1 Markov renewal processes are generalizations of renewal processes. Indeed, let N be the kernel of the Markov renewal process $(Y_n, T_n)_{n\geq 1}$ on E: if E is a single point, then N is the restriction to $\mathbb{R}_+$ of a probability distribution dF on $\bar{\mathbb{R}}_+$ and $(T_n)_{n\geq 1}$ is a renewal process with interarrival distribution dF. Its delay distribution is the initial distribution of the Markov renewal process.

Example 2.2 Define $E = \{e_0, e_1\}$, $P(e_0, e_1) = P(e_1, e_0) = 1$, $P(e_i, e_i) = 0$ $(i \in \{0, 1\})$, $N(e_i, e_j, dv) = P(e_i, e_j)\, 1_{\mathbb{R}_+}(v)\, dF_i(dv)$, where dF_i is a probability distribution on $\bar{\mathbb{R}}_+$ and $1_{\mathbb{R}_+}(v)\, dF_i(v)$ is its restriction to $\mathbb{R}_+$, $\mu_0(dx, dv) = \delta_{e_1}(dx)\, \delta_0(dv)$. Let $(Y_n, T_n)_{n\geq 1}$ be a Markov renewal process with kernel N and initial distribution μ_0. Then $(T_n)_{n\geq 1}$ is an alternating renewal process, that is, $T_n = \sum_{k=1}^{n} \xi_k$, $(\xi_n)_{n\geq 1}$ are independent r.v., $(\xi_{2n+1})_{n\geq 0}$ are identically distributed with common probability distribution dF_1, $(\xi_{2n})_{n\geq 1}$ are identically distributed with common probability distribution dF_0. An alternating renewal process is the easiest way to describe the successive changes of states of an item subjected to failures.

The following results of this section are easily proved.

Proposition 2.4 *Let* $(Y, T) = (Y_n, T_n)_{n\geq 1}$ *be a sequence of r.v. with values in* $(E \times \mathbb{R}_+) \cup \{(\Delta, +\infty)\}$ *such that* $(T_n)_{n\geq 1}$ *is a nondecreasing sequence and* $\mathbb{P}(T_n = \Delta, T_n = +\infty) = \mathbb{P}(T_n = +\infty) = \mathbb{P}(Y_n = \Delta)$.

Given a semi-Markov kernel N on E and a nonnegative measure μ_0 on $E \times \mathbb{R}_+$ such that $\mu_0(E \times \mathbb{R}_+) \leq 1$, the following assertions are equivalent:

1. *the process (Y, T) is a Markov renewal process with kernel N and initial distribution μ_0,*
2. *define $X_0 = (Y_1, T_1)$ and for $n \geq 1$,*

$$X_n = \begin{cases} (Y_{n+1}, T_{n+1} - T_n), & T_n < +\infty, \\ (\Delta, +\infty), & T_n = +\infty; \end{cases}$$

the sequence $(X_n)_{n\geq 0}$ is a Markov chain with initial distribution $\mu_0 + (1 - \mu_0(E \times \mathbb{R}_+))\, \delta_{\Delta,+\infty}$ and probability kernel

$$K_1((x,u), dy, dv) = 1_E(x)\left(N(x, dy, dv) + (1 - N(x, E \times \mathbb{R}_+))\delta_{\Delta,+\infty}(dy, dv)\right) + 1_\Delta(x)\, \delta_{\Delta,+\infty}(dy, dv),$$

3. *for every $n \geq 1$ and Borel function $f : (E \times \mathbb{R}_+)^n \to \mathbb{R}_+$,*

$$\mathbb{E}\left(f(Y_1, T_1, Y_2, T_2 - T_1, \ldots, Y_n, T_n - T_{n-1})\, 1_{\{T_n < +\infty\}}\right) = \int_{(E\times\mathbb{R}_+)^n} f(y_1, v_1, \ldots, y_n, v_n)\, \mu_0(dy_1, dv_1)$$
$$N(y_1, dy_2, dv_2) \ldots N(y_{n-1}, dy_n, dv_n),$$

4. *for every $n \geq 1$ and Borel function $f : (E \times \mathbb{R}_+)^n \to \mathbb{R}_+$,*

$$\mathbb{E}\left(f(Y_1, T_1, Y_2, T_2, \ldots, Y_n, T_n)\, 1_{\{T_n < +\infty\}}\right) = \int_{(E\times\mathbb{R}_+)^n} f(y_1, v_1, y_2, v_1 + v_2, \cdots, y_n, v_1 + v_2 + \cdots + v_n)\, \mu_0(dy_1, dv_1)$$
$$N(y_1, dy_2, dv_2) \cdots N(y_{n-1}, dy_n, dv_n),$$

5. *the process (Y, T) is a Markov chain taking values in $(E \times \mathbb{R}_+) \cup \{(\Delta, +\infty)\}$, with initial distribution $\mu_0 + (1 - \mu_0(E \times \mathbb{R}_+))\, \delta_{\Delta,+\infty}$ and probability kernel*

$$\int \varphi(y, v)\, K_2((x, u), dy\, dv) = 1_E(x) \int_{E\times\mathbb{R}_+} \varphi(y, u + v)\, N(x, dy, dv)$$
$$+ \ (1_E(x)\,(1 - N(x, E \times \mathbb{R}_+)) + 1_\Delta(x))\ \varphi(\Delta, +\infty)$$

for every $(x, u) \in (E \times \mathbb{R}_+) \cup \{(\Delta, +\infty)\}$ and Borel function $\varphi : (E \times \mathbb{R}_+) \cup \{(\Delta, +\infty)\} \to \mathbb{R}_+$.

Notation Assuming that the initial distribution of the Markov renewal process (Y, T) is $N(x, \cdot, \cdot)$ $(x \in E)$, we define $Y_0 = x$, $T_0 = 0$. The process $(Y_n, T_n)_{n\geq 0}$ is then the generalization of a zero-delayed renewal process, and some books consider only this case. It plays an essential role in the theory developed in this book, but we should

not limit ourselves to studying it, as we shall see later.
Define $\mathbb{P}_x$ as the probability measure (on Ω endowed with the σ-field generated by (Y, T)) for which $N(x, \cdot, \cdot)$ is the initial distribution of the Markov renewal process; $\mathbb{E}_x$ is the related expectation. For instance, we get

$$\mathbb{E}_x(\varphi(Y_1, T_1)\, 1_{\{T_1<+\infty\}}) = \int_{E\times\mathbb{R}_+} \varphi(y, v)\, N(x, dy, dv),$$

$$\mathbb{P}_x((Y_1, T_1) = (\Delta, +\infty)) = 1 - N(x, E \times \mathbb{R}_+).$$

Define the function $\bar{F}_x$ as $\bar{F}_x(t) = \mathbb{P}_x(T_1 > t)$.

Corollary 2.1 *Let (Y, T) be a Markov renewal process with initial distribution μ_0 and kernel N. Then*

$$\mathbb{E}(\varphi(Y_n, T_n)\, 1_{\{T_n<+\infty\}}) = \int_{E\times\mathbb{R}_+} \varphi(y, v)\, \mu_0 * N^{*(n-1)}(dy, dv)$$

for every Borel function $\varphi : E \times \mathbb{R}_+ \to \mathbb{R}_+$. In particular,

$$\mathbb{E}_x(\varphi(Y_n, T_n)\, 1_{\{T_n<+\infty\}}) = \int_{E\times\mathbb{R}_+} \varphi(y, v)\, N^{*n}(x, dy, dv).$$

Proposition 2.5 *Given a semi-Markov kernel N on E and a nonnegative measure μ_0 on $E \times \mathbb{R}_+$ such that $\mu_0(E \times \mathbb{R}_+) \le 1$, write their Radon–Nikodym decompositions as*

substochastic

$$N(x, dy, dv) = \pi(x, dy)\, \alpha(x, y; dv), \quad \mu_0(dx, du) = \mu_1(dx)\, \alpha_1(x, du),$$

where π is a kernel on E, α is a probability kernel from E^2 to $\mathbb{R}_+$, μ_1 is a nonnegative measure on E, α_1 is a probability kernel from E to $\mathbb{R}_+$.

Let $(Y, T) = (Y_n, T_n)_{n\ge1}$ be a sequence of r.v. taking values in $(E \times \mathbb{R}_+) \cup \{(\Delta, +\infty)\}$ such that the sequence $(T_n)_{n\ge1}$ is nondecreasing.

The process (Y, T) is a Markov renewal process on E with initial distribution μ_0 and kernel N iff:

1. *the process $Y = (Y_n)_{n\ge1}$ is a Markov chain taking values in $E \cup \{\Delta\}$ with initial distribution μ_1 and probability kernel*

$$P(x, dy) = \begin{cases} \pi(x, dy) + (1 - N(x, E \times \mathbb{R}_+))\, \delta_\Delta(dy), & x \in E, \\ \delta_\Delta(dy), & x = \Delta, \end{cases} \tag{2.5}$$

2. *given the Markov chain Y,*
   - *on $\{Y_1 \in E\}$, the conditional probability distribution of T_1 is $\alpha_1(Y_1, .)$,*

- *for all $n \geq 2$, on $\{Y_n \in E\}$ the conditional probability distribution of $T_n - T_{n-1}$ is $\alpha(Y_{n-1}, Y_n;\ .)$,*
- *for all $n \geq 2$, on $\{Y_n \in E\}$ the r.v. $T_1, T_2 - T_1, \ldots, T_n - T_{n-1}$ are conditionally independent,*

3. *for all $n \geq 1$, $\{Y_n = \Delta\} = \{T_n = +\infty\}$ a.s.*

The Markov chain Y is called the driving chain of the Markov renewal process. Define

$$N_t = \sum_{n \geq 1} 1_{\{T_n \leq t\}}. \tag{2.6}$$

We get

$$T_{N_t} \leq t < T_{N_t+1}, \quad \{N_t = n\} = \{T_n \leq t < T_{n+1}\}, \quad \{T_n \leq t\} = \{N_t \geq n\}.$$

The following lemma, which is a consequence of Proposition 2.4, will be very useful in many computational proofs.

Lemma 2.1 *For all $n \in \mathbb{N}^*$ and every Borel function $f : (E \times \mathbb{R}_+)^n \to \mathbb{R}_+$,*

$$\mathbb{E}\left(1_{\{N_t=n\}} f(Y_1, T_1, \ldots, Y_n, T_n)\right) = \mathbb{E}\left(1_{\{T_n \leq t\}} f(Y_1, T_1, \ldots, Y_n, T_n)\, \bar{F}_{Y_n}(t - T_n)\right).$$

Notation For $(x, u) \in E \times \mathbb{R}_+$ define $N_0^{x,u}$ as follows:

$$\int_{E \times \mathbb{R}_+} \varphi(y, v)\, N_0^{x,u}(dy, dv) = \frac{1}{\bar{F}_x(u)} \int_{E \times \mathbb{R}_+} \varphi(y, v - u)\, 1_{\{v>u\}}\, N(x, dy, dv) \tag{2.7}$$

for all $t > 0$ and all (x, u) such that $\bar{F}_x(u) \neq 0$ and every Borel function $\varphi : E \times \mathbb{R}_+ \to \mathbb{R}_+$.
The measure $N_0^{x,u}$ will appear in the fundamental theorem on changing the observation time (Theorem 2.1, Sect. 2.4) and is of great use in the further developments.

The probabilistic meaning of $N_0^{x,u}$ is

$$\int_{E \times \mathbb{R}_+} \varphi(y, v)\, N_0^{x,u}(dy, dv) = \mathbb{E}_x(\varphi(Y_1, T_1 - u)\, 1_{\{T_1 < +\infty\}} \mid T_1 > u) \tag{2.8}$$

for every Borel function $\varphi : E \times \mathbb{R}_+ \to \mathbb{R}_+$. Thus $N_0^{x,u}$ is the conditional probability distribution under $\mathbb{P}_x$ of $(Y_1, T_1 - u)$ given $T_1 > u$, restricted to $E \times \mathbb{R}_+$.

The construction of a Markov renewal process with kernel N and initial distribution $N_0^{x,u}$ can be achieved by taking (Y_1, T_1) as follows:

1. generate (Y_1, D_1) according to the probability distribution $N(x, dy, dv) + (1 - N(x, E \times \mathbb{R}_+))\, \delta_{\Delta,+\infty}(dy, dv)$,
2. if $D_1 \leq u$, return to 1; otherwise, take $T_1 = D_1 - u$, $Y_1 = X_1$.

Then generate (Y_{n+1}, D_{n+1}) according to the probability distribution $N(Y_n, dy, dv) + (1 - N(Y_n, E \times \mathbb{R}_+)) \delta_{\Delta, +\infty}(dy, dv)$ and take $T_{n+1} = T_n + D_{n+1}$. All drawings are independent.

Lemma 2.2 *Assume that $N_0^{x,u}$ is the initial distribution of the Markov renewal process (Y, T). Then $\mathbb{P}(T_1 > t) = \bar{F}_x(u + t)/\bar{F}_x(u)$, and the conditional probability distribution of $(Y_1, T_1 - t)$ given $T_1 > t$ restricted to $E \times \mathbb{R}_+$ is $N_0^{x,u+t}$.* $\left(\text{assume } \bar{F}_x(u) > 0 \right)$

The assumption $N(x, E \times \{0\}) = 0$ (see Definition 2.3) implies $N_0^{x,0}(dy, dv) = N(x, dy, dv)$.

From now on, we make the assumption

$$\lim_{n \to +\infty} T_n = +\infty \ a.s.$$

throughout this book except in Proposition 2.6 below and in Theorem 8.3, Sect. 8.2 which give sufficient conditions to satisfy it.

The following proposition provides a sufficient condition for the stronger condition $\mathbb{E} \left(\sum_{n \geq 1} 1_{\{T_n \leq t\}} \right) < +\infty$ to hold.

Proposition 2.6 *Let $(Y, T) = (Y_n, T_n)_{n \geq 1}$ be a Markov renewal process with kernel N and initial distribution $N(x_0, \cdot, \cdot)$ for some $x_0 \in E$ or a mixture of such measures. Assume that there exists an r.v. V taking values in $\bar{\mathbb{R}}_+$ such that $\mathbb{P}(V > 0) \neq 0$ and*

$$\forall (x, y) \in E, \ \forall t > 0, \quad \mathbb{P}_x(T_1 > t \mid Y_1 = y) \geq \mathbb{P}(V > t). \tag{2.9}$$

Then $\lim_{n \to +\infty} T_n = +\infty$ a.s. and

$$\mathbb{P}(T_n \leq t) \leq \mathbb{P}(\sum_{k=1}^{n} V_k \leq t), \tag{2.10}$$

where $(V_k)_{k \geq 1}$ are i.i.d. with the same probability distribution as V.

Moreover, for all Borel bounded functions $h_1 : E \to \mathbb{R}_+$ and $h_2 : \mathbb{R}_+ \to \mathbb{R}$, h_2 nonincreasing, we have

$$\mathbb{E} \left(h_1(Y_n) h_2(T_n) 1_{\{T_n < +\infty\}} \right) \leq \mathbb{E} \left(h_1(Y_n) 1_E(Y_n) \right) \mathbb{E} \left(h_2(\sum_{k=1}^{n} V_k) \right). \tag{2.11}$$

Hence

$$\mathbb{E} \left(\sum_{n \geq 1} 1_{\{T_n \leq t\}} \right) \leq e^t \sum_{n \geq 1} (\mathbb{E}(e^{-V}))^n < +\infty.$$

If the initial distribution of the Markov renewal process is not $N(x_0, \cdot, \cdot)$ or a mixture of such measures, then (2.10) and (2.11) hold with $\sum_{k=1}^{n} V_k$ replaced by $\sum_{k=2}^{n} V_k$.

Proof Assume that $N(x_0, ., .)$ is the initial distribution of the Markov renewal process (Y, T). The result for a mixture of such measures follows immediately.

We take the same notation as in Proposition 2.5, Sect. 2.3. Let F_V (resp. $F_{x,y}$) be the cumulative distribution function of V (resp. of the probability distribution $\alpha(x, y; du)$), F_V^{-1} (resp. $F_{x,y}^{-1}$) its left-continuous pseudo-inverse function (see its definition Sect. 3.3). Let $(W_n)_{n\geq 1}$ be a sequence of i.i.d. random variables uniformly distributed on $[0, 1]$, and thus $V_n = F_V^{-1}(W_n)$ (resp. $F_{x,y}^{-1}(W_n)$) are i.i.d. random variables with the same probability distribution as V (resp. with probability distribution $\alpha(x, y; du)$). Finally, let $\tilde{Y} = (\tilde{Y}_n)_{n\geq 0}$ be a Markov chain with probability kernel $P(x, dy) = \pi(x, dy) + (1 - N(x, E \times \mathbb{R}_+))\,\delta_\Delta(dy)$ and initial distribution δ_{x_0}, and which is independent of $(W_n)_{n\geq 1}$. For $n \geq 1$ define $\xi_n = F^{-1}_{\tilde{Y}_{n-1},\tilde{Y}_n}(W_n)$ and $\tilde{T}_n = \xi_1 + \cdots + \xi_n$ if $\tilde{Y}_n \in E$, $\tilde{T}_n = +\infty$ if $\tilde{Y}_n = \Delta$. We deduce from Proposition 2.5, Sect. 2.3 that $(\tilde{Y}_n, \tilde{T}_n)_{n\geq 1}$ is a Markov renewal process with the same probability distribution as (Y, T).

Condition (2.9) is $F_{x,y} \leq F_V$. Therefore, $F_{x,y}^{-1} \geq F_V^{-1}$, and thus for all $k \geq 1$, $\xi_k \geq V_k$. Hence $\tilde{T}_n \geq \sum_{k=1}^n V_k$ holds for all $n \geq 1$. Hence $\mathbb{P}(T_n \leq t) = \mathbb{P}(\tilde{T}_n \leq t) \leq \mathbb{P}(\sum_{k=1}^n V_k \leq t)$. On the other hand, the law of large numbers gives $\mathbb{P}(\lim_n T_n = +\infty) = \mathbb{P}(\lim_n \tilde{T}_n = +\infty) \geq \mathbb{P}(\sum_{k\geq 1} V_k = +\infty) = 1$. Moreover,

$$\mathbb{E}(h_1(Y_n)h_2(T_n)1_{\{T_n<+\infty\}}) = \mathbb{E}(h_1(\tilde{Y}_n)1_E(\tilde{Y}_n)h_2(\tilde{T}_n))$$
$$\leq \mathbb{E}(h_1(\tilde{Y}_n)1_E(\tilde{Y}_n))\,\mathbb{E}(h_2(\sum_{k=1}^n V_k)) = \mathbb{E}(h_1(Y_n)1_E(Y_n))\,\mathbb{E}(h_2(\sum_{k=1}^n V_k)).$$

The last inequality follows from the Markov inequality and the above inequality applied with $h_1 \equiv 1$, $h_2(u) = e^{-u}$.

If the initial distribution of the Markov renewal process is general, we construct $\tilde{T}_n - \tilde{T}_1$ as above by means of $(\xi_k)_{k\geq 2}$ and apply the above reasoning to $T_n - T_1$ instead of T_n. ∎

2.4 Changing the Observation Time

The purpose of this section is to prove Theorem 2.1 below, which gives the properties of the Markov renewal process observed from time t, and to draw immediate consequences such as new definitions.

Following Brémaud [20], the σ-algebra generated by the marked point process $(Y, T) = (Y_n, T_n)_{n\geq 1}$ up to time t is the σ-algebra $\mathcal{F}_t^{(0)}$ generated by the r.v.

$$\sum_{n\geq 1} 1_{\{T_n\leq s\}}1_A(Y_n),\ 0 \leq s \leq t,\ A \in \mathcal{E}.$$

Lemma 2.3 *The σ-algebra $\mathcal{F}_t^{(0)}$ is generated by the r.v. $1_{\{N_t=0\}}$ and*

$$1_{\{N_t=n\}} f_n(Y_1, T_1, \ldots, Y_n, T_n),$$

where $n \in \mathbb{N}^*$ *and* f_n *ranges over the set of nonnegative Borel functions defined on* $(E \times \mathbb{R}_+)^n$.

Proof Define $\mathcal{H}_t$ as the σ-algebra generated by the r.v. $1_{\{N_t=0\}}$ and $1_{\{N_t=n\}} f_n(Y_1, T_1, \ldots, Y_n, T_n)$ $(n \geq 1)$. Then clearly, the r.v. $N_s, s \leq t$ are $\mathcal{F}_t^{(0)}$-measurable. On the other hand, the r.v. Y_k and T_k, $k \leq n$, are $\mathcal{F}_{T_n}^{(0)}$-measurable (see Brémaud [20], Theorem T23 p. 303). Therefore, $f_n(Y_1, T_1, \ldots, Y_n, T_n)\, 1_{\{T_n \leq t\}}$ is $\mathcal{F}_t^{(0)}$-measurable. Consequently, $\mathcal{H}_t \subset \mathcal{F}_t^{(0)}$.

Conversely, given $s \leq t$, $1_{\{T_n \leq s\}} 1_A(Y_n) = \sum_{k \geq n} 1_{\{T_n \leq s\}} 1_A(Y_n) 1_{\{N_t=k\}}$ is $\mathcal{H}_t$-measurable and therefore $\mathcal{F}_t^{(0)} \subset \mathcal{H}_t$. ■

Define

$$\bar{Z}_t = Y_{n+1}, \quad W_t = T_{n+1} - t \quad \text{on } \{N_t = n\}\ (n \geq 0),$$

$$Z_t = Y_n, \quad A_t = t - T_n \quad \text{on } \{N_t = n\}\ (n \geq 1);$$

thus Z_t and A_t are defined only for $t \in [T_1, +\infty[$.

We are now ready to state the main theorem of this chapter.

Theorem 2.1 *Let* $t > 0$ *and* (Y, T) *a Markov renewal process with kernel* N *and initial distribution* $N_0^{x,u}$. *Define* $(Y^t, T^t) = (Y_n^t, T_n^t)_{n \geq 1}$, *the post-t-process that is*

$$\forall n \geq 1, \quad Y_n^t = Y_{N_t+n}, \quad T_n^t = T_{N_t+n} - t,$$

in particular, $Y_1^t = \bar{Z}_t$ *and* $T_1^t = W_t$.

Then given $\mathcal{F}_t^{(0)}$, *the process* (Y^t, T^t) *is a Markov renewal process with kernel* N, *its initial distribution is* $N_0^{Z_t, A_t}$ *on* $\{N_t \geq 1\}$, *and* $N_0^{x,u+t}$ *on* $\{N_t = 0\}$.

Proof According to Proposition 2.4, Sect. 2.3 and Lemma 2.3, we have to check that

$$\begin{aligned} &\mathbb{E}(X\, f(Y_1^t, T_1^t, Y_2^t, T_2^t - T_1^t, \ldots, Y_p^t, T_p^t - T_{p-1}^t)\, 1_{\{T_p^t < +\infty\}}) \\ &= \mathbb{E}\Big(X \int f(y_1, v_1, \ldots, y_p, v_p) N_0^{Z_t, A_t}(dy_1, dv_1) N(y_1, dy_2, dv_2) \ldots N(y_{p-1}, dy_p, dv_p)\Big) \end{aligned} \tag{2.12}$$

for every Borel function $f : (E \times \mathbb{R}_+)^p \to \mathbb{R}_+$ and $X = 1_{\{N_t=n\}}\, g(Y_1, T_1, \ldots Y_n, T_n)$ $(n \geq 1)$, and the same for $X = 1_{\{N_t=0\}}$ and $N_0^{x,u+t}$ instead of $N_0^{Z_t, A_t}$. We get this by applying Proposition 2.4, Sect. 2.3 to the Markov renewal process (Y, T). ■

In the same way, the following result with no assumption on the initial distribution can be proved.

Proposition 2.7 *The process* $(Y_n^t, T_n^t)_{n\geq 1}$ *is a Markov renewal process with kernel* N *and initial distribution the probability distribution of* $(\bar{Z}_t, W_t)$.

Notation Theorem 2.1 suggests the following definitions used throughout this book: when the initial distribution of the Markov renewal process is $N_0^{x,u}$, we take

$$Y_0 = x, \ T_0 = -u, \quad Z_t = x, \ A_t = u + t \text{ on } \{t < T_1\}; \tag{2.13}$$

hence $Y_0 = Z_0 = x$, $A_0 = u = -T_0$. In particular, under $\mathbb{P}_x$ we have $Z_t = x$, $A_t = t$ on $\{t < T_1\}$.

From now on, writing $\mathbb{E}(\cdot \mid Z_0 = x, A_0 = u)$ means that the initial distribution of the process is $N_0^{x,u}$. Consequently, it is assumed that $\bar{F}_x(u) \neq 0$.

The process is said to be zero-delayed if its initial distribution is $N(x, dy, dv)$ or a mixture of such measures. In this case we have $T_0 = 0$, $A_0 = 0$.

From now on, when writing A_t it is assumed that the initial distribution is $N_0^{x,u}$ for some $(x, u) \in E \times \mathbb{R}_+$ such that $\bar{F}_x(u) \neq 0$, or a mixture of such measures and T_0 is defined as $T_0 = -A_0$.

It is equivalent to consider the Markov renewal process $(Y_n, T_n)_{n\geq 0}$ or the process $(Z_t, A_t)_{t\geq 0}$, since

$$Z_t = \begin{cases} Y_0, \ t < T_1, \\ Y_n, \ T_n \leq t < T_{n+1}, \end{cases} \qquad A_t = t - T_n, \ T_n \leq t < T_{n+1}, \tag{2.14}$$

and conversely, $Y_0 = Z_0$, $T_0 = -A_0$,

$$T_1 = \inf\{t > 0 : A_t = 0\}, \ T_{n+1} = \inf\{t > T_n : A_t = 0\}, \ Y_n = Z_{T_n} \ (n \geq 1) \tag{2.15}$$

(understanding $\inf\{\emptyset\} = +\infty$).

Define $\mathcal{F}_t$ as the smallest σ-algebra containing $\mathcal{F}_t^{(0)}$ and the subsets $\{Y_0 \in A, T_0 \leq a\}$, $A \in \mathcal{E}$, $a \leq 0$. From (2.14), (2.15), and Lemma 2.3, Sect. 2.4 it is easily seen that $\mathcal{F}_t$ is the σ-algebra generated by $((Z_s, A_s), s \leq t)$.

The corollary below follows immediately from Theorem 2.1.

Corollary 2.2 *Assume that the initial distribution of the Markov renewal process is* $N_0^{x,u}$ *or a mixture of such measures. Then given* $\mathcal{F}_t$, *the process* (Y^t, T^t) *is a Markov renewal process with kernel* N; *its initial distribution is* $N_0^{Z_t, A_t}$.

2.5 Semi-Markov Processes and CSMPs

Recall the following notation: (Y, T) is a Markov renewal process; we define Z_t and A_t as $Z_t = Y_n$ and $A_t = t - T_n$ when $T_n \leq t < T_{n+1}$, with the convention $Y_0 = x$, $T_0 = -u$ if the initial distribution of (Y, T) is $N_0^{x,u}$.

The process $(Z_t)_{t\geq 0}$ is the continuous-time version of the Markov renewal process, at least when $N(x, \{x\} \times \mathbb{R}_+) = 0$ for all $x \in E$. But $(Z_t)_{t\geq 0}$ is not a Markov process.

To get a Markov process, it needs to be supplemented by another process. It can be $(W_t)_{t\geq 0}$ or $(A_t)_{t\geq 0}$, defined Sect. 2.4, W_t having the advantage of being defined for all $t \geq 0$ regardless of the initial distribution. But we choose $(A_t)_{t\geq 0}$, which is suitable for the definition of PDMPs. The restriction on the initial distributions for A_t to be defined for all t will not be a problem.

Definition 2.5 The process $(Z_t)_{t\geq 0}$ is called the semi-Markov process associated with the Markov renewal process (Y, T).

The process $(Z_t, A_t)_{t\geq 0}$ is called the completed semi-Markov process (CSMP) associated with the Markov renewal process (Y, T). Conversely, (Y, T) is the Markov renewal process associated with the CSMP $(Z, A) = (Z_t, A_t)_{t\geq 0}$.

The CSMP $(Z_t, A_t)_{t\geq 0}$ is said to take values in E, or to be on E, if the random variables Z_t $(t \geq 0)$ take values in E. It is zero-delayed if the associated Markov renewal process is, i.e., $A_0 = 0$. The kernel of the CSMP is the kernel of its associated Markov renewal process.

Theorem 2.1, Sect. 2.4 is actually the Markov property for CSMP. *Assume initial dist. of $N_0^{x,u}$*

Proposition 2.8 *The process $(Z_t, A_t)_{t\geq 0}$ is a time-homogeneous Markov process.*

Proof We must show that $\mathbb{E}(\varphi(Z_{t+s}, A_{t+s}) \mid \mathcal{F}_t) = \mathbb{E}(\varphi(Z_{t+s}, A_{t+s}) \mid Z_t, A_t)$ for all $s, t > 0$ and Borel function $\varphi : E \times \mathbb{R}_+ \to \mathbb{R}_+$. Since $(Z_{t+s}, A_{t+s})_{s\geq 0}$ is the CSMP associated with the post-t-process (Y^t, T^t), Corollary 2.2, at the end of Sect. 2.4 yields $\mathbb{E}(\varphi(Z_{t+s}, A_{t+s}) \mid \mathcal{F}_t) = \psi(Z_t, A_t, s)$ with $\psi(z, u, s) = \mathbb{E}(\varphi(Z_s, A_s) \mid Z_0 = z, A_0 = u)$. The result follows. ∎

The following three examples are based on Example 1.4, Chap. 1 parts 1, 2, and 4, with more details.

Example 2.3 The simplest semi-Markov process is associated with an alternating renewal process (see Example 2.2, Sect. 2.3). It is a classical model in reliability studies. State e_1 is the operating state (or up state), state e_0 is the failure state (or down state), dF_1 is the probability distribution of the duration of trouble-free operation, dF_0 is the probability distribution of repair duration.

If dF_i $(i = 0, 1)$ are not exponential distributions, CSMP must be introduced to get a Markov process.

If dF_i $(i = 0, 1)$ depends, e.g., on a physical phenomena as in Example 3.11, Sect. 3.4 the appropriate model is not a CSMP but a PDMP.

Example 2.4 We use the terminology and notation from the previous example while adding a preventive maintenance: after a failure-free period $a \in \mathbb{R}_+^*$, preventive maintenance is carried out and the probability distribution of its duration is not the same as the probability distribution of the duration of repair. At the end of preventive maintenance and repair, the item is supposed to be as good as new. Here the probability distribution dF_1 is quite general, and we can have $dF_1(\{a\}) \neq 0$. If a failure occurs

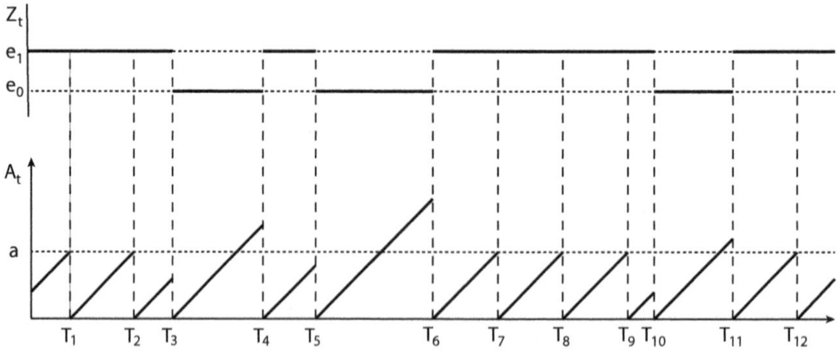

Fig. 2.1 Example 2.4 with instantaneous preventive maintenance

at the same time as the start of preventive maintenance, the probability distribution of refurbishment duration is supposed to be that of preventive maintenance.

First of all, we assume that preventive maintenance is instantaneous. Even if the state of the item does not change when preventive maintenance is performed, the age of the item is reset to 0. The behavior of the item is described by a CSMP $(Z_t, A_t)_{t\geq 0}$ on $E = \{e_0, e_1\}$, A_t is the age of the item when the item is up; it is the time elapsed since the beginning of the repair in progress when the item is down. The kernel N of the CSMP is defined as

$$N(e_1, dy, dv) = 1_{\{v<a\}}\, \delta_{e_0}(dy)\, dF_1(v) + dF_1([a, +\infty])\, \delta_{e_1}(dy)\, \delta_a(dv),$$

$$N(e_0, dy, dv) = 1_{R_+}(v)\, \delta_{e_1}(dy)\, dF_0(v).$$

A trajectory of $(Z_t, A_t)_{t\geq 0}$ is displayed in Fig. 2.1.

Note that the probability distribution of $T_{n+1} - T_n$ given $Y_n = e_1, Y_{n+1} = e_i$ depends on i: it is δ_a if $i = 1$ and $1_{\{v<a\}}\, dF_1(dv)/dF_1(]0, a[)$ if $i = 0$.

We now assume that preventive maintenance is not instantaneous. We denote by e_2 the state of preventive maintenance and by dF_2 the probability distribution of its duration. Then the behavior of the item is described by a CSMP $(Z_t, A_t)_{t\geq 0}$ on $E = \{e_0, e_1, e_2\}$ and kernel N defined as

$$N(e_1, dy, dv) = 1_{\{v<a\}}\, \delta_{e_0}(dy)\, dF_1(v) + dF_1([a, +\infty])\, \delta_{e_2}(dy)\, \delta_a(dv),$$

$$N(e_2, dy, dv) = 1_{R_+}(v)\, \delta_{e_1}(dy)\, dF_2(v),$$

$$N(e_0, dy, dv) = 1_{R_+}(v)\, \delta_{e_1}(dy)\, dF_0(v).$$

Example 2.5 Let us start by recalling the usual reliability terminology. The failure rate (respectively repair rate) of an item, or a component, subject to failure in a given environment is the hazard rate of the probability density function of the duration of trouble -free operation (respectively the duration of repair), which is assumed to have a density. The definition of the hazard rate of a probability density function is given in Appendix A, Definition A.3, Sect. A.3.

Consider an item with two components, each of them being in the up state (state 1) or down state (state 0). The failure rate of component i is λ_i when the other component is up and λ_i' when the other component is down; its repair rate is μ_i. After a repair, a component is as good as new. We are interested in I_t, the state of each component (up or down) at time t, where I_t takes values in $E_0 = (e_1, e_2, e_3, e_4)$ with $e_1 = (1, 1)$, $e_2 = (0, 1)$, $e_3 = (1, 0)$, $e_4 = (0, 0)$. Define T_n $(n \geq 1)$ as the successive jump times of $(I_t)_{t\geq 0}$. If all λ_i, λ_i' and μ_i $(i = 1, 2)$ are constant, then $(I_t)_{t\geq 0}$ is a Markov jump process. Otherwise, $(I_t)_{t\geq 0}$ is not a Markov process. In other words, (I_{T_n}, T_n) is not a Markov renewal process. To get a Markov renewal process, random variables must be added. Let $A_{i,n}$ be the time elapsed at time T_n since the last change of state of component i, i.e., the age of component i if it is up and the time elapsed since the start of its repair if it is down. Define $Y_n = (I_{T_n}, A_{1,n}, A_{2,n}) \in E = E_0 \times \mathbb{R}_+^2$. Then $(Y_n, T_n)_{n\geq 0}$ is a Markov renewal process. Its kernel is obtained from the following result: if U_i $(i = 1, 2)$ are independent nonnegative r.v. with hazard rate h_i $(i = 1, 2)$, respectively, then

$$\mathbb{E}(f(\min(U_1, U_2)\, 1_{\{U_1<U_2\}})) = \int_{\mathbb{R}_+} f(u)\, h_1(u)\, e^{-\int_0^u (h_1(s)+h_2(s))\, ds}\, du.$$

We get

$$N((e_1, x_1, x_2), dy, dv) = e^{-\int_0^v (\lambda_1(x_1+s)+\lambda_2(x_2+s))\, ds}$$
$$\times \big(\lambda_1(x_1 + v)\, \delta_{(e_2,0,x_2+v)}(dy) + \lambda_2(x_2 + v)\, \delta_{(e_3,x_1+v,0)}(dy)\big)\, dv,$$

$$N((e_2, x_1, x_2), dy, dv) = e^{-\int_0^v (\mu_1(x_1+s)+\lambda_2'(x_2+s))\, ds}$$
$$\times \big(\mu_1(x_1 + v)\, \delta_{(e_1,0,x_2+v)}(dy) + \lambda_2'(x_2 + v)\, \delta_{(e_4,x_1+v,0)}(dy)\big)\, dv,$$

$$N((e_3, x_1, x_2), dy, dv) = e^{-\int_0^v (\lambda_1'(x_1+s)+\mu_2(x_2+s))\, ds}$$
$$\times \big(\lambda_1'(x_1 + v)\, \delta_{(e_4,0,x_2+v)}(dy) + \mu_2(x_2 + v)\, \delta_{(e_1,x_1+v,0)}(dy)\big)\, dv,$$

$$N((e_4, x_1, x_2), dy, dv) = e^{-\int_0^v (\mu_1(x_1+s)+\mu_2(x_2+s))\, ds}$$
$$\times \big(\mu_1(x_1 + v)\, \delta_{(e_3,0,x_2+v)}(dy) + \mu_2(x_2 + v)\, \delta_{(e_2,x_1+v,0)}(dy)\big)\, dv.$$

We are interested in the first component I_t of the semi-Markov process associated with (Y, T). Its two other components are not of interest; they are simple tools for modeling. In order to have a process in which all variables have physical meaning, a PDMP model must be used (see Example 3.15, Sect. 3.4).

Results on reliability models based on semi-Markov processes can be found in Cocozza–Thivent and Roussignol [35, 36], Cocozza-Thivent and Eymard [31, 32], and Bloch-Mercier [13].

As is usual for Markov process, we set

$$\mathbb{P}_{x,u}(\,\cdot\,) = \mathbb{P}(\,\cdot \mid Z_0 = x,\, A_0 = u), \quad \mathbb{E}_{x,u}(\,\cdot\,) = \mathbb{E}(\,\cdot \mid Z_0 = x,\, A_0 = u), \tag{2.16}$$

and we have $\mathbb{P}_x = \mathbb{P}_{x,0}$, $\mathbb{E}_x = \mathbb{E}_{x,0}$ (see Sect. 2.3 for the definitions of $\mathbb{P}_x$ and $\mathbb{E}_x$). Thus, under $\mathbb{P}_{x,u}$, $(Y_n, T_n)_{n\geq 1}$ is a Markov renewal process with kernel N and initial distribution $N_0^{x,u}$. Moreover,

$$\mathbb{P}_{x,u}(T_1 \geq v) = \frac{\mathbb{P}_x(T_1 \geq u+v)}{\mathbb{P}_x(T_1 > u)}, \quad \mathbb{P}_{x,u}(T_1 > v) = \frac{\mathbb{P}_x(T_1 > u+v)}{\mathbb{P}_x(T_1 > u)}. \tag{2.17}$$

Define

$$R = \sum_{n\geq 0} N^{*n} = I + \sum_{n\geq 1} N^{*n}. \tag{2.18}$$

The marginal distributions of a CSMP provide nice equations, namely Markov renewal equations, that generalize renewal equations.

Lemma 2.4 *Let $\varphi : E \times \mathbb{R}_+ \to \mathbb{R}_+$ be a Borel function and let $(x, u) \in E \times \mathbb{R}_+$ be such that $\bar{F}_x(u) \neq 0$. Then*

$$\mathbb{E}_x(\varphi(Z_t, A_t)) = \varphi(x,t)\,\bar{F}_x(t) + \int_{E\times\mathbb{R}_+} 1_{\{v\leq t\}}\,\mathbb{E}_y(\varphi(Z_{t-v}, A_{t-v}))\,N(x, dy, dv), \tag{2.19}$$

$$\mathbb{E}_x(\varphi(Z_t, A_t)) = R * g(x,t) \text{ where } g(y,v) = \varphi(y,v)\,\bar{F}_y(v), \tag{2.20}$$

and more generally,

$$\begin{aligned}\mathbb{E}_{x,u}(\varphi(Z_t, A_t)) &= \varphi(x, u+t)\,\frac{\bar{F}_x(u+t)}{\bar{F}_x(u)} \\ &\quad + \int_{E\times\mathbb{R}_+} 1_{\{v\leq t\}}\,\mathbb{E}_y(\varphi(Z_{t-v}, A_{t-v}))\,N_0^{x,u}(dy, dv). \end{aligned}\tag{2.21}$$

Proof By Lemma 2.1, Sect. 2.3

$$\begin{aligned}\mathbb{E}_x(\varphi(Z_t, A_t)) &= \mathbb{E}_x\left(\varphi(Z_t, A_t)\,1_{\{t<T_1\}}\right) + \sum_{n\geq 1} \mathbb{E}_x\left(\varphi(Y_n, t-T_n)\,1_{\{N_t=n\}}\right) \\ &= \varphi(x,t)\,\bar{F}_x(t) + \sum_{n\geq 1} \mathbb{E}_x\left(\varphi(Y_n, t-T_n)\,1_{\{T_n\leq t\}}\,\bar{F}_{Y_n}(t-T_n)\right),\end{aligned}$$

and (2.20) follows according to Corollary 2.1, Sect. 2.3. Then writing $N^{*n} = N * N^{*(n-1)}$, we get (2.19). Similar arguments apply for proving (2.21). ■

Define $f(x,t) = \mathbb{E}_x(\varphi(Z_t, A_t))$, $g(x,t) = \varphi(x,t)\,\bar{F}_x(t)$. Equation (2.19) can be written more consistently as

$$f = g + N * f. \tag{2.22}$$

It is a Markov renewal equation. Some results on such equations are given in Sect. 2.7.

2.6 Other Classical Processes Associated with Markov Renewal Processes

Recall that $\tilde{Z}_t$ and W_t are defined as

$$\tilde{Z}_t = Y_{n+1},\ W_t = T_{n+1} - t, \quad T_n \le t < T_{n+1}.$$

Tedious but straightforward calculations lead to the following result.

Theorem 2.2 *The processes $(Z_t, A_t, \bar{Z}_t, W_t)_{t\geq 0}$ and $(\bar{Z}_t, W_t)_{t\geq 0}$ are time-homogeneous Markov processes and*

$$\begin{aligned} &\mathbb{E}(f(Z_t, A_t, \tilde{Z}_t, W_t) \mid Z_0 = x, A_0 = u, \tilde{Z}_0 = y, W_0 = v) \\ &= f(x, u+t, y, v-t)\, 1_{\{t<v\}} + \mathbb{E}_y(f(Z_{t-v}, A_{t-v}, \tilde{Z}_{t-v}, W_{t-v}))\, 1_{\{v\leq t\}} \end{aligned} \tag{2.23}$$

for all $(x,u) \in E \times \mathbb{R}_+$, all $(y,v) \in (E \times \mathbb{R}_+) \cup \{(\Delta, +\infty)\}$, and every Borel function $f : E \times \mathbb{R}_+ \times (E \times \mathbb{R}_+ \cup \{(\Delta, +\infty)\}) \to \mathbb{R}_+$.

When writing that $(Z_t, A_t, \bar{Z}_t, W_t)_{t\geq 0}$ is a Markov process, it is implicitly assumed that the initial distribution of the Markov renewal process is $N_0^{x,u}$ (or a mixture of such measures) in the first part of the statement. On the other hand, $\bar{Z}_t$ and W_t are defined for every initial distribution, and $(\bar{Z}_t, W_t)_{t\geq 0}$ is a Markov process regardless of the initial distribution.

Semi-Markov processes generalize Markov jump processes on a discrete space. Indeed, let $(I_t)_{t\geq 0}$ be a Markov jump process with a discrete state space E and generator matrix A; define $\lambda(i) = |A(i,i)|$ and denote by p the transition matrix of the embedded Markov chain ($A(i,j) = \lambda(i)\,p(i,j), i \neq j$). We deduce from Proposition 2.5, Sect. 2.3 that $(I_t)_{t\geq 0}$ is a semi-Markov process with kernel $N(i,j,dv) = p(i,j)\,\lambda(i)\,e^{-\lambda(i)v}\,dv$.

The following proposition, which follows immediately from Theorem 3.1, Sect. 3.1 (next chapter), introduces Markov jump process on a general Polish space.

Proposition 2.9 *Given a bounded Borel function $\lambda : E \to \mathbb{R}_+$ and a kernel Q on E, define*

$$N(x, dy, dv) = Q(x; dy)\,\lambda(x)\,e^{-\lambda(x)v}\,dv. \tag{2.24}$$

Then $N_0^{x,u}(\,\cdot\,) = N(x,\,\cdot\,)$ *for all* $(x, u) \in E \times \mathbb{R}_+$, *and the semi-Markov process* $(Z_t)_{t\geq 0}$ *is a Markov process.*

In Proposition 2.9, the function λ is supposed to be bounded to ensure $\lim_{n\to\infty} T_n = +\infty$; weaker conditions may be considered.

Proposition 2.10 *Let* $(Z_t)_{t\geq 0}$ *be a Markov jump process associated with the kernel* N *defined by (2.24) and* $(P_t)_{t\geq 0}$ *its semigroup, i.e.,* $P_t f(x) = \mathbb{E}_x(f(Z_t))$. *Then*

$$\forall x \in E, \quad e^{\lambda(x)t}\,P_t f(x) = f(x) + \int_0^t \left(\int_E P_s f(y)\,Q(x; dy)\right) \lambda(x)\,e^{\lambda(x)s}\,ds \tag{2.25}$$

for every $x \in E$ *and every bounded or nonnegative Borel function* $f : E \to \mathbb{R}$.

Thus, for every $x \in E$ *and bounded Borel function* $f : E \to \mathbb{R}$, *the function* $t \to P_t f(x)$ *is differentiable and*

$$\frac{d}{dt} P_t f(x) = L P_t f(x), \tag{2.26}$$

where the operator L *is defined by* $Lf(x) = \int_E (f(y) - f(x))\,\lambda(x)\,Q(x; dy)$.

Equation (2.26) is the backward Chapman–Kolmogorov equation. Such equations for CSMPS and PDMPS will be set out in Sect. 6.3.

Proof Equation (2.25) comes from (2.19), Sect. 2.5. Equation (2.26) follows by applying Lebesgue's dominated convergence theorem. ∎

2.7 Semiregenerative Processes and Markov Renewal Equations

Given a semi-Markov kernel N on E and a Borel function $g : E \times \mathbb{R}_+ \to \mathbb{R}$, we are interested in the solution f of the Markov renewal equation $f = g + N * f$, i.e.,

$$\forall (x, t) \in E \times \mathbb{R}_+ \quad f(x, t) = g(x, t) + \int_{E\times[0,t]} f(y, t - v)\,N(x, dy, dv).$$

We have already seen such an equation at the end of Sect. 2.5. It appears more generally with semiregenerative processes (see Proposition 2.11 below).

Definition 2.6 A cad-lag process $(X_t)_{t\geq 0}$ is said to be semiregenerative if we can find a zero-delayed Markov renewal process $(Y_n, T_n)_{n\geq 0}$ such that $\lim_n T_n = +\infty$ and for all n, the conditional distribution of $(X_{t+T_n}, T_{n+p})_{t\geq 0, p\geq 0}$ given $(Y_0, Y_1, \ldots, Y_{n-1},$

$Y_n = y, T_1, \dots T_n)$ is the same as the conditional distribution of $(X_t, T_p)_{t\geq 0, p\geq 0}$ given $Y_0 = y$.

The process $(Y_n, T_n)_{n\geq 0}$ is the embedded Markov renewal process, and $T_n (n \geq 1)$ are the regeneration times.

Proposition 2.11 *Let $(X_t)_{t\geq 0}$ be a semiregenerative process taking values in E, and N the kernel of its embedded Markov renewal process (Y, T). Given a bounded or nonnegative Borel function $h : E \to \mathbb{R}$, define*

$$f(x,t) = \mathbb{E}(h(X_t) \mid Y_0 = x), \quad g(x,t) = \mathbb{E}(h(X_t)\, 1_{\{t<T_1\}} \mid Y_0 = x).$$

Then

$$f = g + N * f.$$

Proof We have

$$\begin{aligned}\mathbb{E}(h(X_t) \mid Y_0 = x) &= g(x,t) + \mathbb{E}\left(1_{\{T_1 \leq t\}}\, \mathbb{E}(h(X_{t-T_1+T_1}) \mid Y_0, Y_1, T_1) \mid Y_0 = x\right)\\ &= g(x,t) + \int_{E\times[0,t]} f(y, t-v)\, N(x, dy, dv).\end{aligned}$$

This completes the proof. ■

Proposition 2.12 *Let $g : E \times \mathbb{R}_+ \to \mathbb{R}$ be a Borel function that is bounded on $E \times [0, t]$ for all $t \in \mathbb{R}_+$ and N a semi-Markov kernel such that the associated Markov renewal process $(Y_n, T_n)_{n\geq 1}$ satisfies $\lim_{n\to\infty} T_n = \infty$ $\mathbb{P}_x$ a.s. for all $x \in E$. Then the Markov renewal equation*

$$f = g + N * f. \tag{2.27}$$

has one and only one solution f that is bounded on $E \times [0, t]$ for all $t \in \mathbb{R}_+$; it is $f = R * g$, where $R = \sum_{n\geq 0} N^{*n}$.

Proof Iterating (2.27), we get

$$f = \sum_{k=0}^{n} N^{*k} * g + N^{*(n+1)} * f.$$

It follows from Corollary 2.1, Sect. 2.3 that

$$|N^{*(n+1)} * f(x,t)| \leq \sup_{(y,u)\in E\times[0,t]} |f(y,u)|\, \mathbb{P}_x(T_{n+1} \leq t);$$

hence $|N^{*(n+1)} * f(x,t)|$ tends to 0 as n tends to infinity. This completes the proof. ■

Blackwell's theorem provides asymptotic behavior of the solution of a renewal equation. Similarly, asymptotic theorems for solutions of Markov renewal equations can be obtained, but when E is nondenumerable, it is much more difficult, and the results are substantially more involved.

Many contributions have appeared since the end of the 1970s, and assumptions have been refined all along. Given a Markov renewal process (Y, T), it is assumed that the Markov chain Y is Harris-recurrent (see Definition 8.2, Sect. 8.2). The Markov chain Y then has an invariant measure ν (see Definition 8.1, Chap. 8) that is unique up to a multiplicative constant.

Following Orey [85], the chain Y is Harris-recurrent iff there exist a Borel set $C \subset E$ (called a regeneration set in the terminology of Asmussen [5] Chap. VI Sect. 3), an integer $m \geq 1$, a real number $\beta > 0$, and a probability distribution μ such that

1. $\mu(C) = 1$,
2. for all $x \in E, \mathbb{P}_x(\exists n \geq 1 : Y_n \in C) = 1$,
3. for every Borel set $B \subset E$ and $x \in C, \mathbb{P}_x(Y_m \in B) \geq \beta\,\mu(B)$.

This makes it possible to construct regeneration times τ_p for the Markov chain Y. When the regeneration occurs in a fixed state, the classic result on regenerative processes can be easily applied. But usually it occurs according to μ. Moreover, if the distribution of $T_{n+1} - T_n$ given Y does not depend solely on Y_n but also on Y_{n+1} (as it does for the instantaneous preventive maintenance of Example 2.4, Sect. 2.5), a cycle length depends on the initial conditions for the next cycle (Asmussen [5], Notes page 234). As a matter of fact, the sequence $(T_{\tau_p})_{p \geq 0}$ is only a random walk with stationary, 1-dependent increments (Alsmeyer [3] Lemma 3.2), which makes it very difficult.

As for the classical renewal process, it becomes necessary to distinguish between the arithmetic case and the nonarithmetic one. The concept of "arithmetic" for a Markov renewal process is harder to define than in the case of a renewal process, since there is not one but many interarrival distributions. Such a definition has been provided by Shurenkov [93].

Definition 2.7 Given a Markov renewal process $(Y, T) = (Y_n, T_n)_{n \geq 1}$ such that the Markov chain Y is Harris-recurrent with invariant measure ν, it is called d-arithmetic if $d > 0$ is the maximal number for which there exists a Borel function $\gamma : E \to [0, d[$, called a shift function, such that for all n,

$$\mathbb{P}(T_{n+1} - T_n \in \gamma(x_0) - \gamma(x_1) + d\,\mathbb{Z}) \mid Y_n = x_0, Y_{n+1} = x_1) = 1 \quad \rho - a.s.,$$

where ρ is the probability distribution of (Y_1, Y_2) under $\mathbb{P}_\nu$ when ν is a probability distribution and $\rho(B_1 \times B_2) = \int_{B_1} \mathbb{P}_x(Y_1 \in B_2)\,\nu(dx)$ in the general case.

It is called nonarithmetic if no such d exists.

When every interarrival distribution has an absolutely continuous part, the process (Y, T) is nonarithmetic.

Recall that condition $\lim_{n\to\infty} T_n = +\infty$ is supposed to hold (see Sect. 2.3).

We have chosen to give the results of V. M. Shurenkov (Theorem 2.3 below) and G. Alsmeyer (Theorem 2.4 below) in the nonarithmetic case. The assumptions are slightly different, so it is up to each reader to choose which result is best suited to the model under consideration. We give a corollary on the convergence of the marginal probability distribution of CSMP (Corollary 2.3), and as a consequence of PDMP (Theorem 8.8, Sect. 8.6), based on Shurenkov's result. As an example of the application of Alsmeyer's result, we give the convergence of semiregenerative processes (Corollary 2.4).

The reader interested in the d-arithmetic case may refer directly to Shurenkov [93] and the comment at the beginning of Alsmeyer [4], Sect. 3, or [3].

Theorem 2.3 (Shurenkov [93], Theorem 2)

Let $(Y, T) = (Y_n, T_n)_{n\geq 1}$ be a Markov renewal process on E such that the Markov chain Y is Harris-recurrent with invariant measure ν and $\int_E \mathbb{E}_y(T_1)\,\nu(dy) < +\infty$. Assume that (Y, T) is nonarithmetic.

Let $g : E \times \mathbb{R}_+ \to \mathbb{R}_+$ be a Borel function satisfying

1. $\displaystyle\int_E \sum_{n\geq 0} \sup_{n\leq u<n+1} g(y, u)\,\nu(dy) < +\infty,$
2. $\displaystyle\lim_{\Delta\to 0} \Delta \int_E \sum_{n\geq 0} \left(\sup_{n\Delta\leq u<(n+1)\Delta} g(y, u) - \inf_{n\Delta\leq u<(n+1)\Delta} g(y, u) \right) \nu(dy) = 0,$
3. $\nu(\{y : \sup_{u\geq 0} R * g(y, u) < +\infty\}) > 0.$

Then

$$R * g(x, t) \underset{t\to+\infty}{\longrightarrow} \frac{\displaystyle\int_{E\times\mathbb{R}_+} g(y, u)\,\nu(dy)\,du}{\displaystyle\int_E \mathbb{E}_y(T_1)\,\nu(dy)}$$

holds for ν-almost all $x \in E$.

Corollary 2.3 *Let (Y, T) be a Markov renewal process on E associated with a CSMP (Z, A). Assume that the Markov chain Y is Harris-recurrent with invariant measure ν. Assume that (Y, T) is nonarithmetic and $\int_E \mathbb{E}_y(T_1)\,\nu(dy) < +\infty$.*

(1) Assume $\nu(E) < +\infty$.

Let $\varphi : E \times \mathbb{R}_+ \to \mathbb{R}$ be a continuous bounded function. Then

$$\mathbb{E}_x(\varphi(Z_t, A_t)) \underset{t\to+\infty}{\longrightarrow} \frac{\displaystyle\int_{E\times\mathbb{R}_+} \varphi(y, u)\,\bar{F}_y(u)\,\nu(dy)\,du}{\displaystyle\int_E \mathbb{E}_y(T_1)\,\nu(dy)}$$

for ν-almost all $x \in E$.

(2) Let A be a Borel subset of E such that $\nu(A) < +\infty$. Then for ν-almost all $x \in E$,

$$\mathbb{E}_x(1_A(Z_t)\,\varphi(Z_t, A_t)) \xrightarrow[t\to+\infty]{} \frac{\displaystyle\int_{A\times\mathbb{R}_+} \varphi(y,u)\,\bar{F}_y(u)\,\nu(dy)\,du}{\displaystyle\int_E \mathbb{E}_y(T_1)\,\nu(dy)}$$

for every continuous bounded function $\varphi : E \times \mathbb{R}_+ \to \mathbb{R}$.

(3) Let A *be a Borel subset of* E *such that* $\nu(A) < +\infty$ *and* $a > 0$. *Then*

$$\mathbb{P}_x(Z_t \in A, A_t \le a) \xrightarrow[t\to+\infty]{} \frac{\displaystyle\int_{A\times[0,a]} \bar{F}_y(u)\,\nu(dy)\,du}{\displaystyle\int_E \mathbb{E}_y(T_1)\,\nu(dy)}$$

for ν*-almost all* $x \in E$.

Proof (1) We are going to show the convergence of $\mathbb{E}_x(\varphi(Z_t, A_t))$ for every bounded Lipschitz continuous function $\varphi : E \times \mathbb{R}_+ \to \mathbb{R}_+$. The convergence of the probability distribution of (Z_t, A_t) under $\mathbb{P}_x$ is then given by the portmanteau theorem (Billingsley [11]).

From (2.20), Sect. 2.5 we have $\mathbb{E}_x(\varphi(Z_t, A_t)) = R * g(x,t)$, where $g(y,u) = \varphi(y,u)\bar{F}_y(u)$. So we need to show that this function g satisfies the assumptions of Theorem 2.3.

We have

$$\sum_{n\ge 0} \sup_{n\le u<n+1} g(y,u) \le ||\varphi||_\infty \sum_{n\ge 0} \bar{F}_y(n) \le ||\varphi||_\infty \Big(1 + \int_{\mathbb{R}_+} u\,dF_y(u)\Big),$$

hence
$\int_E \sum_{n\ge 0} \sup_{n\le u<n+1} g(y,u)\,\nu(dy) \le ||\varphi||_\infty(\nu(E) + \int_E \mathbb{E}_y(T_1)\,\nu(dy)) < +\infty$.

For condition 2, let K be the Lipschitz constant of φ. We get

$$\begin{aligned}
&\sup_{n\Delta\le u<(n+1)\Delta} g(y,u) - \inf_{n\Delta\le u<(n+1)\Delta} g(y,u) \\
&\le \sup_{n\Delta\le t_1,t_2<(n+1)\Delta} |\varphi(y,t_1) - \varphi(y,t_2)|\,\bar{F}_y(n\Delta) + ||\varphi||_\infty dF_y(]n\Delta,(n+1)\Delta]) \\
&\le K\Delta\,\bar{F}_y(n\Delta) + ||\varphi||_\infty dF_y(]n\Delta,(n+1)\Delta]);
\end{aligned}$$

hence

$$\begin{aligned}
&\int_E \sum_{n\ge 0}\Big(\sup_{n\Delta\le u<(n+1)\Delta} g(y,u) - \inf_{n\Delta\le u<(n+1)\Delta} g(y,u)\Big)\,\nu(dy) \\
&\le \int_E (K\,\mathbb{E}_y(T_1) + K\Delta + ||\varphi||_\infty)\,\nu(dy).
\end{aligned}$$

Therefore condition 2 holds.

Since $R * g = \varphi$ is bounded, condition 3 holds.

Finally, the result for $\mathbb{E}_{x,u}(\varphi(Z_t, A_t))$ is a consequence of (2.21), Sect. 2.5 when φ has compact support. The result follows, since $\bar{F}_y(u)\, \nu(dy)\, du / \int \mathbb{E}_y(T_1)\, \nu(dy)$ is a probability distribution.

(2) We return to the above proof with $1_A(y)\, \varphi(y, u)$ instead of $\varphi(y, u)$.

(3) We have $g(y, u) = 1_A(y)\, 1_{[0,a]}(u)\, \bar{F}_y(u)$. We get $\sum_{n \geq 0} \sup_{n \leq u < n+1} g(y, u) \leq (a + 1) 1_A(y)$; hence condition 1 of Theorem 2.3 holds. On the other hand,

$$\sum_{n \geq 0} \sup_{n\Delta \leq u < (n+1)\Delta} g(y, u) - \inf_{n\Delta \leq u < (n+1)\Delta} g(y, u)$$
$$\leq 1_A(y) \sum_{n \geq 0 : (n+1)\Delta \leq a} (\bar{F}_y(n\Delta) - \bar{F}_y((n + 1)\Delta) \leq 2\, 1_A(y).$$

Therefore, condition 2 of Theorem 2.3 holds too. ■

Example 2.6 Let $(Z_t, A_t)_{t \geq 0}$ be the CSMP associated with the alternating renewal process described in Example 2.2, Sect. 2.3. The Markov kernel of the Markov chain Y is P, and its stationary probability distribution ν is the uniform distribution on $E = \{e_1, e_0\}$; hence we can take $\nu(e_1) = \nu(e_0) = 1$.

Recall that a measure on $\mathbb{R}_+$ is nonarithmetic if its support is not a subset of some $d\mathbb{N}$, $d > 0$.

The Markov renewal process (Y, T) is nonarithmetic as soon as dF_1 or dF_2 is nonarithmetic, which we assume. Define $\bar{F}_i(u) = dF_i(]u, +\infty])$ $(i = 0, 1)$.

If $\mathbb{E}(dF_i) := \mathbb{E}_{e_i}(T_1) = \int_0^{+\infty} v\, dF_i(v) = \int_0^{+\infty} \bar{F}_i(v)\, dv < +\infty$, then

$$\mathbb{E}_{e_i,u}(\varphi(Z_t, A_t)) \xrightarrow[t \to +\infty]{} \frac{\int_{\mathbb{R}_+} \varphi(e_1, v)\, \bar{F}_1(v)\, dv + \int_{\mathbb{R}_+} \varphi(e_0, v)\, \bar{F}_0(v)\, dv}{\mathbb{E}(dF_1) + \mathbb{E}(dF_0)} \quad (i = 0, 1)$$

for every continuous bounded function $\varphi : E \times \mathbb{R}_+ \to \mathbb{R}$. In particular, $\lim_{t \to \infty} \mathbb{P}(Z_t = e_i) = \mathbb{E}(dF_i)/(\mathbb{E}(dF_1) + \mathbb{E}(dF_0))$.

Example 2.7 Let us go back to Example 2.4, Sect. 2.5 and assume that preventive maintenance is instantaneous. Define $p = dF_1([a, +\infty])$ and, as usual, $\bar{F}_i(u) = dF_i(]u, +\infty])$. We have

$$dF_{e_1}(v) := N(e_1, \mathbb{R}_+, dv) = 1_{\{v < a\}}\, dF_1(v) + p\, \delta_a(dv).$$

The probability kernel P of the Markov chain Y is $P(e_1, e_1) = p$, $P(e_1, e_0) = 1 - p$, $P(e_0, e_1) = 1$, $P(e_0, e_0) = 0$, and the only invariant measure up to a multiplicative constant is $\nu(e_1) = 1$, $\nu(e_0) = 1 - p$. We assume $0 \leq p < 1$, $1_{\{u<a\}} dF_1(u)$ or dF_0 nonarithmetic and $E_{e_0}(T_1) = \mathbb{E}(dF_0) < +\infty$. Then we can apply Corollary 2.3, and we get

$$\mathbb{E}_{e_i}(\varphi(Z_t, A_t)) \xrightarrow[t\to+\infty]{} \frac{\displaystyle\int_{\mathbb{R}_+} \varphi(e_1, u)\, 1_{\{u<a\}} \bar{F}_1(u)\, du + (1-p) \int_{\mathbb{R}_+} \varphi(e_0, u)\, \bar{F}_0(u)\, du}{\displaystyle\int_0^a \bar{F}_1(u)\, du + (1-p)\, \mathbb{E}(dF_0)}$$

($i = 0, 1$) for every continuous bounded function $\varphi : E \times \dot{\mathbb{R}}_+ \to \mathbb{R}$.

In view of Propositions 2.11 and 2.12, Sect. 2.7, for a semiregenerative process $(X_t)_{t\geq 0}$, we have

$$\mathbb{E}(f(X_t) \mid Y_0 = x) = R * g(x, t), \quad g(x, u) = \mathbb{E}(f(X_t)\, 1_{\{u<T_1\}} \mid Y_0 = x). \quad (2.28)$$

When X_t is random for $t \in [0, T_1[$, it is usually not possible to check condition 2 of Theorem 2.3. However, we give below an example in which the process $(X_t)_{t\geq 0}$ is random between the times of semiregeneration we consider, and where we can nevertheless check all the conditions of Theorem 2.3 and understand an approximation commonly used in practice.

Example 2.8 In dependability studies, the availability $D(t)$ of a system at time t is the probability that it is up at that time; MUT (mean up time) is the mean time of trouble-free operation in long time, and MTBF (mean time between failure) is the mean time between failures in long time. When system operation is modeled by an alternating renewal process, Example 2.6 shows that $D(\infty) := \lim_{t\to\infty} D(t) = \lim_{t\to\infty} P(Z_t = e_1) = MUT/MTBF$. This formula is commonly used by practitioners, even when the system consists of several components, in which case its behavior cannot be modeled by an alternating renewal process even if the components are independent (see Example 2.5, Sect. 2.5 with $\lambda_i = \lambda_i'$). One may question its validity. Below, we clarify a general framework and provide some elements of a response.

The behavior of an item subject to failure is described by a stochastic process $(X_t)_{t\geq 0}$ taking values in E. The item can be up or down. Let $(\mathcal{U}, \mathcal{D})$ be the corresponding state space partition: $X_t \in \mathcal{U}$ (respectively $X_t \in \mathcal{D}$) means that the item is up (respectively down) at time t. Assume $X_0 \in \mathcal{U}$, define $S_0 = 0$ and S_n $(n \geq 1)$ as the successive times when X_t returns to $\mathcal{U}$: $S_{n+1} = \inf\{t > S_n : X_t \in \mathcal{U},\ X_s \in \mathcal{D}$ for some $s \in]S_n, t[\}$ $(n \geq 0)$. Assume $S_n < +\infty$ a.s. for every n. Define

$$MUT = \lim_{n\to\infty} \mathbb{E}\left(\int_{S_n}^{S_{n+1}} 1_{\{X_s\in\mathcal{U}\}}\, ds\right), \quad MTBF = \lim_{n\to\infty} \mathbb{E}(S_{n+1} - S_n)$$

when such limits exist.

Assume now that $(X_t)_{t\geq 0}$ is a semiregenerative process with embedded Markov renewal process $(X_{S_n}, S_n)_{n\geq 0}$. Assume that $(X_{S_n})_{n\geq 0}$ is Harris-recurrent with invariant probability distribution ν. Hence the probability distribution of X_{S_n} converges in total variation to ν (Asmussen [5] Chap. VI Theorem 3.6). If it can be proved that $\sup_{x\in E} \mathbb{E}_x(S_1) < +\infty$, then

$$MUT = \int_E \mathbb{P}_y(X_s \in \mathcal{U},\, s < S_1)\, ds\, \nu(dy), \quad MTBF = \int_E \mathbb{E}_y(S_1)\, \nu(dy).$$

In view of (2.28), if we are able to check the assumptions of Theorem 2.3 for $g(y,u) = \mathbb{P}_y(X_u \in \mathcal{U}, S_1 > u)$, we get $\lim_{t\to\infty} \mathbb{P}_x(X_t \in \mathcal{U}) = MUT/MTBF$ for ν-almost all $x \in E$. In the applications,"ν-almost all $x \in E$" is not entirely satisfactory. However, in practice, even without rigorous proof, the approximation of availability $D(t) = \mathbb{P}(X_t \in \mathcal{U})$ by $MUT/MTBF$ seems to make sense when t is "large."

The above assumptions have been checked in Cocozza-Thivent and Roussignol [36] for a class of systems that covers interesting practical cases. As is often the case in reliability studies, $X_t = (I_t, V_t)$, where I_t takes values in a finite state space, $V_t \in \mathbb{R}^k$ for some k, and $\mathcal{U} = \mathcal{U}_0 \times \mathbb{R}^k$ for some $\mathcal{U}_0$ finite. The first difficulty is in proving the Harris recurrence of $(X_{S_n})_{n\geq 0}$. The second is in the assumption $\sup_{x\in E} \mathbb{E}_x(S_1) < +\infty$; it has been overcome thanks to (4.8), Sect. 4.2 which is proved in Cocozza-Thivent and Kalashnikov [34] for the systems under consideration. For condition 2 of Theorem 2.3, the difficulty does not come from $1_{\mathcal{U}}$ but from the jumps of X_t for $t \in]0, S_1]$; their mean number was checked by writing $\mathbb{E}(M_{S_1}) = 0$, where $(M_t)_{t\geq 0}$ is like the martingale of Proposition 7.3, Sect. 7.1. Instead, we could have used Sect. 4.1 and Corollary 5.2, Sect. 5.3.

We now look at G. Alsmeyer's result and give an application to semiregenerative processes.

Theorem 2.4 (Alsmeyer [3, 4]) *Let $(Y,T) = (Y_n, T_n)_{n\geq 1}$ be a Markov renewal process on E such that the Markov chain Y is Harris-recurrent with stationary measure ν. Assume that (Y,T) is nonarithmetic. If $g : E \times \mathbb{R}_+ \to \mathbb{R}$ is any Borel function satisfying*

1. *the function $u \in \mathbb{R}_+ \to g(x,u)$ is Lebesgue-a.e. continuous for ν-almost $x \in E$,*
2. *$\int_E \sum_{n\in\mathbb{N}} \sup_{n\Delta \leq u < (n+1)\Delta} |g(x,u)|\, \nu(dx) < +\infty$ for some $\Delta > 0$,*

then

$$R * g(x,t) \xrightarrow[t\to+\infty]{} \frac{\int_{E\times\mathbb{R}_+} g(y,u)\, \nu(dy)\, du}{\int_E \mathbb{E}_y(T_1)\, \nu(dy)}$$

holds for ν-almost all $x \in E$.

Corollary 2.4 *Let $(X_t)_{t\geq 0}$ be a semiregenerative process on E with embedded semi-Markov process $(Y,T) = (Y_n, T_n)_{n\geq 0}$. Assume that the Markov chain Y is Harris-recurrent and (Y,T) is nonarithmetic.*

Let $f : E \to \mathbb{R}$ be a bounded Borel function such that the function g defined as

$$g(x,u) = \mathbb{E}(f(X_u)\, 1_{\{u<T_1\}} \mid Y_0 = x)$$

satisfies:

- *the function* $u \in \mathbb{R}_+ \to g(x, u)$ *is Lebesgue-a.e. continuous for* ν*-almost* $x \in E$,
- $\int_E \sum_{n\in\mathbb{N}} \sup_{n\Delta \le u < (n+1)\Delta} |g(x, u)|\, \nu(dx) < +\infty$ *for some* $\Delta > 0$.

Then

$$\mathbb{E}(f(X_t) \mid Y_0 = x) \xrightarrow[t\to+\infty]{} \frac{\displaystyle\int_E \mathbb{E}\left(\int_0^{T_1} f(X_v) \mid Y_0 = y)\, dv\right) \nu(dy)}{\displaystyle\int_E \mathbb{E}_y(T_1)\, \nu(dy)}$$

holds for ν*-almost all* $x \in E$.

In case of a regenerative process, the theorem and its proof are much simpler, and convergence of marginal distributions holds for every initial probability distribution (Asmussen [5] Chap. V Theorem 1.2).

Chapter 3
First Steps with PDMPs

In this chapter, we give more general definitions of PDMPs than the usual ones. First of all, we consider a fully general framework, with only assumptions concerning the deterministic behavior between jumps (namely the flow) and a very natural fundamental link between it and the interarrival distributions, which is the probability distributions of the periods between two consecutive jumps. This link will be used extensively in Sect. 8.5 for intensities related to stationary PDMPs, where no other constraints are imposed on interarrival distributions. Then we turn to more classical assumptions by specifying interarrival distributions but keeping a sufficiently broad framework to include many examples without the need for an artificial modification of the state space.

We introduce a PDMP X as a function of a CSMP (Z, A) (defined Sect. 2.5), namely $X_t = \phi(Z_t, A_t)$ $(t \geq 0)$, $\phi : E \times \mathbb{R}_+ \to E$. This means that between two jumps of the Markov renewal process associated with the CSMP, X has a deterministic behavior governed by the flow ϕ. Following Jacobsen [66], we assume that ϕ is a deterministic Markov flow, that is, $\phi(x, 0) = x$, $\phi(x, s + t) = \phi(\phi(x, s), t)$ basically for every $(x, s, t) \in E \times \mathbb{R}_+^2$. Actually, this property is needed only if $\mathbb{P}_x(T_1 > s + t) > 0$, which is why the time $t^*(x)$ is introduced below. This is a refinement that can be omitted at first reading.

In the first section, the kernel N of the CSMP is quite general, and it is proved that condition $N_0^{x,u}(\,\cdot\,) = N(\phi(x, u), \,\cdot\,)$ for "every" $(x, u) \in E \times \mathbb{R}_+$ ensures that $X = \phi(Z, A)$ is a Markov process. This is how we define a PDMP. Inhomogeneous PDMPs are also introduced for the following reason: in the literature, the Markov deterministic flow ϕ is often the flow generated by the solutions of a homogeneous ordinary differential equation (ODE) and we want that inhomogeneous ODEs of the form $z'(t) = \mathbf{v}(t, z(t))$ may be considered. This generalization is not essential; it is intended for readers who are already comfortable with PDMPs.

In Sect. 3.2, we come back to a more classical context: each interarrival distribution is that of the minimum between a random variable with a probability density function and a constant. Unlike in Davis [51], this constant need not be the time for a particle moving according to the flow to reach the boundary of the state space.

C. Cocozza-Thivent, *Markov Renewal and Piecewise Deterministic Processes*, Probability Theory and Stochastic Modelling 100,
https://doi.org/10.1007/978-3-030-70447-6_3

Only a coherence property with respect to the flow is needed. In the examples, this constant is indeed usually the time to reach some subset of the state space, called a boundary even if it is not a subset of its topological boundary, but our presentation focuses on the coherence condition. This constant should not be regarded merely as a way to protect the process from going to "wrong areas." Thanks to the introduction of a partition of the state space, the probability distribution of the time of the next jump, and of the location after the jump, may depend on the area where the PDMP is coming from (see Example 3.10); for the sake of simplicity, it has been formalized only for jumps at a "boundary" because that seems to be what is useful. Jumps on the spot and on a "boundary" are allowed, thus avoiding an artificial enlargement of the state space (see Example 3.8, Sect. 3.2). The partition of the state space may also be introduced to allow some discontinuities, as we will see in Chaps. 6 and 9.

Section 3.4 is devoted to enlightening examples. Its reading is highly recommended for understanding the ins and outs of the chosen model.

Recall that E is a Polish space equipped with its Borel σ-algebra $\mathcal{E}$.

Given a Markov renewal process (Y, T) on E with kernel N, we write

$$N(x, dy, dv) = 1_{\mathbb{R}_+}(v)\, dF_x(v)\, \beta(x, v; dy),$$

where dF_x is a regular conditional probability distribution of $T_{n+1} - T_n$ given $Y_n = x$, $\beta(x, v; dy)$ a regular conditional probability distribution of Y_{n+1} given $Y_n = x$, $T_{n+1} - T_n = v$ $(n \geq 1)$. Thus dF_x is the probability distribution of T_1 under $\mathbb{P}_x$, and $\beta(x, v; dy)$ is a conditional probability distribution of Y_1 given $T_1 = v$ under $\mathbb{P}_x$. The notation $1_{\mathbb{R}_+}(v)\, dF_x(v)$ is for dF_x restricted to $\mathbb{R}_+$; they are the interarrival distributions.

Recall that $dF_x(\{0\}) = 0$ for every $x \in E$, since $N(x, E \times \{0\}) = 0$, and

$$\bar{F}_x(u) = dF_x(]u, +\infty]) = \mathbb{P}_x(T_1 > u).$$

Define

$$t^*(x) = \sup\{u > 0 : \bar{F}_x(u) > 0\},$$

$$\check{E} = \{(x, u) \in E \times \mathbb{R}_+ : \bar{F}_x(u) > 0\} = \{(x, u) \in E \times \mathbb{R}_+ : u < t^*(x)\},$$

$$\hat{E} = \{(x, u) \in E \times \mathbb{R}_+ : u \leq t^*(x)\}.$$

Recall that for $(x, u) \in \check{E}$,

$$\int_{E\times\mathbb{R}_+} \varphi(y, v)\, N_0^{x,u}(dy, dv) = \frac{1}{\bar{F}_x(u)} \int_{E\times\mathbb{R}_+} \varphi(y, v - u)\, 1_{\{v>u\}}\, N(x, dy, dv)$$

for every Borel function $\varphi : E \times \mathbb{R}_+ \to \mathbb{R}_+$. We get

$$N_0^{x,u}(dy, dv) = 1_{\mathbb{R}_+}(v)\, \beta(x, u+v; dy)\, dF_{x,u}(v), \tag{3.1}$$

where $dF_{x,u}$ is the probability distribution of T_1 under $\mathbb{P}_{x,u}$, especially $dF_{x,u}(]t, +\infty]) = \bar{F}_x(u+t)/\bar{F}_x(u)$.

3.1 General PDMPs

A PDMP (piecewise deterministic Markov process) is a Markov process that can be written as $(\phi(Z_t, A_t))_{t\geq 0}$, where $(Z_t, A_t)_{t\geq 0}$ is a CSMP and ϕ a Borel function. A CSMP is the simplest PDMP with general interarrival distributions we can imagine. That is why in the remainder of this book we first establish results for CSMPs and then we deduce those for PDMPs.

The lemma below, which is very general, gives a very natural condition for a function of a Markov process also to be a Markov process.

Lemma 3.1 *Let $(G_1, \mathcal{G}_1)$ and $(G_2, \mathcal{G}_2)$ be two measurable spaces, $\phi : G_1 \to G_2$ a measurable function, and $(X_t)_{t\geq 0}$ a Markov process with values in G_1.*

Assume that for every $x_1 \in G_1$ and $t > 0$, the conditional probability distribution of $\phi(X_t)$ given $X_0 = x_1$ depends only on t and $\phi(x_1)$. More precisely, we assume that for all $t > 0$, there exists a kernel $\tilde{P}_t$ on G_2 such that $\tilde{P}_t(\phi(x_1), \cdot)$ is the probability distribution of $\phi(X_t)$ given $X_0 = x_1$, for all $x_1 \in G_1$.

Then $(\phi(X_t))_{t\geq 0}$ is a Markov process with values in $\phi(G_1)$, and $\tilde{P}_t$ is its semigroup, that is, $\mathbb{E}(f \circ \phi(X_t) \mid \phi(X_0) = x_2) = \int_{G_2} f(y)\tilde{P}_t(x_2, dy)$ for every $x_2 \in \phi(G_1)$ and measurable function $f : \phi(G_1) \to \mathbb{R}_+$.

Proof Given $0 \leq t_1 < t_2 < t_k < t_{k+1}$ and a measurable function $f : G_2 \to \mathbb{R}_+$, we must check that

$$\mathbb{E}(f(\phi(X_{t_{k+1}})) \mid \phi(X_{t_1}), \phi(X_{t_2}), \ldots, \phi(X_{t_k})) = \int_{G_2} f(y)\, \tilde{P}_{t_{k+1}-t_k}(\phi(X_{t_k}), dy).$$

We get this by conditioning first on $(X_s)_{s\leq t_k}$. ∎

In the following proposition we give sufficient (and presumably necessary) conditions for $(\phi(Z_t, A_t))_{t\geq 0}$ to be a Markov process. These conditions relate to ϕ and the CSMP kernel N. Following Jacobsen [66], we assume that $\phi(x, s+t) = \phi(\phi(x, s), t)$, which is a very natural condition. The condition (3.2) below on N is also quite natural.

Theorem 3.1 *Let $(Z_t, A_t)_{t\geq 0}$ be a CSMP with kernel N and $\phi : \hat{E} \to E$ a Borel function such that $v \in [0, t^*(x)] \cap \mathbb{R}_+ \to \phi(x, v)$ is continuous for every $x \in E$. Define $X_t = \phi(Z_t, A_t)$.*

Suppose that

$$\forall (x, u) \in \breve{E}, \quad N_0^{x,u}(\cdot) = N(\phi(x, u), \cdot), \tag{3.2}$$

and

$$\forall (x, s, t) \in E \times \mathbb{R}_+^2, \; s + t \le t^*(x) : \; \phi(x, s + t) = \phi(\phi(x, s), t), \quad \phi(x, 0) = x. \tag{3.3}$$

Then $(X_t)_{t\ge 0}$ *is a Markov process, and its conditional probability distribution given* $X_0 = x_0$ *is the conditional probability distribution of* $(\phi(Z_t, A_t))_{t\ge 0}$ *given* $Z_0 = x$, $A_0 = u$, *for every* x *and* u *satisfying* $\phi(x, u) = x_0$.

Proof According to Lemma 3.1, we must check that the conditional probability distribution of $\phi(Z_t, A_t)$ given $Z_0 = x$, $A_0 = u$ depends only on t and $\phi(x, u)$. This follows directly from (2.21), Sect. 2.5, formulas $\bar{F}_x(u + t)/\bar{F}_x(u) = N_0^{x,u}(E\times]t, +\infty])$, (3.2), and (3.3). ∎

If the conditions of Theorem 3.1 hold, the conditional probability distribution of $(X_t)_{t\ge 0}$ given $X_0 = x_0$ is the conditional probability distribution of $(\phi(Z_t, A_t))_{t\ge 0}$ given $Z_0 = x_0$, $A_0 = 0$; hence it can be assumed that $(Z_t, A_t)_{t\ge 0}$ is a zero-delayed CSMP.

Definition 3.1 A Borel function $\phi : \hat{E} \to E$ is said to be a deterministic (homogeneous) Markov flow on $\hat{E}$ if (3.3) holds and $v \in [0, t^*(x)] \cap \mathbb{R}_+ \to \phi(x, v)$ is continuous for all $x \in E$.

Example 3.1 A function $\phi : \hat{E} \to E$ is a deterministic Markov flow as soon as $E \subset \mathbb{R}^d$ and ϕ is the solution of the ODE

$$\frac{\partial \phi}{\partial t}(x, t) = \mathbf{v}(\phi(x, t)), \; \phi(x, 0) = x,$$

where $\mathbf{v}$ is locally Lipschitz continuous. In this case, the flow is said to be governed by an ODE.

Definition 3.2 A process $(X_t)_{t\ge 0}$ is said to be a piecewise deterministic Markov process (PDMP) on E if there exist a zero-delayed CSMP $(Z_t, A_t)_{t\ge 0}$ on E with kernel N and a deterministic Markov flow on $\hat{E}$ such that $X_t = \phi(Z_t, A_t)$ $(t \ge 0)$ and (3.2) holds.

The PDMP is said to be generated by ϕ and the CSMP $(Z_t, A_t)_{t\ge 0}$ or generated by ϕ and the Markov renewal process $(Y_n, T_n)_{n\ge 0}$ associated with the CSMP $(Z_t, A_t)_{t\ge 0}$ or generated by ϕ and N (note that in this case, only its probability distribution is given).

The CSMP $(Z_t, A_t)_{t\ge 0}$ and the Markov renewal process $(Y_n, T_n)_{n\ge 0}$ are said to be associated with the PDMP $(X_t)_{t\ge 0}$.

The deterministic Markov flow ϕ is said to be the PDMP deterministic flow.

As the name indicates, a PDMP is a Markov process. As usual for a Markov process, the initial distribution of PDMP $(X_t)_{t\geq0}$ is the probability distribution of X_0.

When writing "PDMP generated by ϕ and ...," it is implicitly assumed that ϕ is a deterministic Markov flow on $\hat{E}$. In applications, ϕ is often a deterministic Markov flow on a larger space than $\hat{E}$.

Example 3.2 Let $(T_n)_{n\geq1}$ be a zero-delayed renewal process with $T_0 = 0$, the random variables $T_{n+1} - T_n$ $(n \geq 0)$ are i.i.d. with probability distribution dF. Define $A_t = t - T_n$ when $T_n \leq t < T_{n+1}$ $(n \geq 0)$. Then $(A_t)_{\geq0}$ is a PDMP. Indeed, let $E = \mathbb{R}_+, Y_n = 0$ for all $n \geq 0, \phi(x, t) = x + t$. We have $Z_t = 0$ and $\phi(Z_t, A_t) = A_t$ for all $t \geq 0$. Moreover, $(Y_n, Y_n)_n$ is a zero-delayed Markov renewal process with kernel N defined as $\quad (\overline{T_n}, Y_n) \quad x$ is for value of T_n, $Y_n \geq 0$

$$\int_{E\times\mathbb{R}_+} \varphi(y, v)\, N(x, dy, dv) = \frac{1}{\bar{F}(x)} \int_{\mathbb{R}_+} \varphi(0, v - x)\, 1_{\{v>x\}}\, dF(v),$$

that is, $N(dx, dy, dv) = \delta_0(dy)\, dF_x(v)$, $dF_x(v)$ being the probability distribution of $\xi - x$ given $\xi > x$, where ξ is a random variable with probability distribution dF. We have $N_0^{x,u}(dy, dv) = N(\phi(x, u), dy, dv)$, since $\bar{F}_x(u) = \bar{F}(x + u)/\bar{F}(x)$. Hence $(A_t)_{t\geq0}$ is a PDMP generated by ϕ and N.

More generally, a CSMP is a PDMP as defined in Definition 3.2:

Example 3.3 A CSMP on E with kernel N is a PDMP generated by ϕ and $\tilde{N}$, where
$\quad$ 7 A_t component

$$\phi((x, u), v) = (x, u + v), \quad \tilde{N}((x, u), dydw, dv) = N_0^{x,u}(dy, dv)\, \delta_0(dw).$$

Example 3.4 To operate, a factory needs commodities that are consumed at speed $c \in \mathbb{R}_+$. The amount of commodities at time 0 is $S_0 \in \mathbb{R}_+$. A quantity C_n is delivered at times T_n $(n \geq 1)$. We assume that $(T_n)_{n\geq0}$ is a zero-delayed renewal process with interarrival distribution dF, $(C_n)_{n\geq1}$ are i.i.d. r.v. with probability distribution μ, and S_0, $(C_n)_{n\geq1}$, and $(T_n)_{n\geq1}$ are independent. This example was mentioned in the introduction (Examples 1.1, 1.6, 1.8, 1.11, Chap. 1).

Define S_t as the amount of commodities at time t. For $T_n \leq t < T_{n+1}$, we have $S_t = (S_{T_n} - c(t - T_n))_+$. In addition $S_{T_{n+1}} = (S_{T_n} - c(T_{n+1} - T_n))_+ + C_{n+1}$ $(n \geq 0)$, where $a_+ = \max(a, 0)$. Remember that $T_0 = 0$.

Define $\quad Y_n^{(1)} = S_{T_n} \quad (n \geq 0), \quad N_1(x, dy, dv) = (\delta_{(x-cv)_+} * \mu)(dy)\, dF(v)$, $\phi_1(x, v) = (x - cv)_+$. Then $(Y_n^{(1)}, T_n)_n$ is a Markov renewal process with kernel N_1, ϕ_1 is a deterministic Markov flow, and $S_t = \phi_1(Z_t, A_t)$. However, in general, $(S_t)_{t\geq0}$ is not a PDMP on $E = \mathbb{R}_+$, because "the memory of dF" is not taken into account. Indeed, for every $\varphi : E \times \mathbb{R}_+ \to \mathbb{R}_+$, we have

$$\int_{E\times\mathbb{R}_+} \varphi(y, v)\, N_1(x, dy, dv) = \int_{E\times\mathbb{R}_+} \varphi((x - cv)_+ + y, v)\, \mu(dy)\, dF(v),$$
$$\int_{E\times\mathbb{R}_+} \varphi(y, v)\, (N_1)_0^{x,u}(dy, dv) = \frac{1}{\bar{F}(u)} \int_{E\times\mathbb{R}_+} \varphi((x - cv)_+ + y, v - u)\, 1_{\{v>u\}}\, \mu(dy)\, dF(v),$$

$$\int_{E\times\mathbb{R}_+} \varphi(y,v)\, N_1(\phi_1(x,u), dy, dv) = \int_{E\times\mathbb{R}_+} \varphi((x-c(u+v))_+ + y, v)\, \mu(dy)\, dF(v).$$

The condition $(N_1)_0^{x,u}(dy,dv) = N_1(\phi_1(x,u), dy, dv)$ does not hold, except if $\int_{\mathbb{R}_+} f(v)\, 1_{\{v>u\}}\, dF(v) = \bar{F}(u) \int_{\mathbb{R}_+} f(u+v)\, dF(v)$ for all $u \in \mathbb{R}_+$ such that $\bar{F}(u) > 0$ and every Borel function $f : \mathbb{R}_+ \to \mathbb{R}_+$. Taking $f(v) = 1_{\{v>a+u\}}$, we get $\bar{F}(u+a) = \bar{F}(u)\,\bar{F}(a)$; thus dF is an exponential distribution, which is the only probability distribution on $\mathbb{R}_+$ without memory.

Let us go back to the general case. Define $X_t = (S_t, A_t)$, $E = \mathbb{R}_+^2$,

$$\phi((x_1,x_2),t) = (\phi_1(x_1,t), x_2+t) = ((x_1-ct)_+, x_2+t),$$

$$N((x_1,x_2), dy_1dy_2, dv) = (\delta_{(x_1-cv)_+} * \mu)(dy_1)\, \delta_0(dy_2)\, dF_{x_2}(v), \tag{3.4}$$

where dF_{x_2} is the probability distribution of $\xi - x_2$ given $\xi > x_2$, ξ being a random variable with probability distribution dF. Hence

$$\bar{F}_{(x_1,x_2)}(u) = \bar{F}_{x_2}(u) = \bar{F}(x_2+u)/\bar{F}(x_2). \tag{3.5}$$

We have

$$\int_{E\times\mathbb{R}_+} \varphi(x_1,x_2,v)\, N((x_1,x_2), dy_1dy_2, dv)$$
$$= \frac{1}{\bar{F}(x_2)} \int_{\mathbb{R}_+\times\mathbb{R}_+} 1_{\{v>x_2\}}\, \varphi(\phi_1(x_1, v-x_2) + y_1, 0, v-x_2)\, \mu(dy_1)\, dF(v);$$

hence

$$\int_{E\times\mathbb{R}_+} \varphi(x_1,x_2,v)\, N(\phi((x_1,x_2),u), dy_1dy_2, dv)$$
$$= \frac{1}{\bar{F}(x_2+u)} \int_{\mathbb{R}_+\times\mathbb{R}_+} 1_{\{v>x_2+u\}}\, \varphi(\phi_1(x_1, v-x_2) + y_1, 0, v-x_2-u)\, \mu(dy_1)\, dF(v),$$

$$\int_{E\times\mathbb{R}_+} \varphi(x_1,x_2,v)\, N_0^{(x_1,x_2),u}(dy_1dy_2, dv)$$
$$= \frac{1}{\bar{F}_{x_2}(u)} \frac{1}{\bar{F}(x_2)} \int_{\mathbb{R}_+\times\mathbb{R}_+} 1_{\{v-u>x_2\}}\, \varphi(\phi_1(x_1, v-x_2) + y_1, 0, v-u-x_2)\, \mu(dy_1)\, dF(v).$$

Applying (3.5), we get $N_0^{(x_1,x_2),u}(dy_1dy_2,dv) = N(\phi((x_1,x_2),u), dy_1dy_2, dv)$; hence $(S_t, A_t)_{t\ge 0}$ is a PDMP generated by ϕ and N.

If we are interested in the periods during which the stock is empty, it can be interesting to introduce jumps when the stock becomes empty that is what we do for assessing loss of production with a numerical scheme (Sect. 9.5).

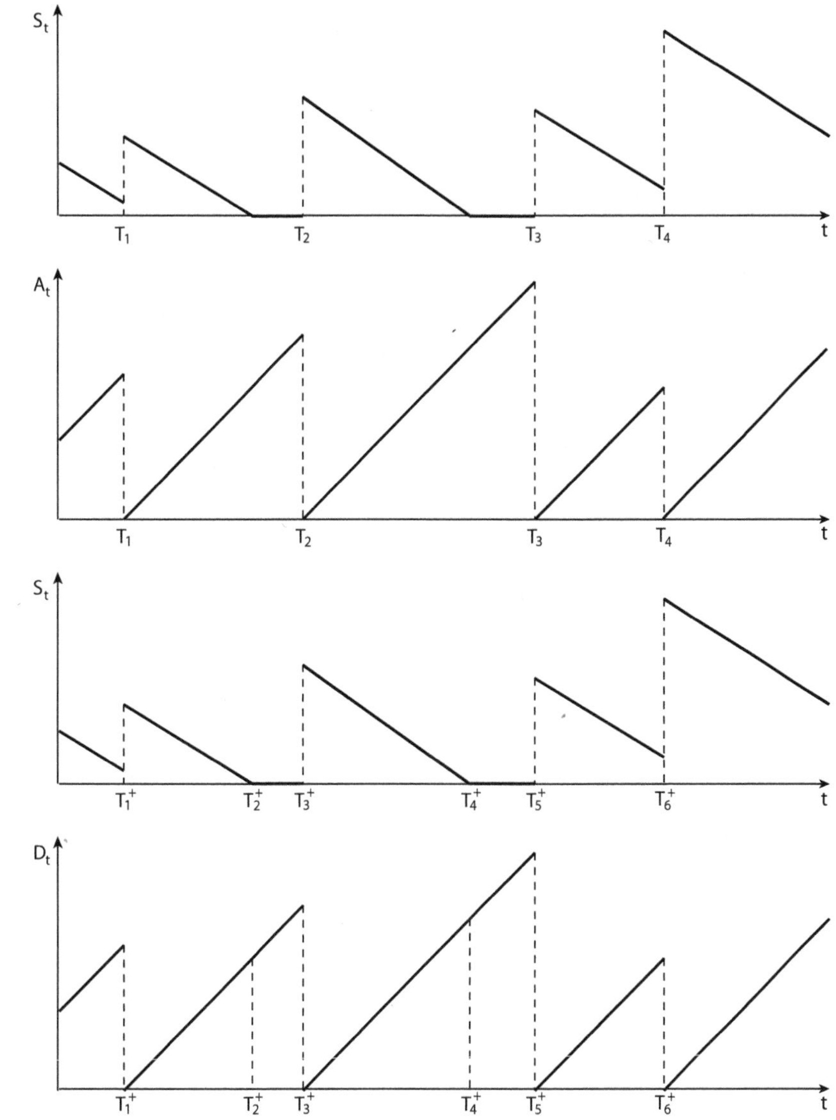

Fig. 3.1 Stock models of Example 3.4: $(T_k)_{k\geq 1}$ are the jumps corresponding to N, $(T_k^+)_{k\geq 1}$ are the jumps corresponding to N_+

When adding jumps when the stock becomes empty, the times since the last delivery must be kept in mind. This is no longer A_t, the time elapsed since the last jump. We call D_t the time elapsed since the last delivery. The models without and with additional jumps are displayed in Fig. 3.1.

Process $(S_t, D_t)_{t\geq 0}$ is a PDMP generated by ϕ defined as above and the Markov renewal kernel N_+ defined as

$$N_+((x_1, x_2), dy_1 dy_2, dv) = 1_{\{v\leq x_1/c\}} (\delta_{x_1-cv} * \mu)(dy_1)\, \delta_0(dy_2)\, dF_{x_2}(v) \qquad (3.6)$$
$$+ \bar{F}_{x_2}(x_1/c)\, \delta_0(dy_1)\, \delta_{x_2+x_1/c}(dy_2)\, \delta_{x_1/c}(dv), \quad x_1 > 0,$$
$$N_+((0, x_2), dy_1 dy_2, dv) = \mu(dy_1)\, \delta_0(dy_2)\, dF_{x_2}(v). \qquad (3.7)$$

When $c = 0$ and $(T_n)_{n\geq 0}$ is a Poisson process, $(S_t)_{t\geq 0}$ is the well-known compound Poisson process. It is extensively revisited in Rolski et al. [88], Chap. 11.

When $c = 1$, $(S_t)_{t\geq 0}$ is the virtual waiting time of a GI/G/1 queue. More precisely, in a GI/G/1 queue, one customer arrives at each time T_n $(n \geq 1)$ in a queue with one server, and its service time is C_n. The assumptions are as above. In [5], S. Asmussen gives prominence to the study of such queues (Chaps. III, VIII, IX). The virtual waiting time, denoted by V_t in [5], is the waiting time of a virtual customer arriving just after time t. It is also known as the residual work, that is, the total time the server has to work to clear the system from that moment on if no new customer arrives afterward.

Definition 3.2 is more general than the one given in Davis [51]. A PDMP defined by M.H.A. Davis is a particular case of what we call a parametrized PDMP (see Sect. 3.2).

In Davis [51], jumps on site are not allowed. This is not the case here. As a consequence, several Markov renewal processes, and as a result several CSMPs, may be associated with the same PDMP. This is what happens in Example 3.4, since $A_t = D_t$ when the kernel of the Markov renewal process is N. The example below provides another illustration.

Example 3.5 Let $(S_n)_{n\geq 0}$ be an ordinary Poisson process that is $S_0 = 0$, and $S_{n+1} - S_n$, $n \geq 0$, are i.i.d. random variables with the same Poisson probability distribution. On the other hand, let $(B_n)_{n\geq 1}$ be i.i.d. Bernoulli distributed random variables and define

$$Z_0 = 0, \quad Z_{n+1} = \begin{cases} 1 - Z_n, & B_n = 1, \\ Z_n, & B_n = 0 \end{cases} \quad (n \geq 0), \quad X_t^1 = Z_n, \; S_n \leq t < S_{n+1}.$$

Then $(X_t^{(1)})_{t\geq 0}$ is a PDMP generated by the deterministic Markov flow $\phi(x, u) = x$ for all $u \in \mathbb{R}_+$ and the zero-delayed Markov renewal process $(Z_n, S_n)_{n\geq 0}$.

Now erase the Poisson process $(S_n)_{n\geq 0}$ using the sequence $(B_n)_{n\geq 1}$, i.e., S_n $(n \geq 1)$ is recorded if $B_n = 1$ and erased if $B_n = 0$. Denote by $(T_n)_{n\geq 1}$ the recorded times and $T_0 = 0$. Then $(T_n)_{n\geq 0}$ is a Poisson process. Define

$$Y_0 = 0, \quad Y_{n+1} = 1 - Y_n \; (n \geq 0), \quad X_t^{(2)} = Y_n \text{ if } T_n \leq t < T_{n+1}.$$

Then $(X_t^{(2)})_{t\geq 0}$ is a PDMP generated by the deterministic Markov flow $\phi(x,u)=x$ for all $u\in\mathbb{R}_+$ and the zero-delayed Markov renewal process $(Y_n,T_n)_{n\geq 0}$. Moreover, $X_t^{(1)}=X_t^{(2)}$ for all $t\geq 0$.

More generally, let $N(x,dy,dv)=Q(x;dy)\,\lambda(x)\,e^{-\lambda(x)v}dv$ be a proper semi-Markov kernel of a Markov jump process on E (see Proposition 2.9, Sect. 2.6) such that $\sup_{x\in E}\lambda(x)\leq\lambda<+\infty$. Define

$$N'(x,dy,dv)=Q'(x;dy)\,\lambda\,e^{-\lambda v}\,dv,\quad Q'(x;dy)=(1-\frac{\lambda(x)}{\lambda})\,\delta_x(dy)+\frac{\lambda(x)}{\lambda}\,Q(x;dy)$$

and $\phi(x,u)=u$ for all $u\in\mathbb{R}_+$. One can construct a Markov jump process generated by ϕ and N and also by ϕ and N' (see Cocozza-Thivent [29] Sect. 8.5).

Proposition 3.1 *Let $(X_t)_{t\geq 0}$ be the PDMP generated by ϕ and the zero-delayed Markov renewal process $(Y_k,T_k)_{k\geq 0}$ with kernel N. Denote by $(Z_t,A_t)_{t\geq 0}$ the CSMP associated with $(Y_k,T_k)_{k\geq 0}$. Define the post-t-point process $(Y^t,T^t)=(Y_k^t,T_k^t)_{k\geq 1}$ as in Theorem 2.7, Sect. 2.4, namely*

$$Y_k^t=Y_{n+k},\ T_k^t=T_{n+k}-t,\ \ T_n\leq t<T_{n+1}\ (n\geq 0,k\geq 1).$$

Then, given $(Z_s,A_s)_{s\leq t}$, (Y^t,T^t) is a Markov renewal process with kernel N and initial distribution $N(X_t,\cdot,\cdot)$. In particular, given $(X_s)_{s\leq t}$, (Y^t,T^t) is a Markov renewal process with kernel N and initial distribution $N(X_t,\cdot,\cdot)$.

Proof The result is a straightforward consequence of Corollary 2.2, Sect. 2.4 and (3.2), Sect. 3.1 since $N_0^{Z_t,A_t}(\cdot,\cdot)=N(\phi(Z_t,A_t),\cdot,\cdot)=N(X_t,\cdot,\cdot)$. ■

In Chap. 6 we will look at the process $(X_t,t)_{t\geq 0}$. The following proposition asserts that it is also a Markov process and that the same is true for the process $(X_t,A_t)_{t\geq 0}$.

Proposition 3.2 *Let $(X_t)_{t\geq 0}$ be the PDMP on E generated by ϕ and the CSMP $(Z_t,A_t)_{t\geq 0}$ with kernel N. Define the deterministic Markov flow $\bar\phi$ on $E\times\mathbb{R}_+$ as $\bar\phi((x,c),t)=(\phi(x,t),c+t)$.*

Then $(X_t,t)_{t\geq 0}$ (respectively $(X_t,A_t)_{t\geq 0}$) is a PDMP generated by $\bar\phi$ and $N^{(1)}$ (respectively $\bar\phi$ and $N^{(2)}$) defined as

$$N^{(1)}((x,c),dx_1dc_1,dv)=N(x,dx_1,dv)\,\delta_{c+v}(dc_1),$$

$$N^{(2)}((x,c),dx_1dc_1,dv)=N(x,dx_1,dv)\,\delta_0(dc_1).$$

The proof is straightforward. Indeed, let $(Y_n,T_n)_{n\geq 0}$ be a Markov renewal process associated with $(X_t)_{t\geq 0}$. Then $(X_t,t)=\bar\phi(Z_t^{(1)},A_t)$, where $(Z_t^{(1)},A_t)_{t\geq 0}$ is the CSMP associated with the Markov renewal process $((Y_n,T_n),T_n)_{n\geq 0}$ with kernel $N^{(1)}$. Similarly, $(X_t,A_t)=\bar\phi(Z_t^{(2)},A_t)$, where $(Z_t^{(2)},A_t)_{t\geq 0}$ is the CSMP associated with the Markov renewal process $((Y_n,0),T_n)_{n\geq 0}$ with kernel $N^{(2)}$.

Inhomogeneous PDMPs

This subparagraph is a generalization that is not essential for what follows. It may be omitted on first reading.

Suppose now that we are concerned with a deterministic flow solution of the inhomogeneous ODE $z'(t) = \mathbf{v}(t, z(t))$, where $\mathbf{v}$ is a locally Lipschitz continuous function. Define $\phi(s, x, s+t)$ as the value at time $s+t$ of the ODE solution satisfying $z(s) = x$. Then ϕ is an inhomogeneous flow as defined below.

Definition 3.3 A function $\phi : \mathbb{R}_+ \times E \times \mathbb{R}_+ \to E$ is said to be an (a priori) inhomogeneous deterministic Markov flow on E if for all $(x, u, s, t) \in E \times \mathbb{R}_+^3$

$$\phi(u, x, u) = x, \ \phi(u, x, u+s+t) = \phi(u+s, \phi(u, x, u+s), u+s+t). \quad (3.8)$$

Given an inhomogeneous deterministic Markov flow ϕ on E, we define the homogeneous deterministic Markov flow $\widetilde{\phi}$ on $\bar{E} = \mathbb{R}_+ \times E$ as

$$\widetilde{\phi}((s, x), t) = (s+t, \phi(s, x, s+t)). \quad (3.9)$$

Lemma 3.2 *Given a kernel $\bar{N}$ from $\bar{E}$ to $E \times \mathbb{R}_+$ such that $\bar{N}((s, x), E \times \{0\}) = 0$ and $\bar{N}((s, x), E \times \mathbb{R}_+) \le 1$ for all $(s, x) \in \bar{E}$, we define the semi-Markov kernel $\widetilde{N}$ on $\bar{E}$ as follows:*

$$\widetilde{N}((s, x), ds_1 dx_1, dv) = \bar{N}((s, x), dx_1, dv)\, \delta_{s+v}(ds_1). \quad (3.10)$$

Then a zero-delayed Markov renewal process with kernel $\widetilde{N}$ can be written as $(\widetilde{Y}_n, T_n)_{n\ge 0}$, where $\widetilde{Y}_n = (S_0 + T_n, Y_n)$ a.s., S_0 and Y_n $(n \ge 0)$ are random variables with values in $\mathbb{R}_+$ and E, respectively, $T_0 = 0$, and $(T_n)_{n\ge 1}$ is an increasing sequence of positive random variables.

Let $(\widetilde{Z}_t, A_t)_{t\ge 0}$ be the associated CSMP. Then $\widetilde{Z}_t = (S_0 + T_{N_t}, Z_t)$ where $N_t = \sum_{k\ge 1} 1_{\{T_k \le t\}}$ and $Z_t = Y_k$ on $\{N_t = k\}$.

Let Y_n, T_n $(n \ge 0)$ be as in Lemma 3.2. We have

$$\mathbb{E}(\varphi(Y_{n+1}, T_{n+1} - T_n)|\, S_0, Y_0, Y_1, T_1, \cdots Y_{n-1}, T_{n-1}, S_0 + T_n = s, Y_n = x)$$
$$= \int_{E\times\mathbb{R}_+} \varphi(y, v)\, \bar{N}((s, x), dy, dv),$$

for every Borel function $\varphi : E \times \mathbb{R}_+ \to \mathbb{R}_+$; hence $(Y_n, T_n)_{n\ge 0}$ is not a Markov renewal process except if $\bar{N}((s, x), dy, dv)$ does not depend on s.

Define $\widetilde{X}_t = \widetilde{\phi}(\widetilde{Z}_t, A_t)$, where $\widetilde{\phi}$ is as in (3.9). We get $\widetilde{X}_t = (S_0 + t, X_t)$ where $X_t = \phi(S_0 + T_n, Y_n, S_0 + t)$ on $\{N_t = n\}$.

As a consequence of Theorem 3.1 we get the following result.

Proposition 3.3 *Given the previous notation, assume*

$$\widetilde{N}_0^{(s,x),u}(ds_1dx_1, dv) = \widetilde{N}(\widetilde{\phi}((s,x),u), ds_1dx_1, dv) \tag{3.11}$$

for all $(s, x, u) \in \bar{E} \times \mathbb{R}_+$ *such that* $\widetilde{N}((s,x), \bar{E} \times [0,u]) < 1$. *Then* $(S_0 + t, X_t)$ *is a PDMP. We call* $(X_t)_{t\geq 0}$ *an inhomogeneous PDMP on* E *generated by* ϕ *and* $\bar{N}$.

Example 3.6 Consider an inhomogeneous Poisson process $(T_n)_{n\geq 1}$ with rate λ, this means that the hazard rate of the conditional distribution of $T_{n+1} - T_n$ given $T_1, \ldots, T_n$ is $\lambda(T_n + \cdot)$ for all $n \geq 0$ (with the convention $T_0 = 0$). Given a real r.v. V, define μ_{x+V} as the probability distribution of $x + V$ $(x \in \mathbb{R})$ and

$$\phi(s, x, s+t) = x \quad \bar{N}((s,x), dx_1, dv) = \mu_{x+V}(dx_1)\, \lambda(s+v)\, e^{-\int_0^v \lambda(s+w)\, dw}\, dv.$$

Condition (3.11) is fulfilled, and the inhomogeneous PDMP $(X_t)_{t\geq 0}$ generated by ϕ and $\bar{N}$, and such that $X_0 = 0$, is a reward process. More precisely, $(X_t)_{t\geq 0}$ has the same distribution as $(R_t)_{t\geq 0}$, where $R_t = \sum_{i=1}^n V_i$ if $T_n \leq t < T_{n+1}$ and V_i are i.i.d. with same probability distribution as V.

For every $(s_0, x_0) \in E \times \mathbb{R}_+$ define $R((s_0, x_0), \cdot)$ as the intensity of the marked point process $(Y_n, T_n)_{n\geq 0}$ given $\widetilde{Y}_0 = (s_0, x_0)$, that is,

$$\int_{E\times\mathbb{R}_+} \varphi(x, v)\, R((s_0, x_0), dx, dv) = \sum_{n\geq 0} \mathbb{E}(\varphi(Y_n, T_n)\, 1_{\{T_n < +\infty\}} \mid \widetilde{Y}_0 = (s_0, x_0))$$

for every Borel function $\varphi : E \times \mathbb{R}_+ \to \mathbb{R}_+$. Define $\widetilde{R} = \sum_{k\geq 0} \widetilde{N}^{*k}$. We have

$$\int_{\mathbb{R}_+\times E\times\mathbb{R}_+} \psi(s, x, v)\widetilde{R}((s_0, x_0), dsdx, dv) = \int_{E\times\mathbb{R}_+} \psi(s_0 + v, x, v)\, R((s_0, x_0), dx, dv) \tag{3.12}$$

for every Borel function $\psi : \mathbb{R}_+ \times E \times \mathbb{R}_+ \to \mathbb{R}_+$.

3.2 Parametrized CSMPs and PDMPs

In this section, we consider PDMPs for which each interarrival distribution is like the probability distribution of $\min(T, \alpha)$, where $\alpha \in \bar{\mathbb{R}}_+^*$ and T is a positive random variable with a probability density function f. So it is as in Davis [51], but conditions on α and on the jumps' locations will be less restrictive. To get nice formulas in the following chapters, it is better to introduce the hazard rate λ of T, that is, to write $f(t) = \lambda(t)\, e^{-\int_0^t \lambda(w)\, dw}$. Roughly speaking, $\mathbb{P}(T \in (t, t + \Delta t) \mid T > t) = \lambda(t)\, \Delta t + o(\Delta t)$ (see Appendix A, Sect. A.3).

The study of a CSMP, and consequently of a PDMP, each interarrival distribution of which is the mixture of a probability density function and a countable number of

Dirac measures, can be reduced to the study a simple CSMP (see Example 3.8, Sect. 3.2, and Appendix B, Propositions B.1 and B.2).

Throughout this section, $\ell : E \times \mathbb{R}_+ \to \mathbb{R}_+$ and $\alpha : E \to]0, +\infty]$ are Borel functions, dF_x ($x \in E$) is the probability distribution of the minimum between $\alpha(x)$ and an r.v. with hazard rate $\ell(x, \cdot)$, that is,

$$1_{\mathbb{R}_+}(v)\, dF_x(v) = \ell(x, v)\, e^{-\int_0^v \ell(x,w)\, dw}\, 1_{\{v < \alpha(x)\}}\, dv$$
$$+ 1_{\{\alpha(x) < +\infty\}}\, e^{-\int_0^{\alpha(x)} \ell(x,w)\, dw}\, \delta_{\alpha(x)}(dv),$$

and β is a probability kernel from $\hat{E}$ to E.

Definition 3.4 With the above notation, a simple CSMP on E with parameters ℓ, α, β is a CSMP on E with kernel $N(x, dy, dv) = 1_{\mathbb{R}_+}(v)\, \beta(x, v; dy)\, dF_x(v)$.

We have

$$t^*(x) = \alpha(x) \wedge t_\ell^*(x) \quad \text{where} \quad t_\ell^*(x) = \sup\{v > 0 : \int_0^v \ell(x, w)\, dw < +\infty\},$$

and $P_{x_0}\left(Y_1 \in dy, \; T_1 - u \in dv \mid T_1 > u \right)$

$$N_0^{x,u}(dy, dv) = \Big(\ell(x, u + v)\, e^{-\int_0^v \ell(x,u+w)\, dw}\, 1_{\{v < \alpha(x) - u\}} \tag{3.13}$$
$$+ 1_{\{\alpha(x) < +\infty\}}\, e^{-\int_0^{\alpha(x)-u} \ell(x,u+w)\, dw}\, \delta_{\alpha(x)-u}(dv) \Big)\, \beta(x, u + v; dy).$$

Example 3.7 Let us go back to Example 2.4, Sect. 2.5 with instantaneous preventive maintenance. Assume that dF_1 and dF_0 have densities. Denote by λ (respectively μ) the failure rate (respectively the repair rate) of the item, that is, the hazard rate of dF_1 (respectively of dF_0). The CSMP corresponding to this modeling has parameters ℓ, α, β with

$$\ell(e_1, v) = \lambda(v), \quad \ell(e_0, v) = \mu(v), \quad \alpha(e_1) = a, \quad \alpha(e_0) = +\infty,$$

$$\beta(e_1, v; dy) = \begin{cases} \delta_{e_0}(dy), & v < a, \\ \delta_{e_1}(dy), & v = a, \end{cases} \quad \beta(e_0, v; dy) = \delta_{e_1}(dy).$$

Thanks to Theorem 3.1, we give sufficient conditions that ℓ and α must satisfy so that $\phi(Z_t, A_t)_{t \geq 0}$ is a Markov process. From this, we derive a more general definition of parametrized PDMPs than usual, which allows for easier modeling.

From Theorem 3.1, Sect. 3.1 and (3.13) we get the following lemma.

Lemma 3.3 Let $(Z_t, A_t)_{t \geq 0}$ be a simple CSMP on E with parameters ℓ, α, β and $\phi : \hat{E} \to E$ a Borel function. Assume that

1. ϕ is a deterministic Markov flow on $\hat{E}$,

2. *for all* $(x, u) \in \breve{E}$,

$$1_{\{u+v<t^*(x)\}}\ell(x, u+v)\,\beta(x, u+v; dy)\,dv = 1_{\{u+v<t^*(x)\}}\ell(\phi(x,u), v)\,\beta(\phi(x,u), v; dy)\,dv,$$

3. *for all* $(x, u) \in \breve{E}$,

   a. $\alpha(\phi(x, u)) = \alpha(x) - u$,
   b. $\beta(x, \alpha(x); dy) = \beta(\phi(x, u), \alpha(\phi(x, u)); dy)$.

Define $X_t = \phi(Z_t, A_t)$. *Then the process* $(X_t)_{t\geq 0}$ *is a PDMP generated by* ϕ *and* $(Z_t, A_t)_{t\geq 0}$.

Note that if we assume that conditions 1 and 2 of Lemma 3.3 hold and $u \in [0, t^*(x)[\to \ell(x, u)$ is right continuous for all $x \in E$, then $\ell(x, u) = \lambda(\phi(x, u))$ with $\lambda(x) = \ell(x, 0)$. Similarly, it is not a great restriction to assume as well that $\beta(x, u; dy) = Q(\phi(x, u); dy)$ for all $u < \alpha(x)$ and some probability kernel Q on E in place of Condition 2 of Lemma 3.3.

On the other hand, assume that conditions 1 and 3(a) of Lemma 3.3 hold. Let $(E_i)_{i\in I}$ be a countable partition of E such that for all $i \in I, x \in E_i, \phi(x, u) \in E_i$ for all $u < \alpha(x)$ and $\beta(x, \alpha(x); dy) = q_i(\phi(x, \alpha(x)); dy)$ for some q_i. Then condition 3(b) of Lemma 3.3 holds, since $\phi(\phi(x, u), \alpha(\phi(x, u))) = \phi(x, \alpha(x))$.

In view of these remarks we give the following definition for a parametrized PDMP.

Definition 3.5 Let $\phi : E \times \mathbb{R}_+ \to E$, $\lambda : E \to \mathbb{R}_+$, $\alpha : E \to]0, +\infty]$ be Borel functions, dF_x ($x \in E$) the probability distribution of the minimum between $\alpha(x)$ and an r.v. with hazard rate $\lambda(\phi(x, \cdot)$, that is,

$$1_{\mathbb{R}_+}(v)\,dF_x(v) = \lambda(\phi(x, v))\,e^{-\int_0^v \lambda(\phi(x,w))\,dw}\,1_{\{v<\alpha(x)\}}\,dv$$
$$+ 1_{\{\alpha(x)<+\infty\}}\,e^{-\int_0^{\alpha(x)} \lambda(\phi((x,w))\,dw}\,\delta_{\alpha(x)}(dv).$$

Assume that ϕ is a deterministic Markov flow on $\hat{E}$ and

$$\forall (x, u) \in \breve{E} : \quad \alpha(\phi(x, u)) = \alpha(x) - u. \tag{3.14}$$

Let $(E_i)_{i\in I}$ be a countable partition of E or of a subset of E. For $i \in I$, define

$$\Gamma_i = \{\phi(x, \alpha(x)) : x \in E_i, \alpha(x) < +\infty\} \tag{3.15}$$

and assume that

1. $\forall i \in I,\ \forall x \in E_i,\ \forall u < \alpha(x) : \phi(x, u) \in E_i$,
2. $\forall i \in I,\ E_i \cap \Gamma_i = \emptyset$.

Let Q be a probability kernel on E and q_i a probability kernel from Γ_i^* to E, where

$$\Gamma_i^* = \{\phi(x, \alpha(x)) : x \in E_i,\ \alpha(x) < +\infty,\ \int_0^{\alpha(x)} \lambda(\phi(x, u))\, du < +\infty\}.$$

Define

$$\beta(x, v; dy) = \begin{cases} Q(\phi(x, v); dy), & v < \alpha(x), \\ q_i(\phi(x, \alpha(x); dy), & x \in E_i, v = \alpha(x)\ (i \in I). \end{cases}$$

A PDMP with parameters $\phi, \lambda, Q, \alpha, (E_i, q_i)_{i \in I}$ is a PDMP with values in $\cup_{i \in I} E_i$, generated by ϕ and N, where for all $i \in I, x \in E_i, N(x, dy, dv) = 1_{\mathbb{R}_+}(v)\, \beta(x, v; dy)\, dF_x(v)$ that is

$$N(x, dy, dv) = \lambda(\phi(x, v))\, e^{-\int_0^v \lambda(\phi(x,s))\, ds}\, 1_{\{v < \alpha(x)\}}\, Q(\phi(x, v); dy)\, dv \qquad (3.16)$$
$$+ 1_{\{\alpha(x) < +\infty\}}\, e^{-\int_0^{\alpha(x)} \lambda(\phi(x,s))\, ds}\, q_i(\phi(x, \alpha(x)); dy)\, \delta_{\alpha(x)}(dv).$$

If I is a singleton, the PDMP parameters are $\phi, \lambda, Q, \alpha, q$.

If $\alpha(x) = +\infty$ for every $x \in E$, the PDMP is said to be without Dirac measure or without boundary; its parameters are ϕ, λ, Q.

Remark 3.1 Define $\underline{E}$ as the subset where the PDMP has values. In view of (3.16):

1. $\lambda(x)$ has to be specified for all $x \in \underline{E}$,
2. $Q(x; dy)$ has to be specified for all $x \in \underline{E}$ such that $\lambda(x) \neq 0$,
3. because of the Lebesgue measure dv, $1_{\mathbb{R}_+^*}(v)$ can be added in the first term of the second member of (3.16),
4. $q_i(x; dy)$ has to be specified for $x \in \Gamma_i^*$ $(i \in I)$. Usually, $q_i(x; dy)$ is given for all $x \in \Gamma_i$.

Remark 3.2 Given a PDMP with parameters $\phi, \lambda, Q, \alpha, (E_i, q_i)_{i \in I}$, $x \in E_i$, $u \in [0, \alpha(x)]$, then $u < \alpha(x)$ iff $\phi(x, u) \in E_i$ and $u = \alpha(x)$ iff $\phi(x, u) \in \Gamma_i$.

Usually,

$$\alpha(x) = \inf\{u > 0 : \phi(x, u) \in B\} \qquad (3.17)$$

for some Borel set $B \subset E$. Then (3.14) holds as soon as ϕ is a deterministic Markov flow on $\hat{E}$. Moreover, if B is an open set, we deduce from $\{x : \alpha(x) < t\} = \cup_{\{x \in \mathbb{Q}, s < t\}} \phi(x, s) \in B$ that α is a Borel function. Furthermore, if B is closed, then α is also a Borel function as soon as ϕ is continuous (see Davis [51] Lemma (27.1)).

When $\alpha(x) < +\infty$ for some $x \in E$, we shall say that we have a PDMP with boundary or, more correctly, a PDMP with Dirac measures.

The word "boundary" and the terminology "without boundary" are somewhat improper in our setting, but they are commonly used. It comes from the definition of PDMPs in Davis [51], in which α is defined as in (3.17), where B is a subset of $\partial \underline{E}$, the topological boundary of $\underline{E}$. We will use the term boundary for B as soon as $\alpha(x) = \inf\{u > 0 : \phi(x, u) \in B\} < +\infty$ for some $x \in E$, even if B is not a subset of the topological boundary of the PDMP state space (see Example 3.8 below). In that case, the parameter α may be replaced by the boundary B or $(B_i)_{i\in I}$, since it is often more descriptive. We can say that it is a PDMP with parameters ϕ, λ, Q, B, $(E_i, q_i)_{i\in I}$ or ϕ, λ, Q, $(E_i, B_i, q_i)_{i\in I}$. It is with parameters ϕ, λ, Q, B, q when I is a singleton.

Introducing subsets $E_i (i \in I)$ may appear to be an unnecessary complication. Actually, they might appear naturally in the description of the model (see Example 3.9, Sect. 3.4) or give a better understanding of the model (see Example 3.8 below). They also make it possible to take into account where the process is coming from, keeping the natural state space (see Example 3.10, Sect. 3.4). In a more fundamental way, in some cases, even if it is not needed for the model description, it is required to get comprehensive results (see comments before and after Theorem 6.1, Sect. 6.2).

Example 3.8 Let $(\alpha_i)_{1\le i\le n}$ be real numbers such that $\alpha_i < \alpha_{i+1}$. Define $\alpha_0 = -\infty$, $\alpha_{n+1} = +\infty$. A particle is moving on $\mathbb{R}$ with speed 1. When it reaches point α_i $(1 \le i \le n)$, with probability $p_i \in]0, 1]$ it jumps to another point of $\mathbb{R}$ chosen with probability distribution μ_i, and with probability $1 - p_i$, it continues with the same speed. The particle's motion can be described by a PDMP $(X_t)_{t\ge 0}$ with parameters $\phi, \lambda, Q, \alpha, q$, where $\phi(x, t) = x + t$, $\lambda = 0$, $\alpha(x) = \alpha_j - x$ when $\alpha_{j-1} \le x < \alpha_j$ $(1 \le j \le n + 1)$, $q(\alpha_i; dy) = p_i\,\mu_i(dy) + (1 - p_i)\,\delta_{\alpha_i}(dy)$ $(1 \le i \le n)$. We can also say that it is a PDMP with parameters $\phi, \lambda, Q, \cup_{i=1}^n\{\alpha_i\}, q$. Note that it is fundamental to have a strict inequality in (3.17), so that $\alpha(x) > 0$. Indeed, in the current example, $\inf\{t \ge 0 : \phi(x, t) \in \cup_{i=1}^n\{\alpha_i\}\} = 0$ for $x \in \cup_{i=1}^n\{\alpha_i\}$. On the other hand, on-site jumps avoid introducing a hybrid artificial space.

In order to have a model that satisfies (nearly) the assumptions made in Davis [51], it is necessary to introduce a state space as $\{1\}\times]-\infty, \alpha_1[\cup_{i=1}^n\{i\} \times [\alpha_{i-1}, \alpha_i[\cup\{n+1\} \times [\alpha_n, +\infty[$. We write "mainly" because the intervals are not open subsets of $\mathbb{R}$ as in [51], but we can tell that it doesn't matter, because they are closed on the left and the particle moves to the right.

Our model described above is the simplest one, but it actually has two drawbacks related to discontinuities at points α_i. The first is that it leads to writing Chapman–Kolmogorov equations for continuous functions, including at points α_i, which is not enough (see comments before Theorem 6.1, Sect. 6.2). This is not surprising, because the model with a hybrid state space shows that there is no reason to consider continuous functions at these points. The second is that this model does not satisfy the assumptions for the convergence of the numerical scheme of Chap. 9, Sect. 9.4, since α is not continuous. The solution is to keep the idea that is in the definition of the hybrid space while keeping $\mathbb{R}$ as state space. We introduce the partition $(E_i)_{1\le i\le n+1}$ defined as $E_1 =]-\infty, \alpha_1[$, $E_i = [\alpha_{i-1}, \alpha_i[$ $(2 \le i \le n)$, $E_{n+1} = [\alpha_n, +\infty[$, $\Gamma_i = \{\alpha_i\}$ $(1 \le i \le n)$, $\Gamma_{n+1} = \emptyset$. Then $(X_t)_{t\ge 0}$ is a PDMP with parameters $\phi, \lambda, Q, (E_i, \Gamma_i, q)_{1\le i\le n+1}$. The assumptions of the Chapman–Kolmogorov equations for such models require only that for each i, a test function can be extended into an absolutely continuous function over $E_i \cup \Gamma_i$. For the convergence of the numerical scheme in Chap. 9, the condition on α required only that each $\alpha_i(x) = \inf\{t > 0 : x + t \in \Gamma_i\}$ $(x \in E_i)$ can be extended into a continuous function on $E_i \cup \Gamma_i$.

Let us note that $(X_t)_{t \geq 0}$ is also a PDMP generated by ϕ and N, where

$$N(x, dy, dv) = \beta(x, v; dy)\, 1_{\mathbb{R}_+}(v)\, dF_x(v), \quad \beta(x, v; dy) = q(\phi(x, v); dy),$$

q is as above, and

$$1_{\mathbb{R}_+}(v)\, dF_x(v) =$$
$$\begin{cases} p_1 \delta_{\alpha_1 - x}(dv) + \sum_{j=2}^{n} \prod_{k=1}^{j-1}(1 - p_k)\, p_j\, \delta_{\alpha_j - x}(dv), & x < \alpha_1 \\ p_i\, \delta_{\alpha_i - x}(dv) + \sum_{j=i+1}^{n} \prod_{k=i}^{j-1}(1 - p_k)\, p_j\, \delta_{\alpha_j - x}(dv), & \alpha_{i-1} \leq x < \alpha_i, (2 \leq i \leq n) \\ 0, & x > \alpha_n. \end{cases}$$

So our model allows us to have interarrival distributions that are mixtures of Dirac measures. When $\lambda \neq 0$, the interarrival distributions are mixtures of a probability density function and Dirac measures. The interested reader will find more details on this subject in Appendix B.

Terminology:
We saw earlier Sect. 3.1 that several CSMPs can be associated with the same PDMP. In the case of a PDMP with parameters $\phi, \lambda, Q, \alpha, (E_i, q_i)_{i \in I}$, unless otherwise specified, the associated CSMP is assumed to have parameters ℓ, α, β defined as

$$\forall x \in E : \ell(x, u) = \lambda(\phi(x, u)),$$

$$\forall i \in I, \forall x \in E_i : \beta(x, u; dy) = \begin{cases} Q(\phi(x, u); dy), & u < \alpha(x) \leq +\infty, \\ q_i(\phi(x, \alpha(x)); dy), & u = \alpha(x) < +\infty. \end{cases}$$

The Markov renewal process associated with this CSMP is called the Markov renewal process associated with the parametrized PDMP.

Let us end this section by summarizing the differences between our definition of parametrized PDMPs and the definition of PDMPs given by M.H.A. Davis and M. Jacobsen, respectively.

In Davis [51], a PDMP has values in a hybrid space as $\underline{E} = \bigcup_{i \in I} \{i\} \times E_i$, where I is countable and E_i is an open subset of $\mathbb{R}^{d_i}$. On each E_i, ϕ is a deterministic Markov flow governed by an ODE. The function α is given by (3.17), where B is a subset of the topological boundary of $\underline{E}$. On-site jumps are not allowed.

In [66], no boundary is introduced, but M. Jacobsen does not require the flow to be governed by an ODE. It is assumed to be only a deterministic Markov flow. Moreover, more results than in this book are explicitly given for non-time-homogeneous flows and PDMPs.

3.3 Simulation

Consider a piecewise deterministic process $(X_t)_{t\geq 0}$ generated by ϕ and N that is $X_t = \phi(Z_t, A_t)$, where $(Z_t, A_t)_{t\geq 0}$ is a CSMP with kernel N. Write $N(x, dy, dv) = 1_{\mathbb{R}_+}(v)\,\beta(x, v; dy)\,dF_x(v)$.

Assume $X_0 = x_0$. A sample path of $(X_t)_{t\geq 0}$ is obtained as follows:

1. generate t_1 from the probability distribution dF_{x_0},
2. set $X_t(\omega) = \phi(x_0, t)$ for $t < t_1$,
3. generate y_1 from the probability distribution $\beta(x_0, t_1; dy)$, set $X_{t_1}(\omega) = y_1$,
4. generate ξ_2 from the probability distribution dF_{y_1}, set $t_2 = t_1 + \xi_2$,
5. set $X_t(\omega) = \phi(y_1, t - t_1)$ for $t_1 < t < t_2$,
6. generate y_2 from the probability distribution $\beta(y_1, \xi_2; dy)$, set $X_{t_2}(\omega) = y_2$, and so on:
7. generate ξ_{n+1} from the probability distribution dF_{y_n}, set $t_{n+1} = t_n + \xi_{n+1}$,
8. set $X_t(\omega) = \phi(y_n, t - t_n)$ for $t_n \leq t < t_{n+1}$,
9. generate y_{n+1} from the probability distribution $\beta(y_n, \xi_{n+1}; dy)$, set $X_{t_{n+1}}(\omega) = y_{n+1}$

   ...

All drawings are independent.

In parametrized PDMPs, the hazard rates of the interarrival distributions are part of the parameters. Hence, instead of simulating a random variable with density from its pdf, it is better to simulate it from its hazard rate h. Let $H(t) = \int_0^t h(s)\,ds$ and H^{-1} its left-continuous pseudoinverse function defined as $H^{-1}(t) = \inf\{u : H(u) \geq t\}$. We have $H^{-1}(s) > t$ iff $s > H(t)$. Let U be a random variable with uniform probability distribution on $[0, 1]$ and $T = H^{-1}(-\log U)$. Then $P(T > t) = \mathbb{P}(-\log U > H(t)) = \mathbb{P}(U < e^{-H(t)}) = e^{-\int_0^t h(s)\,ds}$. Therefore, T has hazard rate h.

Assume $(X_t)_{t\geq 0}$ is a PDMP with parameter ϕ, λ, Q, α, $(E_i, q_i)_{i\in I}$, that is, for all $i \in I$, $x \in E_i$,

$$N(x, dy, dv) = \lambda(\phi(x, v))\,e^{-\int_0^v \lambda(\phi(x,s))\,ds}\,1_{\{v<\alpha(x)\}}\,Q(\phi(x, v); dy)\,dv \\ + 1_{\{\alpha(x)<+\infty\}}\,e^{-\int_0^{\alpha(x)} \lambda(\phi(x,s))\,ds}\,q_i(\phi(x, \alpha(x)); dy)\,\delta_{\alpha(x)}(dv).$$

Let $(U_n)_{n\geq 1}$ be i.i.d. random variables with uniform probability distribution on $[0, 1]$. Then:

- Step 7 of the above simulation is: define $H(t) = \int_0^t \lambda(\phi(y_n, s))\,ds$, $\xi_{n+1} = H^{-1}(-\log U_{n+1})$, set $t_{n+1} = t_n + \xi_{n+1}$ if $\xi_{n+1} < \alpha(y_n)$, $t_{n+1} = t_n + \alpha(y_n)$ if $\xi_{n+1} \geq \alpha(y_n)$.
- Step 9 is: if $\xi_{n+1} < \alpha(y_n)$ then generate y_{n+1} from the probability distribution $Q(\phi(y_n, \xi_{n+1}); dy)$; if $\xi_{n+1} \geq \alpha(y_n)$ and $y_n \in E_i$ then generate y_{n+1} from the probability distribution $q_i(\phi(y_n, \alpha(y_n)); dy)$; set $X_{t_{n+1}}(\omega) = y_{n+1}$.

De Saporta, Dufour, and Zhang [53] is devoted to PDMP simulations.

3.4 Some Examples

As stated in the preface, PDMPs are known to arise in various application areas. As we have seen in some examples, they may result from the addition of supplementary variables to get a Markov process.
The examples of this section have been chosen for their instructive interest in the context of this book. For more examples, the reader is invited to refer to the references given in the preface.

Example 3.9 A particle is moving on $E =]-\infty, 0] \cup [1, +\infty[$ with speed 1 on $]-\infty, 0]$ and -1 on $[1, +\infty[$. When the particle hits the point $\{0\}$ (respectively the point $\{1\}$), it jumps immediately to a new point of $\underline{E} =]-\infty, 0[\cup]1, +\infty[$ chosen with probability distribution $q(0; \,.\,)$ (respectively $q(1; \,.\,)$). Define X_t as the position of the particle at time t. Then $(X_t)_{t\geq 0}$ is a PDMP with parameters $\phi, \lambda, Q, \alpha, q$, where

$$\phi(x,t) = \begin{cases} x+t, \; x<0, \\ x-t, \; x>1, \end{cases} \quad \lambda = 0, \; \alpha(x) = \begin{cases} |x|, \quad\; x<0, \\ x-1, \; x>1, \end{cases}$$

where Q need not be specified, since $\lambda = 0$.

It is a PDMP with boundary $B = \{0, 1\}$. The partition $(E_1 =]-\infty, 0[, E_2 =]1, +\infty[)$ of the state space appears also in a natural way. We have $\Gamma_1 = \{0\}$, $\Gamma_2 = \{1\}$. Define $q_1(\,.\,) = q(0; \,.\,)$, $q_2(\,.\,) = q(1; \,.\,)$, $(X_t)_{t\geq 0}$ is a PDMP with parameters $\phi, \lambda, Q, (E_i, \Gamma_i, q_i)_{i=1,2}$.

In the following example, the partition allows for the process to have behavior at the boundary that depends on where the process is coming from.

Example 3.10 A particle is moving in the plane with speed $(1, 0)$ when it is on $A = \mathbb{R} \times \{0\}$, with speed $(0, -1)$ when it is on $\mathbb{R}\times]0, +\infty[$, and with speed $(0, 1)$ when it is on $\mathbb{R}\times]-\infty, 0[$. When it reaches A from above (respectively below), with probability p_1 (respectively p_2) it continues with speed $(1, 0)$, and with probability $1 - p_1$ (respectively $1 - p_2$), it jumps to a point of $\mathbb{R}^2$ chosen with probability distribution μ_1 (respectively μ_2). The particle's motion is described by a PDMP with boundary. Its parameters are $\phi, \lambda, Q, (E_i, \Gamma_i, q_i)_{1\leq i\leq 3}$ defined as

$$\phi((x_1,x_2),t) = \begin{cases} (x_1, x_2 - \frac{x_2}{|x_2|}\, t), \quad x_2 \neq 0, t < |x_2|, \\ (x_1 + t - |x_2|, 0), \; t \geq |x_2|, \end{cases} \qquad \lambda = 0,$$

$$E_1 = \mathbb{R} \times \mathbb{R}_+^*, \quad E_2 = \mathbb{R} \times \mathbb{R}_-^*, \quad E_3 = A, \quad \Gamma_1 = \Gamma_2 = A, \quad \Gamma_3 = \emptyset,$$

$$q_1((x_1, 0); dy_1 dy_2) = p_1\, \delta_{(x_1,0)}(dy_1 dy_2) + (1 - p_1)\, \mu_1(dy_1 dy_2),$$

$$q_2((x_1, 0); dy_1 dy_2) = p_2\, \delta_{(x_1,0)}(dy_1 dy_2) + (1 - p_2)\, \mu_2(dy_1 dy_2).$$

Q is not specified since $\lambda = 0$.

Example 3.11 Let us return to Example 1.4 part 3, Chap. 1 in a slightly different context. We assume that the failure rate $\bar{\lambda}$ of the equipment depends on both its age and its temperature and that its repair rate depends on the duration since the beginning of the repair in progress. When the item is up (respectively down), its temperature is governed by the ODE $z'(t) = \mathbf{v_1}(z(t))$ (respectively $z'(t) = \mathbf{v_0}(z(t))$). The repair rate does not depend on the temperature, but the temperature continues to change during repairs, and we need to be aware of it during operating periods. The model that makes the most physical sense is therefore the one that integrates the tracking of the temperature at all times.

Denote by $I_t \in \{e_1, e_0\}$ the state of the item at time t, where $I_t = e_1$ (respectively $I_t = e_0$) means that the item is up (respectively down) at time t. Define τ_t as the temperature of the item at time t, A_t as the time elapsed at time t since the last change of state and $X_t = (I_t, A_t, \tau_t) \in E = \{e_1, e_0\} \times \mathbb{R}_+ \times \mathbb{R}$. Then $(X_t)_{t\geq 0}$ is a PDMP (without boundary) with parameters ϕ, λ, Q, where

$$\phi((e_i, x_1, x_2), t) = (e_i, x_1 + t, \varphi_{i,x_1,x_2}(t)), \ \frac{d}{dt}\varphi_{i,x_1,x_2} = \mathbf{v_i} \circ \varphi_{i,x_1,x_2}, \ \varphi_{i,x_1,x_2}(0) = x_2,$$

$$\lambda(e_1, x_1, x_2) = \bar{\lambda}(x_1, x_2), \quad \lambda(e_0, x_1, x_2) = \mu(x_1),$$

$$Q((e_i, x_1, x_2); dj dy_1 dy_2) = \delta_{e_{1-i}}(dj)\, \delta_0(dy_1)\, \delta_{x_2}(dy_2).$$

As already seen, several models could be considered for the same situation. Using a more general deterministic Markov flow than a flow governed by an ODE can avoid introducing boundaries. Furthermore, the choice of a model has to be made according to the information we want to get.

Example 3.12 Let us go back to the factory that needs commodities (Example 3.4, Sect. 3.1). Let us recall that in the first model, $X_t = (S_t, A_t)$, where S_t is the stock size at time t (or the virtual waiting time of a GI/G/1 queue when $c = 1$). Assume now that dF has hazard rate $\bar{\lambda}$. Then $(X_t)_{t\geq 0}$ is a PDMP without boundary, taking values in $E = \mathbb{R}_+^2$, with parameters ϕ, λ, Q defined as

$$\phi((x_1, x_2), t) = ((x_1 - ct)_+, x_2 + t),$$

$$\lambda(x_1, x_2) = \bar{\lambda}(x_2), \quad Q((x_1, x_2); dy_1 dy_2) = \delta_{x_1} * \mu(dy_1)\, \delta_0(dy_2).$$

To get a flow governed by an ODE, it is necessary to make a distinction between the stock being empty (state e_0) or not empty (state e_1) and to introduce a boundary. The state space becomes a hybrid one, namely $\{e_0\} \times \mathbb{R}_+ \cup \{e_1\} \times \mathbb{R}_+^* \times \mathbb{R}_+$, and the boundary is $\Gamma = \{e_1\} \times \{0\} \times \mathbb{R}_+$. It is a subset of the topological boundary of the state space.

Even if the model without boundary is a good description, a boundary may be introduced to get more information, as mentioned in Example 3.4, Sect. 3.1. We keep the same space $E = \mathbb{R}_+^2$, introduce the partition (E_1, E_0) of E with $E_1 =$

$\mathbb{R}_+^* \times \mathbb{R}_+$, $E_0 = \{0\} \times \mathbb{R}_+$, and the boundary $\Gamma_1 = \{0\} \times \mathbb{R}_+$. We have $\Gamma_0 = \emptyset$. Parameters ϕ, λ, Q are as above, and $q_1((0, x_2); dy_1 dy_2) = \delta_0(dy_1)\,\delta_{x_2}(dy_2)$. The PDMP $(S_t, D_t)_{t\geq 0}$ has parameters $\phi, \lambda, Q, E_0, E_1, q_1$, and there is no q_0, since $\Gamma_0 = \emptyset$.

The mean number of times the factory must be stopped because the stock becomes empty during the time interval $[0, t]$ is then given by the boundary measure σ defined in Proposition 5.5, Sect. 5.3 (see Example 5.11, Sect. 5.3 and numerical experiments in Sect. 9.5).

Example 3.13 A production unit is required to produce gas at a nominal flow rate c. This unit can be up or down. When the unit is down, the production is achieved by taking the gas from a tank with capacity M as long as it is not empty. When the unit is up and the tank is full, the unit produces gas at nominal flow rate c. When the unit is up and the tank is not full, the unit produces gas at flow rate $c_1 > c$ in order to refill the tank with the complementary production $c_1 - c$. The failure rate of the unit is $\bar{\lambda}$, and its repair rate μ. After a repair, the unit is as good as new.

Define $\eta_t \in \{e_1, e_0\}$ as the state of the unit at time t: $\eta_t = e_1$ (respectively $\eta_t = e_0$) if the unit is up (respectively down) at time t, $X_{1,t} \in [0, M]$ as the volume of gas in the tank at time t, $X_{2,t} \in \mathbb{R}_+$ as the time elapsed (at time t) since the last unit state change, and $X_t = (\eta_t, X_{1,t}, X_{2,t})$. Then $(X_t)_{t\geq 0}$ is a PDMP without boundary with state space $E = \{e_1, e_0\} \times [0, M] \times \mathbb{R}_+$ and parameters ϕ, λ, Q given, for $(x_1, x_2) \in [0, M] \times \mathbb{R}_+$, by

$$\phi((e_1, x_1, x_2), t) = (e_1, (x_1 + (c_1 - c)\,t) \wedge M, x_2 + t),$$

$$\phi((e_0, x_1, x_2), t) = (e_0, (x_1 - ct)_+, x_2 + t),$$

$$\lambda(e_1, x_1, x_2) = \bar{\lambda}(x_2), \quad \lambda(e_0, x_1, x_2) = \mu(x_2),$$

$$Q((e_1, x_1, x_2); dj dy_1 dy_2) = \delta_{e_0}(dj)\,\delta_{x_1}(dy_1)\,\delta_0(dy_2),$$

$$Q((e_0, x_1, x_2); dj dy_1 dy_2) = \delta_{e_1}(dj)\,\delta_{x_1}(dy_1)\,\delta_0(dy_2).$$

Assume now that we are interested in the ability to deliver gas; for example, we want to assess the expectation $LP(t)$ of the number of times there is a loss of production i.e., no delivery, during the time interval $[0, t]$. The number of times there is no delivery is the number of times the tank becomes empty. It is not reachable in the above model, since the times when the tank becomes empty are not identified as jump times. To add them, we have to introduce the boundary $\{e_0\} \times \{0\} \times \mathbb{R}_+$ to the model. To do this, we define the process $(X_t)_{t\geq 0}$ as a PDMP with parameters $\phi, \lambda, Q, (E_i, \Gamma_i, q_i)_{i\in\{1,2,3\}}$, for instance as follows:

$$E_1 = \{e_1\} \times [0, M] \times \mathbb{R}_+, \quad E_2 = \{e_0\} \times]0, M] \times \mathbb{R}_+, \quad E_3 = \{e_0\} \times \{0\} \times \mathbb{R}_+,$$

$$\Gamma_1 = \emptyset, \quad \Gamma_2 = E_3, \quad \Gamma_3 = \emptyset, \quad q((e_0, 0, x_2); dj dy_1 dy_2) = \delta_{e_0}(dj)\,\delta_0(dy_1)\,\delta_{x_2}(dy_2).$$

Then, as in Example 3.12, $LP(t)$ is given by the boundary measure σ defined in the next chapter.

In some cases, boundaries cannot be avoided. This is the case in Example 3.14 below and when there is preventive maintenance (Examples 3.7, Sect. 3.2 and 3.15 hereunder).

Example 3.14 Let us return to Example 1.9, Chap. 1. A tank is required to deliver water at the flow rate $c \in \mathbb{R}_+^*$. When the level in the tank is under h_{min}, a pump fills up the tank at flow rate $c_1 > c$ if the pump is up, and the pump is switched off when the tank is full i.e., when its level reaches h_{max}. The pump failure rate is 0 when it is switched off; it is $\bar{\lambda}$ and depends on its age when the pump is up. After a repair, the pump is as good as new. The pump repair rate is a function μ of the time elapsed since the start of repair. Let $I_t \in \{e_a, e_1, e_0\}$ be the pump state at time t, $I_t = e_1$ means that the pump is up, $I_t = e_a$ that it is switched off, and $I_t = e_0$ that it is down. Denote by H_t the tank level at time t, D_t the age of the pump if it is up or switched off and the time elapsed since the start of repair if it is down. Define $X_t = (I_t, H_t, D_t)$. Then $(X_t)_{t \geq 0}$ is a PDMP on $E = \{e_a\} \times [h_{min}, h_{max}] \times \mathbb{R}_+ \cup \{e_0\} \times [0, h_{max}[\times \mathbb{R}_+ \cup \{e_1\} \times [0, h_{max}] \times \mathbb{R}_+$ with boundary. It has values in $\underline{E} = \{e_a\} \times]h_{min}, h_{max}] \times \mathbb{R}_+ \cup \{e_0\} \times [0, h_{max}[\times \mathbb{R}_+ \cup \{e_1\} \times [0, h_{max}[\times \mathbb{R}_+$. Its parameters $\phi, \lambda, Q, \Gamma, q$ are defined as

$$\phi((e_a, x_1, x_2), t) = (e_a, x_1 - ct, x_2),$$

$$\phi((e_1, x_1, x_2), t) = (e_1, x_1 + (c_1 - c)t, x_2 + t),$$

$$\phi((e_0, x_1, x_2), t) = (e_0, (x_1 - ct)_+, x_2 + t)$$

$$\lambda(e_a, x_1, x_2) = 0, \quad \lambda(e_1, x_1, x_2) = \bar{\lambda}(x_2), \quad \lambda(e_0, x_1, x_2) = \mu(x_2),$$

$$Q((e_1, x_1, x_2); dj dy_1 dy_2) = \delta_{e_0}(dj)\, \delta_{x_1}(dy_1)\, \delta_0(dy_2),$$

$$Q((e_0, x_1, x_2); dj dy_1 dy_2) = 1_{[0, h_{min}]}(x_1)\, \delta_{e_1}(dj)\, \delta_{x_1}(dy_1)\, \delta_0(dy_2),$$

$$Q((e_0, x_1, x_2); dj dy_1 dy_2) = 1_{]h_{min}, h_{max}[}(x_1)\, \delta_{e_a}(dj)\, \delta_{x_1}(dy_1)\, \delta_0(dy_2),$$

$$\Gamma = \{(e_a, h_{min})\} \times \mathbb{R}_+, \cup \{(e_1, h_{max})\} \times \mathbb{R}_+,$$

$$q((e_a, h_{min}, x_2); dj dy_1 dy_2) = \delta_{e_1}(dj)\, \delta_{h_{min}}(dy_1)\, \delta_{x_2}(dy_2),$$

$$q((e_1, h_{max}, x_2); dj dy_1 dy_2) = \delta_{e_a}(dj)\, \delta_{h_{max}}(dy_1)\, \delta_{x_2}(dy_2),$$

$Q((e_a, x_1, x_2);\ .)$ need not be specified since $\lambda(e_a, x_1, x_2) = 0$.

Note that because of the boundary, the values of $\phi((e_a, x_1, x_2), t)$ for $t > x_1/c$ and of $\phi((e_1, x_1, x_2), t)$ for $t > (M - x_1)/(c_1 - c)$ are irrelevant.

The boundary is essential here to take into account the transition from pump state e_a to e_1, because in the first case, the pump failure rate is 0, and in the second case, it is $\bar{\lambda}$.

Example 3.15 Let us go back to Example 2.5, Sect. 2.5. We are modeling this time with a PDMP to have a continuous-time model in which all variables have a physical meaning. We slightly complicate the context by adding preventive maintenance: when component i reaches age a_i ($i = 1, 2$), it is instantaneously replaced by a new component (age 0). Example 2.5 corresponds to $a_1 = a_2 = +\infty$. Define $X_{i,t}$ ($i = 1, 2$) as the age of component i at time t if this component is up and the time elapsed since the start of its repair when it is down. Then $X_t = (I_t, X_{1,t}, X_{2,t})$ is a PDMP with values in $\underline{E} = \cup_{i=1}^4 E_i$ with $E_1 = \{e_1\} \times [0, a_1[\times[0, a_2[$, $E_2 = \{e_2\} \times \mathbb{R}_+ \times [0, a_2[$, $E_3 = \{e_3\} \times [0, a_1[\times\mathbb{R}_+$, $E_4 = \{e_4\} \times \mathbb{R}_+^2$. We can take $E = \{e_1, e_2, e_3, e_4\} \times \mathbb{R}_+^2$ or, to be more accurate, $E = \{e_1\} \times [0, a_1] \times [0, a_2] \cup \{e_2\} \times \mathbb{R}_+ \times [0, a_2] \cup \{e_3\} \times [0, a_1] \times \mathbb{R}_+ \cup \{e_4\} \times \mathbb{R}_+^2$. The PDMP parameters are $\phi, \lambda, Q, \alpha, q$:

$$\phi((e_i, x_1, x_2), t) = (e_i, x_1 + t, x_2 + t),$$

$$\lambda(e_1, x_1, x_2) = \lambda_1(x_1) + \lambda_2(x_2),\ \ \lambda(e_2, x_1, x_2) = \mu_1(x_1) + \lambda_2'(x_2),$$

$$\lambda(e_3, x_1, x_2) = \lambda_1'(x_1) + \mu_2(x_2),\ \ \lambda(e_4, x_1, x_2) = \mu_1(x_1) + \mu_2(x_2),$$

$$Q((e_1, x_1, x_2); dy) = \frac{1}{\lambda_1(x_1) + \lambda_2(x_2)} \left(\lambda_1(x_1)\, \delta_{(e_2,0,x_2)}(dy) + \lambda_2(x_2)\, \delta_{(e_3,x_1,0)}(dy)\right),$$

$$Q((e_2, x_1, x_2); dy) = \frac{1}{\mu_1(x_1) + \lambda_2'(x_2)} \left(\mu_1(x_1)\, \delta_{(e_1,0,x_2)}(dy) + \lambda_2'(x_2)\, \delta_{(e_4,x_1,0)}(dy)\right),$$

$$Q((e_3, x_1, x_2); dy) = \frac{1}{\lambda_1'(x_1) + \mu_2(x_2)} \left(\lambda_1'(x_1)\, \delta_{(e_4,0,x_2)}(dy) + \mu_2(x_2)\, \delta_{(e_1,x_1,0)}(dy)\right),$$

$$Q((e_4, x_1, x_2); dy) = \frac{1}{\mu_1(x_1) + \mu_2(x_2)} \left(\mu_1(x_1)\, \delta_{(e_3,0,x_2)}(dy) + \mu_2(x_2)\, \delta_{(e_2,x_1,0)}(dy)\right),$$

$$\alpha(e_1, x_1, x_2) = (a_1 - x_1) \wedge (a_2 - x_2), \quad \alpha(e_2, x_1, x_2) = a_2 - x_2,$$

$$\alpha(e_3, x_1, x_2) = a_1 - x_1, \quad \alpha(e_4, x_1, x_2) = +\infty,$$

$$q((e_1, x_1, x_2); dy) = 1_{\{x_1=a_1, x_2<a_2\}}\delta_{(e_1,0,x_2)}(dy) + 1_{\{x_1<a_1, x_2=a_2\}}\delta_{(e_1,x_1,0)}(dy) + 1_{\{x_1=a_1, x_2=a_2\}}\delta_{(e_1,0,0)}(dy),$$

$$q((e_2, x_1, a_2); dy) = \delta_{(e_2,x_1,0)}(dy), \quad q((e_3, a_1, x_2); dy) = \delta_{(e_3,0,x_2)}(dy).$$

Chapter 4
Hitting Time Distribution

This chapter provides methods to assess the probability distribution of a hitting time. It is motivated by applications in predictive reliability in which the probability that a piece of equipment is in operation throughout a given period of time has to be evaluated. It is an opportunity to look at new PDMPs that are obtained from the initial model. We give two methods to derive the probability that the hitting time τ of B is strictly greater than t.

The first is well known. It consists in introducing the killed process or the process stopped at time τ. The probability that the hitting time is greater than t is the probability that the new process is not in B at time t. The purpose of the first section is to show that the killed process and the stopped process are PDMPs and to give their characteristics.

The second method is much less well known. It consists in some way in decomposing the initial process. The probability that the hitting time is greater than t is the expectation of a functional of a new process obtained, roughly speaking, by removing the jumps that lead to B. It is a method derived from results of R. L. Dobrushin and promoted by V. Kalashnikov. We do not rely on their work; we get the decomposition from the basic definition of a PDMP.

This chapter is not useful for an understanding of the following chapters.

4.1 Killed and Stopped PDMPs

Given a PDMP $(X_t)_{t\geq 0}$ on E and a Borel set $B \subset E$, we want to get the distribution function of the first hitting time of B, namely

$$\tau_B = \inf\{t > 0 : X_t \in B\}. \tag{4.1}$$

C. Cocozza-Thivent, *Markov Renewal and Piecewise Deterministic Processes*, Probability Theory and Stochastic Modelling 100,
https://doi.org/10.1007/978-3-030-70447-6_4

Note that when B is closed, $X_{\tau_B} \in B$, since $(X_t)_{t\geq 0}$ is right-continuous. When B is not closed, we can have $X_{\tau_B} \notin B$ (see Example 4.2 below).

We assume $\tau_B > 0$ a.s. We are interested in $\mathbb{P}(\tau_B > t)$.

Example 4.1 In reliability studies, the reliability at time t of a piece of equipment subject to failure is the probability that this equipment is operating for the entire period of time $[0, t]$. Assume that $(X_t)_{t\geq 0}$ is such that the equipment is operating at time t iff $X_t \in \mathcal{U}$, where $\mathcal{U}$ is an open subset of E. Define $\mathcal{D} = \mathcal{U}^c$; then the reliability at time t is $\mathbb{P}(\tau_{\mathcal{D}} > t)$.

Let $(\widetilde{X}_t)_{t\geq 0}$ be the process killed at time τ_B and sent to a cemetery $\Delta_B \notin E \cup \{\Delta\}$, then $\mathbb{P}(\tau_B > t) = \mathbb{P}(\widetilde{X}_t \in E)$. We will start with its study. Then we will turn our attention to the stopped process $(X_{t\wedge\tau_B})_{t\geq 0}$. When B is closed, we have $\{\tau_B > t\} = \{X_{t\wedge\tau_B} \notin B\}$. When B is not closed, we may have $\mathbb{P}(\tau_B > t) \neq \mathbb{P}(X_{t\wedge\tau_B} \notin B)$, as in the following example.

Example 4.2 Return to Example 3.8, Sect. 3.2 with $n = 1$, $\alpha_1 = 0$, $\mu_1 = 1_{[-1,0[}$, $X_0 = -2$, and take $B =]1, +\infty[$; then $X_{\tau_B} = 1 \notin B$, and hence $\mathbb{P}(X_{t\wedge\tau_B} \notin B) = 1$ for all $t \in \mathbb{R}_+$, while $\mathbb{P}(\tau_B > 3) = p_1$.

When B is not closed, B must be replaced by B_+ defined as follows:

$$\alpha_B(x) = \inf\{t \geq 0 : \phi(x, t) \in B\},$$

$$B_+ = \{x \in E : \alpha_B(x) = 0\}, \quad B_+^c = E \setminus B_+ = \{x \in E : \alpha_B(x) > 0\},$$

If B is closed then $B = B_+$.

We assume that α_B is a Borel function. This holds in particular if B is an open set or a closed set. (see Sect. 3.2).

We have $X_{\tau_B} \in B_+$ and $\{\tau_B > t\} = \{X_{t\wedge\tau_B} \notin B_+\}$.

The following technical lemma will be useful later on.

Lemma 4.1 *Let $(Y_n, T_n)_{n\geq 0}$ be a zero-delayed Markov renewal process, $(X_t)_{t\geq 0}$ a PDMP generated by ϕ and $(Y_n, T_n)_{n\geq 0}$, and τ_B as in (4.1). The events $\{\tau_B \geq T_n\}$ and $\{\tau_B > T_n\}$ belong to the σ-field generated by $Y_0, Y_1, T_1, \cdots, Y_n, T_n$. Moreover, τ_B is a stopping time for the filtration $(\mathcal{F}_t)_{t\geq 0}$ generated by the Markov renewal process (Y, T), (defined Sect. 2.5), that is, $\{\tau_B > t\} \in \mathcal{F}_t$ for all t.*

Proof The two first results follow from $\{\tau_B \geq T_n\} = \bigcap_{i=0}^{n-1}\{\alpha_B(Y_i) \geq T_{i+1} - T_i\}$ and $\{\tau_B > T_n\} = \{\tau_B \geq T_n\} \cap \{\alpha_B(Y_n) > 0\}$.

The third result follows from

$$\{\tau_B > t\} = \cup_{k\geq 0}\left(\{N_t = k\} \cap \{\tau_B > T_k\} \cap \{\alpha_B(Y_k) > t - T_k\}\right),$$

the first result, and Lemma 2.3, Sect. 2.4. ■

Killed Processes

Let $\Delta_B \notin E \cup \{\Delta\}$. The killed process, defined in Proposition 4.1 below, is the process obtained by sending the initial process to the cemetery Δ_B at time τ_B. Its associated Markov renewal process is $(\widetilde{Y}_k, \widetilde{T}_k)_{k\geq 0}$ defined at the beginning of Proposition 4.1.

The interest of the killed process is that its marginal distributions make it possible to calculate the cumulative distribution function of τ_B. On the other hand, it does not allow one to obtain the state of the PDMP at time τ_B.

Proposition 4.1 *Let $(Y_k, T_k)_{k\geq 0}$ be a zero-delayed Markov renewal process with kernel N such that $Y_0 \in B_+^c$.*

Define $(\widetilde{Y}_k, \widetilde{T}_k)_{k\geq 0}$ as follows:

- *on $\{\tau_B < +\infty\} \cap \{T_{k_0-1} < \tau_B \leq T_{k_0}\}$ $(k_0 \geq 1)$,*
  - *for $0 \leq k < k_0$, $\widetilde{T}_k = T_k$, $\widetilde{Y}_k = Y_k$,*
  - *$\widetilde{T}_{k_0} = \tau_B$, $\widetilde{Y}_{k_0} = \Delta_B$,*
  - *for $k > k_0$, $\widetilde{T}_k = +\infty$, $\widetilde{Y}_k = \Delta$,*
- *on $\{\tau_B = +\infty\}$, $\widetilde{T}_k = T_k$ and $\widetilde{Y}_k = Y_k$ for any $k \geq 0$.*

Define the renewal kernel $\widetilde{N}$ on $\widetilde{E} = B_+^c \cup \{\Delta_B\}$ as follows:

- *for $x \in B_+^c$,*

$$\begin{aligned}\widetilde{N}(x, dy, dv) &= 1_{\{v\leq\alpha_B(x)\}}\, 1_{\{\alpha_B(y)>0\}}\, N(x, dy, dv) \\ &\quad + 1_{\{v\leq\alpha_B(x)\}}\, \delta_{\Delta_B}(dy) \int_{\{z\in B_+\}} N(x, dz, dv) \\ &\quad + 1_{\{\alpha_B(x)<+\infty\}}\, \mathbb{P}_x(T_1 > \alpha_B(x))\, \delta_{\Delta_B}(dy)\, \delta_{\alpha_B(x)}(dv),\end{aligned} \tag{4.2}$$

- *Δ_B is an absorbing state, that is, $\widetilde{N}(\Delta_B, (B_+^c \cup \{\Delta_B\}) \times \mathbb{R}_+) = 0$.*

Then $(\widetilde{Y}_k, \widetilde{T}_k)_{k\geq 0}$ is a Markov renewal process on $B_+^c \cup \{\Delta_B\}$ with kernel $\widetilde{N}$.

Let $(X_t)_{t\geq 0}$ be a PDMP generated by ϕ and $(Y_k, T_k)_{k\geq 0}$. Define $\widetilde{\phi}_B : (B_+^c \cup \{\Delta_B\}) \times \mathbb{R}_+ \to E$ as

$$\widetilde{\phi}_B(x, t) = \phi(x, t),\ (x, t) \in B_+^c \times \mathbb{R}_+,\quad \widetilde{\phi}_B(\Delta_B, t) = \Delta_B,\ \forall t \in \mathbb{R}_+, \tag{4.3}$$

and $(\widetilde{X}_t)_{t\geq 0}$ as

$$\widetilde{X}_t = \begin{cases} X_t, & t < \tau_B, \\ \Delta_B, & t \geq \tau_B. \end{cases}$$

Then $(\widetilde{X}_t)_{t\geq 0}$ is a PDMP generated by $\widetilde{\phi}_B$ and $(\widetilde{Y}_k, \widetilde{T}_k)_{k\geq 0}$, and

$$\mathbb{P}(\tau_B > t) = \mathbb{P}(\widetilde{X}_t \in B_+^c).$$

Proof Let $k \geq 1$ and $f_1 : B_+^c \times (B_+^c \times \mathbb{R}_+)^k \to \mathbb{R}_+$, $f_2 : \widetilde{E} \times \mathbb{R}_+ \to \mathbb{R}_+$ be Borel functions. The proof of the following result:

$$\mathbb{E}\Big(f_1(\widetilde{Y}_0, \widetilde{Y}_1, \widetilde{T}_1, \widetilde{Y}_2, \widetilde{T}_2 - \widetilde{T}_1, \dots, \widetilde{Y}_k, \widetilde{T}_k - \widetilde{T}_{k-1})\, f_2(\widetilde{Y}_{k+1}, \widetilde{T}_{k+1} - \widetilde{T}_k)\, 1_{\{\widetilde{T}_{k+1} < +\infty\}}\Big)$$
$$= \mathbb{E}\bigg(f_1(\widetilde{Y}_0, \widetilde{Y}_1, \widetilde{T}_1, \widetilde{Y}_2, \widetilde{T}_2 - \widetilde{T}_1, \dots, \widetilde{Y}_k, \widetilde{T}_k - \widetilde{T}_{k-1})\, 1_{\{\widetilde{T}_k < +\infty\}} \times \int f_2(y, v)\, \widetilde{N}(\widetilde{Y}_k, dy, dv)\bigg). \tag{4.4}$$

proceeds in four steps.

First we notice that if f_1 is replaced by $1_{\{\tau_B \le T_k\}}\, f_1$, then the two members of (4.4) are equal to 0, since on $\{\tau_B \le T_k\}$ we get $\widetilde{T}_{k+1} = +\infty$ and $\widetilde{Y}_k = \Delta_B$.

Then it can be proved that

$$\mathbb{E}\Big(1_{\Omega_i}\, f_1(\widetilde{Y}_0, \widetilde{Y}_1, \widetilde{T}_1, \widetilde{Y}_2, \widetilde{T}_2 - \widetilde{T}_1, \dots, \widetilde{Y}_k, \widetilde{T}_k - \widetilde{T}_{k-1})\, f_2(\widetilde{Y}_{k+1}, \widetilde{T}_{k+1} - \widetilde{T}_k)\, 1_{\{\widetilde{T}_{k+1} < +\infty\}}\Big)$$
$$= \mathbb{E}\left(1_{\{T_k < \tau_B\}}\, f_1(Y_0, Y_1, T_1, Y_2, T_2 - T_1, \dots, Y_k, T_k - T_{k-1}) \int f_2(z, v)\, N_i(Y_k, dz, dv)\right)$$

$(i = 1, 2, 3)$, where

$$\Omega_1 = \{\tau_B > T_{k+1}\} = \{\tau_B > T_k\} \cap \{\alpha_B(Y_k) > T_{k+1} - T_k\} \cap \{\alpha_B(Y_{k+1}) > 0\},$$

$$N_1(x, dy, dv) = 1_{\{v \le \alpha_B(x)\}}\, 1_{\{\alpha_B(y) > 0\}}\, N(x, dy, dv),$$

$$\Omega_2 = \{\tau_B = T_{k+1}\} = \{\tau_B > T_k\} \cap \{\alpha_B(Y_k) \ge T_{k+1} - T_k\} \cap \{\alpha_B(Y_{k+1}) = 0\},$$

$$N_2(x, dy, dv) = \delta_{\Delta_B}(dy)\, 1_{\{v \le \alpha_B(x)\}} \int_{\{z \in E\}} 1_{\{\alpha_B(z) = 0\}}\, N(x, dz, dv),$$

$$\Omega_3 = \{T_k < \tau_B < T_{k+1}\} = \{\tau_B > T_k\} \cap \{\alpha_B(Y_k) < T_{k+1} - T_k\},$$

$$N_3(x, dy, dv) = 1_{\{\alpha_B(x) < +\infty\}}\, \mathbb{P}_x(T_1 > \alpha_B(x))\, \delta_{\Delta_B}(dy)\, \delta_{\alpha_B(x)}(dv).$$

Summing the above results, we get (4.4), since on $\{\tau_B > T_k\}$ we have $\widetilde{Y}_j = Y_j$ and $\widetilde{T}_j = T_j$ for $j \le k$.

To conclude the proof it remains only to check that $\widetilde{X}_t = \widetilde{\phi}_B(\widetilde{Y}_k, t - \widetilde{T}_k)$ for $\widetilde{T}_k \le t < \widetilde{T}_{k+1}$ and $\widetilde{N}(\widetilde{\phi}(x, u), dy, dv) = \widetilde{N}_0^{x,u}(dy, dv)$ for $x \in B_+^c$. The first point is straightforward, while the second is a consequence of $\alpha_B(\phi(x, u)) = \alpha_B(x) - u$ $(u < \alpha_B(x))$ and $N(\phi(x, u)), dy, dv) = N_0^{x,u}(dy, dv)$. ■

Corollary 4.1 *The notation is as in Proposition 4.1. Assume that $(X_t)_{t \ge 0}$ is a PDMP with parameters $\phi, \lambda, Q, \alpha, (E_i, q_i)_{i \in I}$ and $X_0 \in B_+^c$. Then $(\widetilde{X}_t)_{t \ge 0}$ is a PDMP on $B_+^c \cup \{\Delta_B\}$ with parameters $\widetilde{\phi}_B, \lambda 1_{B_+^c}, \widetilde{Q}, \widetilde{\alpha}, (\widetilde{E}_i, \widetilde{q}_i)_{i \in I}$ defined as follows:*

$$\widetilde{Q}(x; dy) = Q(x; B_+)\, \delta_{\Delta_B}(dy) + 1_{\{\alpha_B(y) > 0\}}\, Q(x; dy) \;\; (x \in B_+^c), \quad \widetilde{E}_i = E_i \cap B_+^c,$$

$$\widetilde{\alpha}(x) = \begin{cases} \alpha(x) \wedge \alpha_B(x), & x \in B_+^c, \\ +\infty, & x = \Delta_B, \end{cases}$$

$$\widetilde{\Gamma}_i = \{\widetilde{\phi}_B(x, \widetilde{\alpha}(x)) : x \in \widetilde{E}_i\},$$

$$\widetilde{q}_i(x; dy) = \begin{cases} 1_{\{\alpha_B(y)>0\}}\, q_i(x; dy) + q_i(x; B_+)\, \delta_{\Delta_B}(dy), & x \in \widetilde{\Gamma}_i \cap E_i^c, \\ \delta_{\Delta_B}(dy), & x \in \widetilde{\Gamma}_i \cap E_i, \end{cases}$$

$\widetilde{Q}(\Delta_B; dy)$ (respectively $q(\Delta_B; dy)$) has not to be defined, since $(\lambda 1_E)(\Delta_B) = 0$ (respectively $\alpha(\Delta_B) = +\infty$).

By returning to the notation $\Gamma_i = \{\phi(x, \alpha(x)) : x \in E_i\}$, we get $\widetilde{\Gamma}_i \cap E_i^c = \widetilde{\Gamma}_i \cap \Gamma_i$.

Stopped Processes

A stopped process is a process stopped at time τ_B, namely $(X_{t\wedge\tau_B})_{t\geq 0}$. We are going to show that it is a PDMP and give its parameters when $(X_t)_{t\geq 0}$ is a parametrized PDMP.

We have $\mathbb{P}(\tau_B > t) = \mathbb{P}(X_{t\wedge\tau_B} \notin B_+)$. Moreover, on $\{\tau_B < +\infty\}$, $\lim_{t\to+\infty} X_{t\wedge\tau_B} = X_{\tau_B}$, which makes it possible to obtain the probability distribution of X_{τ_B}.

Define

$$\phi_B(x, t) = \phi(x, t \wedge \alpha_B(x)).$$

When ϕ is a deterministic Markov flow, it is the same for ϕ_B.

Proposition 4.2 *Let $(Y_k, T_k)_{k\geq 0}$ be a zero-delayed Markov renewal process with kernel N such that $Y_0 \in B_+^c$.*

Define $\bar{Y}_k$ and $\bar{T}_k$ $(k \geq 0)$ as follows:

- *if $T_k \leq \tau_B$, then $\bar{T}_k = T_k$ and $\bar{Y}_k = Y_k$,*
- *if $T_k > \tau_B$, then $\bar{T}_k = +\infty$ and $\bar{Y}_k = \Delta$.*

Then $(\bar{Y}, \bar{T}) = (\bar{Y}_k, \bar{T}_k)_{k\geq 0}$ is a Markov renewal process with kernel

$$\bar{N}(x, dy, dv) = 1_{\{v\leq\alpha_B(x)\}}\, N(x, dy, dv).$$

Let $(X_t)_{t\geq 0}$ be a PDMP generated by ϕ and $(Y_k, T_k)_{k\geq 0}$, and $(\bar{Z}_t, \bar{A}_t)_{t\geq 0}$ the CSMP associated with the Markov renewal process $(\bar{Y}, \bar{T})$. Then $X_{t\wedge\tau_B} = \phi_B(\bar{Z}_t, \bar{A}_t)$, and $(X_{t\wedge\tau_B})_{t\geq 0}$ is a PDMP generated by ϕ_B and $\bar{N}$.

Proof We have to check Formula (4.4), Sect. 4.1 where $\widetilde{Y}_j$ $(j \geq 0)$ is replaced by $\bar{Y}_j$, $\widetilde{T}_j$ $(j \geq 1)$ by $\bar{T}_j$, and $\widetilde{N}$ by $\bar{N}$. This is straightforward using $\{T_{k+1} \leq \tau_B\} = \{T_k \leq \tau_B\} \cap \{\alpha_B(Y_k) \geq T_{k+1} - T_k\}$ and Lemma 4.1, Sect. 4.1.

Clearly $X_{t\wedge\tau_B} = \phi_B(\bar{Z}_t, \bar{A}_t)$. To prove that $(X_{t\wedge\tau_B})_{t\geq 0}$ is a PDMP generated by ϕ_B and $\bar{N}$, all that is left is to check that $\bar{N}_0^{x,u}(dy, dv) = \bar{N}(\phi_B(x, u), dy, dv)$, which is quite easy. ∎

Corollary 4.2 *Assume that $(X_t)_{t\geq 0}$ has parameters $\phi, \lambda, Q, \alpha, (E_i, q_i)_{i\in I}$. Define*

$$\forall x \in E : \ \bar{\alpha}(x) = \begin{cases} \alpha(x), & \alpha(x) \leq \alpha_B(x), \\ +\infty, & \alpha(x) > \alpha_B(x). \end{cases}$$

Then $(X_{t\wedge\tau_B})_{t\geq 0}$ is a PDMP with parameters $\phi_B, \lambda 1_{B_+^c}, Q, \bar{\alpha}, (E_i, q_i)_{i\in I}$.

The proof is a consequence of Proposition 4.2 and the equivalence between $\alpha(x) \leq \alpha_B(x)$, $\alpha(x) < +\infty$ and $\bar{\alpha}(x) < +\infty$.

At first glance it may seem surprising that in Proposition 4.1 we have $\widetilde{\phi}(x,t) = \phi(x,t)$ for $x \in B_+^c$ instead of $\widetilde{\phi}(x,t) = \phi_B(x,t)$. In fact, both are suitable, because we have $\mathbb{P}_x(\widetilde{T}_1 > \alpha_B(x)) = 0$ for all $x \in B_+^c$; hence the values of $\widetilde{\phi}(x,t)$ for $x > \alpha_B(x)$ do not matter. Also, in Corollary 4.1 we could replace $\lambda 1_{B_+^c}$ by $\lambda 1_E$ in the parameters of $(\widetilde{X}_t)_{t\geq 0}$. Indeed, for $x \in B_+^c$, $v < \widetilde{\alpha}(x)$ we have $v < \alpha_B(x)$, and hence $\alpha_B(\phi(x,v)) = \alpha_B(x) - v > 0$. Therefore $\phi(x,v) \in B_+^c$. But writing $\lambda 1_{B_+^c}$ makes more sense, since this parameter does not have to be considered on B_+.

Note that because of the jumps, the process $(X_{t\wedge\tau_B})_{t\geq 0}$ has values not only in $B_+^c \cup \{\phi(x, \alpha_B(x)) : x \in B_+^c\}$. Indeed, let $(X_t)_{t\geq 0}$ be a PDMP on $\mathbb{R}_+$ with parameters $\phi(x,t) = x+t$, $\lambda = 0$, $\Gamma = \{1\}$, $q(1;]2,+\infty[) = 1$; assume $X_0 = 0$ and $B = [2,+\infty[$. Then $B_+^c \cup \{\phi(x,\alpha_B(x)) : x \in B_+^c\} = [0,2]$ while $\tau_B = 1$ and $\mathbb{P}(X_{\tau_B} \in]2,+\infty[) = q(1;]2,+\infty[) = 1$.

Examples

We give below three examples. The two first show the interest of the above results in reliability studies; in these examples B is closed. The third is an illustrative example in which B is not closed.

Example 4.3 Let us go back to Example 3.15, Sect. 3.4 and its notation. Assume that the two components of the item are in active redundancy, that is, the item is up if at least one component is up. Let us denote by $\mathcal{U}$ (respectively $\mathcal{D}$) the up states (respectively the down states) of the item. We have $\mathcal{U} = \cup_{i=1}^3 E_i$, $\mathcal{D} = E_4$. The reliability of the item at time t is the probability $R(t)$ that the item is up during the whole time interval $[0,t]$. We take $B = \mathcal{D}$, and then $R(t)$ is the probability that the killed or the stopped process is in $\mathcal{U}$ at time t.

The killed process is a PDMP on $\mathcal{U} \cup \Delta_{\mathcal{D}}$ with parameters $\widetilde{\phi}, \lambda 1_{\mathcal{U}}, \widetilde{Q}, \widetilde{\alpha}, q$ where

$$\widetilde{\phi}(x,t) = \begin{cases} \phi(x,t), & (x,t) \in \mathcal{U} \times \mathbb{R}_+, \\ \Delta_{\mathcal{D}}, & (x,t) \in \{\Delta_{\mathcal{D}}\} \times \mathbb{R}_+, \end{cases} \qquad \widetilde{Q}((e_1,x_1,x_2);dy) = Q((e_1,x_1,x_2);dy),$$

$$\widetilde{Q}((e_2,x_1,x_2);dy) = \frac{1}{\mu_1(x_1) + \lambda_2'(x_2))}\left(\mu_1(x_1)\,\delta_{(e_1,0,x_2)}(dy) + \lambda_2'(x_2)\,\delta_{\Delta_{\mathcal{D}}}(dy)\right),$$

$$\widetilde{Q}((e_3,x_1,x_2);dy) = \frac{1}{\lambda_1'(x_1) + \mu_2(x_2)}\left(\lambda_1'(x_1)\,\delta_{\Delta_{\mathcal{D}}}(dy) + \mu_2(x_2)\,\delta_{(e_1,x_1,0)}(dy)\right),$$

$$\widetilde{\alpha}(e_i,x_1,x_2) = \alpha(e_i,x_1,x_2) \ \ (i = 1,2,3), \quad \widetilde{\alpha}(\Delta_{\mathcal{D}}) = +\infty.$$

The stopped process is a PDMP on $\mathcal{U} \cup \{e_4\} \times \{0\} \times \mathbb{R}_+ \cup \{e_4\} \times \mathbb{R}_+ \times \{0\}$ with parameters $\bar{\phi}, \lambda 1_{\mathcal{U}}, Q, \bar{\alpha}, q$, where

$$\bar{\phi}(x,t) = \begin{cases} \phi(x,t), & (x,t) \in \mathcal{U} \times \mathbb{R}_+, \\ x, & (x,t) \in \mathcal{D} \times \mathbb{R}_+, \end{cases} \quad \bar{\alpha}(x) = \begin{cases} \alpha(x), & x \in \mathcal{U}, \\ +\infty, & x \in \mathcal{D}. \end{cases}$$

Example 4.4 Return to Example 3.14, Sect. 3.4 and its notation. Assume that the tank is not empty at the initial time. We want to compute the reliability of the equipment, i.e., the probability $R(t)$ that the water is delivered throughout the time period $[0,t]$. We have $R(t) = \mathbb{P}(\tau_B > t)$, where $B = \{e_0\} \times \{0\} \times \mathbb{R}_+$. The set B is closed, and hence $B = B_+$ and

$$B^c = \{e_0\} \times \mathbb{R}_+^* \times \mathbb{R}_+ \cup \{e_1\} \times [0, h_{max}] \times \mathbb{R}_+ \cup \{e_a\} \times [h_{min}, h_{max}] \times \mathbb{R}_+.$$

The killed process $(\widetilde{X}_t)_{t\geq 0}$ is a PDMP on $B^c \cup \{\Delta_B\}$ with boundary and parameters $\widetilde{\phi}_B, \lambda 1_{B^c}, Q, \Gamma \cup B, \tilde{q}$, where $\widetilde{\phi}_B$ is as in (4.3), Sect. 4.1 and $\tilde{q}(x;dy) = 1_\Gamma(x)\, q(x;dy) + 1_B(x)\, \delta_{\Delta_B}(dy)$. We have $R(t) = \mathbb{P}(\widetilde{X}_t \in B^c)$.

Why is Q in the PDMP parameters of the killed process and not $\widetilde{Q}$? Because we can consider that $Q(x;B) = 0$. Indeed, it is the quantity $\lambda(\phi(x,v))\, Q(\phi(x,v);dy)\, dv$ that comes into play (see (3.16), Sect. 3.2). But $Q(\phi(x,v);B) \neq 0$ gives $x = (e_1, 0, x_2)$ and $v = 0$, hence $Q(\phi(x,v);B)\, dv = 0$.

We have $\phi_B(x,t) = \phi(x,t)$ if $x \in \{e_1\} \times [0, h_{max}] \times \mathbb{R}_+ \cup \{e_a\} \times [h_{min}, h_{max}] \times \mathbb{R}_+$,

$$\phi_B(x,t) = \begin{cases} (e_0, x_1 - ct, x_2 + t), & x = (e_0, x_1, x_2), t < x_1/c, \\ (e_0, 0, x_2 + x_1/c) & x = (e_0, x_1, x_2), t \geq x_1/c. \end{cases}$$

The stopped process $(X_{t\wedge\tau_B})_{t\geq 0}$ is a PDMP on E with boundary and parameters $\phi_B, \lambda 1_{B^c}, Q, \Gamma, q$. We have $R(t) = \mathbb{P}(X_{t\wedge\tau_B} \in B^c)$.

Example 4.5 We return to Example 3.10, Sect. 3.4 and take $B = \mathbb{R} \times \mathbb{R}_-^* \cup \mathbb{R}_+^* \times \{0\}$. Then $B_+ = B \cup \{(0,0)\}$, and hence $B_+^c = \mathbb{R}_-^* \times \mathbb{R}_+ \cup \mathbb{R}_+ \times \mathbb{R}_+^*$.

Recall that $\lambda = 0$. The killed process $(\widetilde{X}_t)_{t\geq 0}$ is a PDMP on $B_+^c \cup \{\Delta_B\}$ with boundary and parameters $\widetilde{\phi}_B, \lambda, Q, A, \tilde{q}$, where $\widetilde{\phi}_B$ is given in (4.3), Sect. 4.1; Q need not be specified, since $\lambda = 0$; and $\tilde{q}$ is defined as follows:

- if $x_1 < 0$,

$$\tilde{q}((x_1,0);dy) = p_1\, \delta_{(x_1,0)}(dy) + (1-p_1)\, 1_{B_+^c}(y)\, \mu_1(dy) \\ + (1-p_1)\, \mu_1(B_+)\, \delta_{\Delta_B}(dy);$$

- if $x_1 \geq 0$, $\tilde{q}((x_1,0);dy) = (1-p_1)\, 1_{B_+^c}(y)\, \mu_1(dy) + (p_1 + (1-p_1)\, \mu_1(B_+))\, \delta_{\Delta_B}(dy)$.

We have $\mathbb{P}(\tau_B > t) = \mathbb{P}(\widetilde{X}_t \notin B)$.

The stopped process $(X_{t\wedge\tau_B})_{t\geq 0}$ is a PDMP on $\mathbb{R}^2$ with boundary and parameters $\phi_B, \lambda 1_{B^c}, Q, A, q_1$ with

$$
\begin{aligned}
&x_1 < 0, x_2 \geq 0, \quad \phi_B((x_1,x_2),t) = \begin{cases} (x_1, x_2 - t), & t < x_2, \\ (x_1 + t - x_2, 0), & x_2 \leq t < x_2 - x_1, \\ (0,0), & t \geq x_2 - x_1, \end{cases} \\
&x_1 \geq 0, x_2 > 0, \quad \phi_B((x_1,x_2),t) = (x_1, (x_2 - t)_+), \\
&(x_1, x_2) \in B, \qquad \phi_B((x_1,x_2),t) = (x_1, x_2).
\end{aligned}
$$

Since $\lambda 1_{B^c} = 0$, Q need not be specified. We have $\mathbb{P}(\tau_B > t) = \mathbb{P}(X_{t\wedge\tau_B} \notin B_+)$.

Let us note that Corollary 4.2 states that $(X_{t\wedge\tau_B})_{t\geq 0}$ is a PDMP with parameters $\phi_B, \lambda 1_{B^c}, Q, \bar{\alpha}, (E_i, q_i)_{i\in\{1,2,3\}}$, but the complication given by the partition $(E_i)_{i\in\{1,2,3\}}$ is meaningless here.

4.2 Process Decomposition

In order to have notation more adapted to the method of this section, we do not keep that of the previous section; we take $A = B^c$, and below, α_B becomes α_ϕ.

Let $X = (X_t)_{t\geq 0}$ be a process with values in E. Define $\tau = \inf\{t > 0, X_t \in A^c\}$. The decomposition of (X, τ) consists in writing $\mathbb{P}(\tau > t)$ as the expectation of a functional of another process X^0. The computation of $\mathbb{P}(\tau > t)$ can then be done by a Monte Carlo simulation of the process X^0. The interest of this method compared to a Monte Carlo simulation on X is that it is a variance reduction method. It is particularly efficient when $\mathbb{P}(\tau > t)$ is close to 1 and when a good precision on $\mathbb{P}(\tau \leq t)$ is desired, which is the case when τ is the failure rate of a safety equipment.

Let $\gamma : E_0 \to \mathbb{R}_+$ be a Borel function. We say that (X, τ) can be decomposed into (X^0, γ) if for all $m \geq 1$, $f_i : E \to \mathbb{R}_+$, and $t_i \leq t$ $(1 \leq i \leq m)$ we have

$$
\mathbb{E}\left(f_1(X_{t_1}) \cdots f_m(X_{t_m})\, 1_{\{\tau > t\}}\right) = \mathbb{E}\left(f_1(X^0_{t_1}) \cdots f_m(X^0_{t_m})\, e^{-\int_0^t \gamma(X^0(s))\,ds}\right). \tag{4.5}
$$

This concept was developed by V. Kalashnikov from an idea of R.L. Dobrushin. Basically, assume that X is a Markov process with generator L, and we can write Lf as $Lf(x) = \int f(y)\, L(x, dy) = L_0 f(x) + L_1 f(x)$, where $L_0 f(x)$ does not depend on the values $f(y)$, $y \in A^c$, when $x \in A$, and L_1 is such that $L_1 f(x) = \int_{A^c} f(y)\, L(x, dy) - f(x)\, L(x, A^c)$. Then (X, τ) can be decomposed into (X^0, γ), where X^0 has generator L_0 and $\gamma(x) = L(x, A^c)$.

Such a decomposition was used in Cocozza-Thivent and Kalashnikov [34] to solve a very practical problem: compare a failure rate in reliability studies to a commonly used approximation. It has been applied to PDMP models to compute equipment reliability and also to get quantitative estimates in queues. Examples can be found in Kalashnikov [73] and Lorton [82].

In this section the decomposition is not obtained through the generator. It is directly addressed from the process construction. This allows for less-restrictive assumptions.

For the sake of simplicity, at first our emphasis is $\mathbb{P}(\tau > t)$.

Proposition 4.3 *Given N and N_0 two semi-Markov kernels on E, let $(X_t)_{t\geq 0}$ (respectively $(X_t^0)_{t\geq 0}$) a PDMP generated by ϕ and N (respectively N_0), $dF_x(v) = N(x, E, dv)$, $\bar{F}_x(v) = 1 - N(x, E \times [0, t]) = dF_x(]v, +\infty])$, and the same for dF_x^0 and $\bar{F}_x^0$ with N_0 instead of N. Let $A \subset E$ a Borel set, define*

$$\tau = \inf\{t > 0 : X_t \in A^c\}, \quad \alpha_\phi(x) = \inf\{t > 0 : \phi(x,t) \in A^c\},$$

$$\tau^0 = \inf\{t > 0, X_t^0 \in A^c\}.$$

Assume $\mathbb{P}(X_0 \in A) = 1$, $X_0 = X_0^0$ a.s. and that there exists $\gamma : A \to \mathbb{R}_+$ such that

$$\forall v < \alpha_\phi(x), \; e^{-\int_0^v \gamma(\phi(x,s))\,ds}\, \bar{F}_x^0(v) = \bar{F}_x(v), \tag{4.6}$$

$$e^{-\int_0^v \gamma(\phi(x,s))\,ds}\, 1_A(y)\, 1_{\{v \leq \alpha_\phi(x)\}}\, N_0(x, dy, dv) = 1_A(y)\, 1_{\{v \leq \alpha_\phi(x)\}}\, N(x, dy, dv). \tag{4.7}$$

Then

$$\mathbb{P}(\tau > t) = \mathbb{E}\left(1_{\{t<\tau_0\}}\, e^{-\int_0^t \gamma(X_s^0)\,ds}\right). \tag{4.8}$$

Remark 4.1 In Proposition 4.3, X^0 may have values in E and not only in A if $\phi(x,t) \in A^c$ for some x and t such that $\bar{F}_x(t) > 0$, which explains $1_{\{t<\tau_0\}}$ in (4.8). If $\alpha_\phi(x) \geq t^*(x)$ for all x, then $\mathbb{P}(\tau > t) = \mathbb{E}\left(e^{-\int_0^t \gamma(X_s^0)\,ds}\right)$.

Proof Let $(Y_n, T_n)_{n\geq 0}$ be a Markov renewal process associated with X, and $(Y_n^0, T_n^0)_{n\geq 0}$ a Markov renewal process associated with X^0. We have

$$\mathbb{P}_x(\tau > t) = \sum_{n\geq 1} a_n,$$

$$a_n = \mathbb{E}_x\left(\prod_{k=1}^n f(Y_{k-1}, T_k - T_{k-1}, Y_k)\, g(Y_n, T_{n+1} - T_n, T_n))\right),$$

$$f(x,v,y) = 1_A(y)\, 1_{\{v \leq \alpha_\phi(x)\}}, \quad g(x,v,w) = 1_{\{0 \leq t-w < \alpha_\phi(x)\}}\, 1_{\{t < v+w\}},$$

$$\mathbb{E}_x\left(1_{\{t<\tau^0\}}\, e^{-\int_0^t \gamma(X_s^0)\,ds}\right) = \sum_{n\geq 1} a_n^0,$$

$$a_n^0 = \mathbb{E}_x\left(\prod_{k=1}^n f_0(Y_{k-1}^0, T_k^0 - T_{k-1}^0, Y_k^0)\, g_0(Y_n^0, T_{n+1}^0 - T_n^0, T_n^0)\right),$$

$$f_0(x,v,y) = f(x,v,y)\, e^{-\int_0^v \gamma(\phi(s,s))\,ds}, \quad g_0(x,v,w) = g(x,v,w)\, e^{-\int_0^{t-w} \gamma(\phi(x,s))\,ds}.$$

Conditioning on $(Y_k, T_k)_{1\le k\le n}$, we get

$$a_n = \mathbb{E}\left(\prod_{k=1}^{n} f(Y_{k-1}, T_k - T_{k-1}, Y_k)\,\varphi_1(Y_n, T_n)\right), \quad \varphi_1(y, w) = \int_{\mathbb{R}_+} g(y, v, w)\,dF_y(v),$$

then conditioning on $(Y_k, T_k)_{1\le k\le n-1}$ yields

$$a_n = \mathbb{E}_x\left(\prod_{k=1}^{n-1} f(Y_k, T_k - T_{k-1}, Y_k)\,\varphi_2(Y_{n-1}, T_{n-1})\right),$$

$$\varphi_2(y, w) = \int_{E\times\mathbb{R}_+} f(y, v, z)\,\varphi_1(z, v + w)\,N(y, dz, dv),$$

and by induction,

$$a_n = \mathbb{E}_x\left(\prod_{k=1}^{n-p} f(Y_{k-1}, T_k - T_{k-1}, Y_k)\,\varphi_{p+1}(Y_{n-p}, T_{n-p})\right),$$

$$\varphi_{p+1}(y, w) = \int_{E\times\mathbb{R}_+} \varphi_p(z, v + w)\,f(y, v, z)\,N(y, dz, dv) \quad (1 \le p \le n).$$

Hence

$$a_n = \mathbb{E}_x(f(Y_0, T_1, Y_1)\,\varphi_n(Y_1, T_1)) = \int_{E\times\mathbb{R}_+} 1_{\{v\le t\}}\varphi_n(y, v)\,f(x, v, y)\,N(x, dy, dv).$$

Therefore,

$$\begin{aligned} a_n^0 &= \int_{E\times\mathbb{R}_+} 1_{\{v\le t\}}\varphi_n^0(y, v)\,f_0(x, v, y)\,N_0(x, dy, dv), \\ &= \int_{E\times\mathbb{R}_+} 1_{\{v\le t\}}\varphi_n^0(y, v)\,f(x, v, y)\,e^{-\int_0^v \gamma(\phi(x,s))\,ds}\,N_0(x, dy, dv), \end{aligned}$$

$$\varphi_1^0(y, w) = \int_{\mathbb{R}_+} g_0(y, v, w)\,dF_y^0(v) = \int_{\mathbb{R}_+} g(y, v, w)\,e^{-\int_0^{t-w} \gamma(\phi(y,s))\,ds}dF_y^0(v),$$

and for $2 \le p \le n$,

$$\begin{aligned} \varphi_{p+1}(y, w) &= \int_{E\times\mathbb{R}_+} \varphi_p^0(z, v + w)\,f_0(y, v, z)\,N_0(y, dz, dv) \\ &= \int_{E\times\mathbb{R}_+} \varphi_p^0(z, v + w)\,f(y, v, z)\,e^{-\int_0^v \gamma(\phi(y,s))\,ds}\,N_0(y, dz, dv). \end{aligned}$$

Condition (4.6) gives $a_0^0 = a_0$, and then $a_n^0 = a_n$ for $n \ge 1$ comes from Condition (4.7). ∎

Corollary 4.3 *Let X a PDMP with parameters λ, $Q, \alpha, (E_i, q_i)_{i\in I}$, $A \subset E$, and $\tau = \inf\{t : X_t \in A^c\}$. Define*

$$\gamma(x) = \lambda(x)\, Q(x;\, A^c), \quad \lambda_0(x) = \lambda(x)\, Q(x;\, A), \quad Q_0(x;\, dy) = 1_A(y)\, \frac{Q(x;\, dy)}{Q(x;\, A)},$$

X^0 as a PDMP with parameters λ_0, $Q_0, \alpha, (E_i, q_i)_{i\in I}$. Assume $X_0 = X_0^0$ a.s. and $\mathbb{P}(X_0 \in A) = 1$. Then

$$\mathbb{P}(\tau > t) = \mathbb{E}\left(1_{\{t<\tau_0\}}\, e^{-\int_0^t \gamma(X_s^0)\, ds}\right).$$

If $\phi(x, t) \in A$ for all $x \in A$, $t < t^*(x)$, and $q_i(x;\, A) = 1$ for $x \in \Gamma_i^* \cap A$, $i \in I$, then X^0 takes values in A and $\mathbb{P}(\tau > t) = \mathbb{E}\left(e^{-\int_0^t \gamma(X_s^0)\, ds}\right)$.

Note that Q_0 need not be defined when $Q(x;\, A) = 0$, since $\lambda_0(x) = 0$.

Under the assumptions of Proposition 4.3, and therefore under those of Corollary 4.3, it can be seen that (4.5) holds. In fact, $f(X_s)$ $(s \le t)$ is introduced by writing $f(X_s) = \sum_m 1_{\{T_m \le s < T_{m+1}\}}\, f(\phi(Y_m, s - T_m))$, a_n is replaced by $a_{n,m}$, and then the principle and the arguments are the same as in the proof of Proposition 4.2, but explicit calculations are heavy and boring.

For a better understanding of how to get X^0 from X, let us look at two examples. In all of them, X^0 takes values in A. They are models of predictive reliability.

We give parameters λ and Q (respectively λ_0, Q_0) through a diagram that provides the transition rates $\Upsilon(x;\, dy) := \lambda(x)\, Q(x;\, dy)$. Parameters λ and Q are derived by writing $\lambda(x) = \Upsilon(x;\, E)$, $Q(x;\, dy) = \Upsilon(x;\, dy)/\lambda(x)$. Actually, the state space is hybrid; it is a subset of $E_0 \times \mathbb{R}^d$, where E_0 is finite. Hence $x = (e_i, x_1, \ldots, x_d)$. The first example is like Example 2.5, Sect. 2.5 but with three components instead of two and modeling by a PDMP. The second example is with two components and instantaneous preventive maintenance.

Example 4.6 An equipment is composed of three stochastically independent components. Each component is as in Example 2.5, Sect. 2.5. The sate of a component is 1 when the component is up and 0 when it is down. The failure rate of component i is λ_i, its repair rate is μ_i $(1 \le i \le 3)$.

The item is modeled by a parametrized PDMP with values in $E_0 \times \mathbb{R}^3$, where

$$E_0 = \{(1, 1, 1), (0, 1, 1), (1, 0, 1), (1, 1, 0), (0, 0, 1), (0, 1, 0), (1, 0, 0), (0, 0, 0)\}.$$

Its parameters λ, Q are derived from the diagram of Fig. 4.1, as explained below.

Firstly, λ_i and μ_i are for $\lambda_i(x_i)$ and $\mu_i(x_i)$ respectively. Secondly, when there is an arrow from $e_i = (\eta_1, \eta_2, \eta_3) \in \{1, 0\}^3$ to e_j, then $e_j = (1 - \eta_1, \eta_2, \eta_3)$ or $e_j = (\eta_1, 1 - \eta_2, \eta_3)$ or $e_j = (\eta_1, \eta_2, 1 - \eta_3)$. Thirdly, when the label is κ, this means that

$$\Upsilon((e_i, x_1, x_2, x_3);\, \{e_j\}, dy_1 dy_1 dy_3) = \kappa(x_1, x_2, x_3)\, \mu_{i,j,x_1,x_2,x_3}(dy_1 dy_2 dy_3)$$

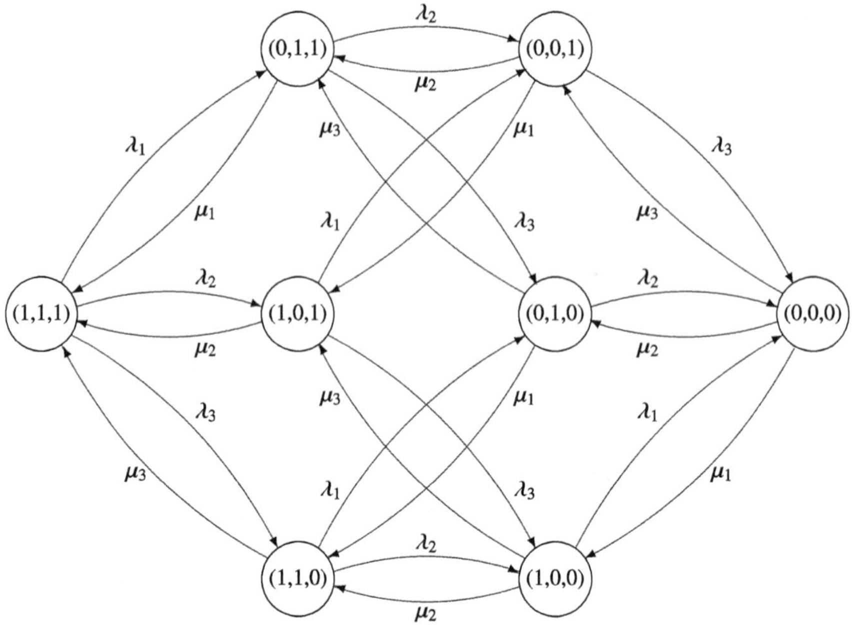

Fig. 4.1 Transition rates for the piece of equipment with three independent components of Example 4.6

with

$$\mu_{i,j,x_1,x_2,x_3}(dy_1dy_2dy_3) = \begin{cases} \delta_0(dy_1)\,\delta_{x_2}(dy_2)\,\delta_{x_3}(dy_3), & e_j = (1-\eta_1, \eta_2, \eta_3), \\ \delta_{x_1}(dy_1)\,\delta_0(dy_2)\,\delta_{x_3}(dy_3), & e_j = (\eta_1, 1-\eta_2, \eta_3), \\ \delta_{x_1}(dy_1)\,\delta_{x_2}(dy_2)\,\delta_0(dy_3), & e_j = (\eta_1, \eta_2, 1-\eta_3). \end{cases}$$

Therefore, $\lambda(e_i, x_1, x_2, x_3)$ is equal to the sum of the labels of the arrows that start at e_i, and $Q((e_i, x_1, x_2, x_3); \{e_j\}, dy_1dy_2dy_3)$ is $\mu_{i,j,x_1,x_2,x_3}(dy_1dy_2dy_3)$ impacted with a coefficient that is the label of the arrow from e_i to e_j divided by $\lambda(e_i, x_1, x_2, x_3)$. For example, $\lambda((0,1,1), x_1, x_2, x_3) = \mu_1(x_1) + \lambda_2(x_2) + \lambda_3(x_3)$,

$$Q(((0,1,1), x_1, x_2, x_3); (0,0,1), dy_1dy_2dy_3) \\ = \frac{\lambda_2(x_2)}{\mu_1(x_1) + \lambda_2(x_2) + \lambda_3(x_3)}\, \delta_{x_1}(dy_1)\,\delta_0(dy_2)\,\delta_{x_3}(dy_3).$$

When failure and repair rates do not depend on time, the diagram is the usual diagram of the Markov jump process, taking values in E_0, that models the equipment. It gives the transition rates that are the nondiagonal terms of the generator matrix (Çinlar [27], Chap. 8, Cocozza-Thivent [29], Chap. 8).

The equipment reliability at time t, i.e., the probability that the equipment is operating throughout the whole period $[0, t]$, is $R(t) = \mathbb{P}(\tau > t) = \mathbb{E}(e^{-\int_0^t \gamma(X^0(s))\,ds})$, where τ, γ, X^0 are as in Corollary 4.3 above, where A depends on how the equipment operates.

We first assume that the equipment is a 2-out-of-3 redundant system, that is, the equipment is operating if and only if at least two components out of three are up. We have $A = \{(1, 1, 1)\ (0, 1, 1), (1, 0, 1), (1, 1, 0)\} \times \mathbb{R}^3$. The transition rates of X^0 and γ are given in Fig. 4.2.

Then the three components are assumed to be connected in parallel, that is, the equipment is operating when at least one component is up. We have $A^c = \{(0, 0, 0)\} \times \mathbb{R}_+^3$. The transition rates of X^0 and γ are given in Fig. 4.3.

Example 4.7 Consider the PDMP $(X_t)_{t\geq 0}$ from Example 3.15, Sect. 3.4. It models an equipment made of two components, each of which can be in state up (state 1) or down (state 0). The failure rate of component i is λ_i when the other component is up and λ_i' when the other component is down. The repair rate of component i is μ_i. An instantaneous preventive maintenance is carried out on component i when it reaches the age of a_i. Define $e_1 = (1, 1)$, $e_2 = (0, 1)$, $e_3 = (1, 0)$, $e_4 = (0, 0)$. The PDMP takes values in $e_1 \times [0, a_1] \times [0, a_2] \cup \{e_2\} \times \mathbb{R}_+ \times [0, a_2] \cup \{e_3\} \times [0, a_1] \times \mathbb{R}_+ \cup \{e_4\} \times \mathbb{R}_+^2$.

The two components are assumed to be in parallel, that is, the equipment is operating if and only if at least one of the components is up; hence its set of operating states is $A = \{e_1, e_2, e_3\} \times \mathbb{R}_+^2$. The equipment reliability at time t is $R(t) = \mathbb{P}(\tau > t) = \mathbb{E}(e^{-\int_0^t \gamma(X_s^0)\, ds})$, where $\tau = \inf\{t : X_t \in A^c\}$ and X^0 is as in Corollary 4.3.

In Fig. 4.4, the transition rates of X between states e_i and e_j $(i, j \in \{1, 2, 3, 4\}, i \neq j)$ are displayed. The meaning of an arrow from e_i to e_j and of its label is as in Example 4.6. It provides parameters λ and Q as in that example. An arrow from e_i to e_i is for an instantaneous preventive maintenance that creates a jump from (e_i, x_1, x_2) to either $(e_i, 0, x_2)$, $(e_i, x_1, 0)$, or $(e_i, 0, 0)$, depending on i, x_1, x_2.

The transition rates of X^0 are given in Fig. 4.5. Write $X^0 = (I^0, X_1^0, X_2^0)$. We get

$$R(t) = \mathbb{E}\left(\exp\left(-\int_0^t (1_{\{I_s^0=(0,1)\}}\lambda_2'(X_{2,s}^0) + 1_{\{I_s^0=(1,0)\}}\lambda_1'(X_{1,s}^0))\, ds\right)\right).$$

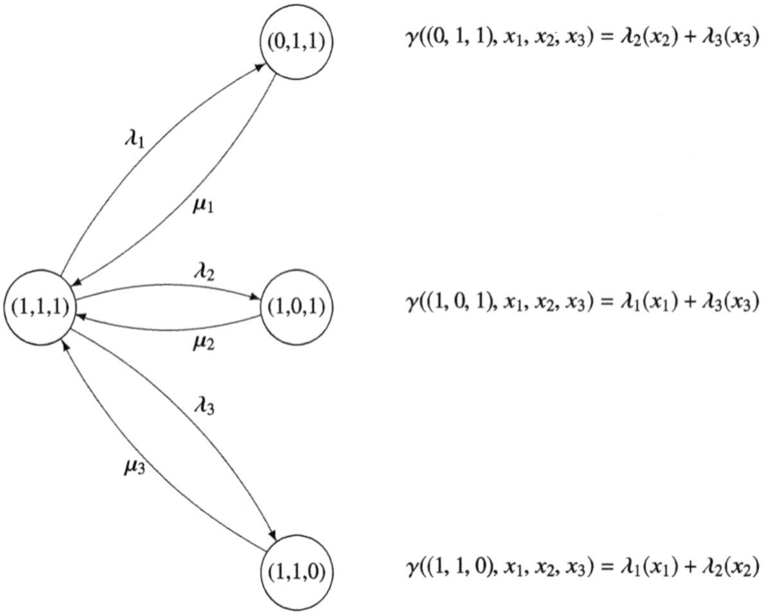

$$\gamma((0, 1, 1), x_1, x_2, x_3) = \lambda_2(x_2) + \lambda_3(x_3)$$

$$\gamma((1, 0, 1), x_1, x_2, x_3) = \lambda_1(x_1) + \lambda_3(x_3)$$

$$\gamma((1, 1, 0), x_1, x_2, x_3) = \lambda_1(x_1) + \lambda_2(x_2)$$

Fig. 4.2 Transition rates of X^0 for the 2-out-of-3 system of Example 4.6

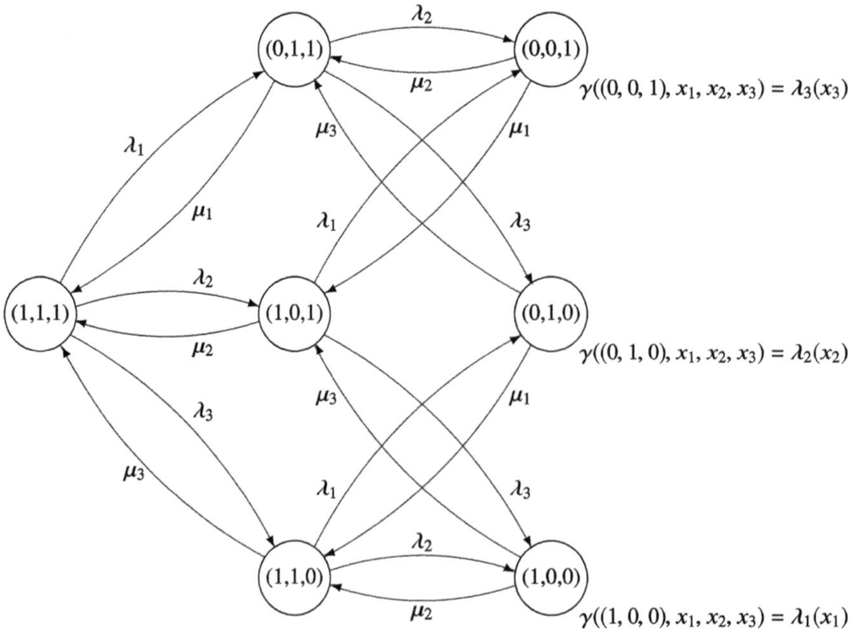

$$\gamma((0, 0, 1), x_1, x_2, x_3) = \lambda_3(x_3)$$

$$\gamma((0, 1, 0), x_1, x_2, x_3) = \lambda_2(x_2)$$

$$\gamma((1, 0, 0), x_1, x_2, x_3) = \lambda_1(x_1)$$

Fig. 4.3 Transition rates of X^0 for the parallel system of Example 4.6

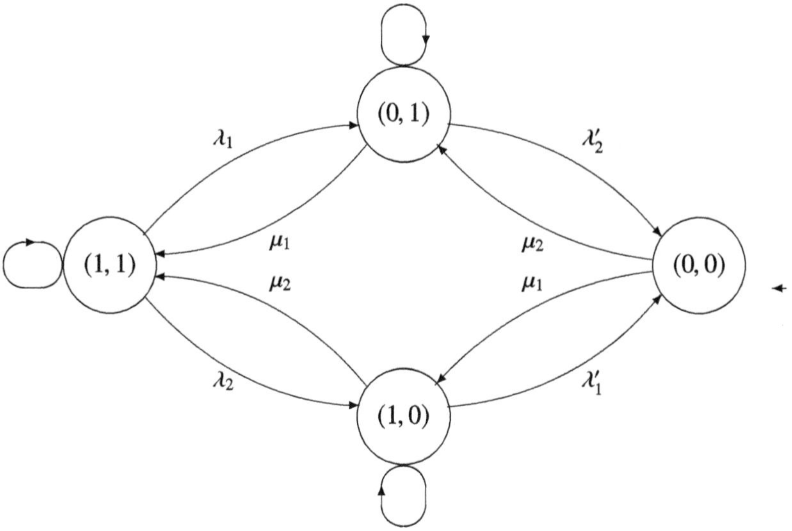

Fig. 4.4 Transition rates for X of Example 4.7

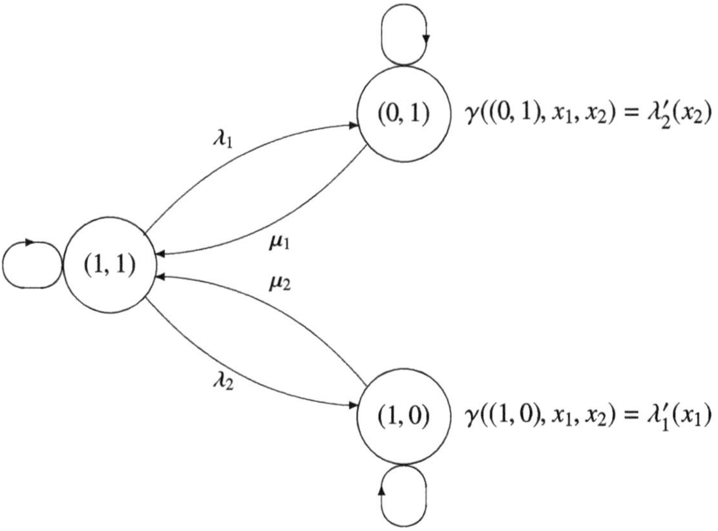

Fig. 4.5 Transition rates for X^0 of Example 4.7

Chapter 5
Intensity of Some Marked Point Processes

In this chapter we introduce the intensities of marked point processes that will be useful for the future. They give the mean number of times the marks are in a given set during a given time period. They are both tools for proofs and aids for the understanding of formulas, and they provide interesting performance indicators in applications, as we will see in examples.

Kolmogorov equations can be obtained through a martingale approach (see Chap. 7). An alternative approach is presented in Chap. 6 based on the intensities of marked point processes. It is more in line with the spirit of this book, yields greater insight into the formulas, and no prerequisites are required.

This chapter introduces these intensities. The structure of the renewal kernel is progressively specified and the consequence on the intensities revealed.

In the classical case in which the interarrival distributions are those of the minimum between a random variable with a probability density function and a positive constant, special attention should be paid to two marked point processes: the first one linked to the "purely random" jumps and the second one to the jumps caused by the constants (namely the Dirac measures in the kernel); their respective intensities are investigated. For PDMPs, the intensity σ of the process formed by the jumps caused by the Dirac measures marked by the state of the PDMP just before these jumps is essential in the generalized Kolmogorov equations and is interesting in itself in the applications; it is called the boundary measure. It can be approximated by numerical schemes (see Chap. 9).

In Davis [51], as for most of our results, it is assumed that the expectation of the number of jumps up to time t is finite for all $t > 0$. When there exists a Dirac measure in some interarrival distributions, it should be noted that in Corollary 5.1, end of Sect. 5.2, a more practical condition than the usual one is given for this assumption to hold.

C. Cocozza-Thivent, *Markov Renewal and Piecewise Deterministic Processes*, Probability Theory and Stochastic Modelling 100,
https://doi.org/10.1007/978-3-030-70447-6_5

Let us recall the framework and the notation.

The marked point process $(Y, T) = (Y_n, T_n)_{n\geq 1}$ is a Markov renewal process with values in a Polish space E equipped with its Borel σ-algebra. Its kernel N satisfies $N(x, E \times \{0\}) = 0$ for all $x \in E$. It is assumed that $\lim_{n\to\infty} T_n = +\infty$. We write

$$N(x, dy, dv) = 1_{\mathbb{R}_+}(v)\,\beta(x, v; dy)\,dF_x(v),$$

where dF_x is a regular conditional probability distribution of $T_{n+1} - T_n$ given $Y_n = x$ and $\beta(x, v; dy)$ is a regular conditional probability distribution of Y_{n+1} given $Y_n = x$, $T_{n+1} - T_n = v$ $(n \geq 1)$. We defined

$$\bar{F}_x(u) = dF_x(]u, +\infty]), \quad t^*(x) = \sup\{u \in \mathbb{R}_+ : \bar{F}_x(u) \neq 0\},$$

$$\breve{E} = \{(x, u) \in E \times \mathbb{R}_+ : \bar{F}_x(u) \neq 0\} = \{(x, u) \in E \times \mathbb{R}_+ : u < t^*(x)\},$$

$$\hat{E} = \{(x, u) \in E \times \mathbb{R}_+ : u \leq t^*(x)\}.$$

The initial distribution of the Markov renewal process (Y, T) is assumed to be $N_0^{x,u}$ defined Sect. 2.3 ($(x, u) \in \breve{E}$) or a mixture of such measures.

The CSMP associated with (Y, T) is denoted by $(Z_t, A_t)_{t\geq 0}$. It is a Markov process, and we write

$$\mathbb{P}_{x,u}(\,\cdot\,) = \mathbb{P}(\,\cdot \mid Z_0 = x, A_0 = u), \quad \mathbb{E}_{x,u}(\,\cdot\,) = \mathbb{E}(\,\cdot \mid Z_0 = x, A_0 = u),$$

$$\mathbb{P}_x = \mathbb{P}_{x,0}, \quad \mathbb{E}_x = \mathbb{E}_{x,0};$$

$dF_{x,u}$ is the probability distribution of T_1 under $\mathbb{P}_{x,u}$, i.e., $dF_{x,u}(v) = \int_{\{y\in E\}} N_0^{x,u}(dy, dv)$.

Definition 5.1 Given a marked point process $(\eta_k, S_k)_{k\geq 1}$ with values in E, its intensity is the nonnegative measure ρ on $E \times \mathbb{R}_+$ defined as

$$\rho(B \times [0, t]) = \mathbb{E}\left(\sum_{k\geq 1} 1_B(\eta_k)\, 1_{\{S_k \leq t\}}\right)$$

for every Borel set $B \subset E$ and all $t \geq 0$, or equivalently,

$$\int \varphi\, d\rho = \sum_{k\geq 1} \mathbb{E}(\varphi(\eta_k, S_k)\, 1_{\{S_k < +\infty\}})$$

for every Borel function $\varphi : E \times \mathbb{R}_+ \to \mathbb{R}_+$.

Example 5.1 Let $(T_n)_{n\geq 1}$ be a zero-delayed renewal process with interarrival distribution dF and $(Y_n)_{n\geq 1}$ i.i.d. random variables with probability distribution ν. Then the intensity of $(Y_n, T_n)_{n\geq 1}$ is $\rho(dy, dv) = \nu(dy) \sum_{n\geq 1} dF^{*n}(dv)$.

If $(T_n)_{n\geq 1}$ is a Poisson process with parameter λ, we have

$$\sum_{n\geq 1} dF^{*n}(dv) = \sum_{n\geq 1} \frac{\lambda^n}{(n-1)!}\, v^{n-1}\, e^{-\lambda v}\, dv = \lambda\, dv.$$

Example 5.2 We return to Example 3.11, Sect. 3.4. Let ρ be the intensity of the Markov renewal process associated with PDMP $(X_t)_{t\geq 0}$, (see the terminology Sect. 3.2). Then the mean number of failures before time t that occur when the temperature is above a is $\rho(\{e_0\} \times \{0\} \times [a, +\infty[\times[0, t]) = \rho(\{e_0\} \times \mathbb{R}_+ \times [a, +\infty[\times[0, t])$.

Example 5.3 We take again Example 2.4, Sect. 2.5 assuming that preventive maintenance is not instantaneous. Let ρ be the intensity of the Markov renewal process associated with the CSMP. Then the mean number of failures before time t is $\rho(\{e_0\} \times [0, t])$, and the mean number of preventive maintenances started before time t is $\rho(\{e_2\} \times [0, t])$.

When preventive maintenance is considered instantaneous, the intensity of the Markov renewal process does not distinguish between preventive maintenance and repair times. To separate preventive maintenance from repair, a different intensity must be used (see Example 5.4 below). In Example 5.6, Sect. 5.2 we will see another way to address this problem when dF_i has a density ($i = 0, 1$).

5.1 Two Intensities Related to Markov Renewal Processes

Let us start by defining the intensities that will interest us and by giving some immediate formulas.

Notation Denote by $\rho^{x,u}$ the intensity of the Markov renewal process $(Y_n, T_n)_{n\geq 1}$ under $\mathbb{P}_{x,u}$, and by $\bar{\rho}^{x,u}$ the intensity of the marked point process $((Y_0, A_0 + T_1, T_1), (Y_k, T_{k+1} - T_k, T_{k+1})_{k\geq 1})$ under $\mathbb{P}_{x,u}$, that is,

$$\int_{E\times\mathbb{R}_+} \varphi(y, v)\, \rho^{x,u}(dy, dv) = \sum_{k\geq 1} \mathbb{E}_{x,u}(\varphi(Y_k, T_k)\, 1_{\{T_k<+\infty\}}),$$

$$\int_{E\times\mathbb{R}_+^2} \psi(y, v, w)\, \bar{\rho}^{x,u}(dy, dv, dw) = \mathbb{E}_{x,u}(\psi(x, u + T_1, T_1)\, 1_{\{T_1<+\infty\}})$$

$$+ \sum_{k\geq 1} \mathbb{E}_{x,u}(\psi(Y_k, T_{k+1} - T_k, T_{k+1}) 1_{\{T_{k+1}<+\infty\}}) \tag{5.1}$$

$$= \int_{\mathbb{R}_+} \psi(x, u + v, v)\, dF_{x,u}(v) + \int_{E\times\mathbb{R}_+^2} \psi(y, v_1, v + v_1)\, dF_y(v_1)\, \rho^{x,u}(dy, dv) \tag{5.2}$$

for all Borel functions $\varphi : E \times \mathbb{R}_+ \to \mathbb{R}_+$ and $\psi : E \times \mathbb{R}_+^2 \to \mathbb{R}_+$.

Example 5.4 We return to Example 2.4, Sect. 2.5. We assume that the item is new at time 0. The mean number of failures up to time t is $\rho(e_1, \{e_0\} \times [0, t])$, and also $\bar{\rho}^{e_1,0}(\{e_1\} \times [0, a[\times[0, t])$. The mean number of preventive maintenances up to time t is $\bar{\rho}^{e_1,0}(\{e_1\} \times \{a\} \times [0, t])$.

Define

$$\rho(x, dy, dv) = \rho^{x,0}(dy, dv), \quad \bar{\rho}(x, dy, dv, dw) = \bar{\rho}^{x,0}(dy, dv, dw).$$

Recall that R is defined as $R = \sum_{n\geq 0} N^{*n}$ (see (2.18), Sect. 2.5). It is called *potentiel* in Jacod [67]. We have

$$R(x, dy, dv) = \delta_{x,0}(dy, dv) + \rho(x, dy, dv), \tag{5.3}$$

$$\rho^{x,u} = N_0^{x,u} * R, \tag{5.4}$$

$$\mathbb{E}_{x,u}(\sum_{n\geq 1} 1_{\{T_n \leq t\}}) = \rho^{x,u}(E \times [0, t]) = \bar{\rho}^{x,u}(E \times \mathbb{R}_+ \times [0, t]).$$

From (5.2) we get

$$\int_{E\times\mathbb{R}_+^2} \psi(y, v, w)\, \bar{\rho}(x, dy, dv, dw) = \int_{E\times\mathbb{R}_+^2} \psi(y, v_1, v + v_1)\, dF_y(v_1)\, R(x, dy, dv).$$

Example 5.5 Let us consider the Markov renewal process $(Y_n, T_n)_{n\geq 1}$ related to the alternating renewal process in Example 1.4 Part 1, Chap. 1. The intensity of $(Y_n, T_n)_{n\geq 1}$ is

$$\rho(e_1, dy, dv) = \delta_{e_1}(dy) \sum_{n\geq 1} dF_1^{*n} * dF_0^{*n}\,(dv) + \delta_{e_0}(dy) \sum_{n\geq 0} dF_1^{*(n+1)} * dF_0^{*n}\,(dv).$$

When it is a model for an item subject to failure and the item is up (state e_1) and new at time 0, the mean number of failures up to time t is $\rho(e_1, \{e_0\} \times [0, t]) = \sum_{n\geq 0} dF_1^{*(n+1)} * dF_0^{*n}\,([0, t])$, the mean number of repairs completed before time t is $\rho(e_1, \{e_1\} \times [0, t]) = \sum_{n\geq 1} dF_1^{*n} * dF_0^{*n}\,([0, t])$. The mean number of repairs completed before time t of duration less than a is $\bar{\rho}(e_1, \{e_0\} \times [0, a] \times [0, t])$.

When $dF_i(v) = 1_{\mathbb{R}_+}(v) \frac{\lambda^{\gamma_i}}{\Gamma(\gamma_i)} v^{\gamma_i - 1} e^{-\lambda v}\, dv$ (Γ-probability distribution with parameters $(\gamma_i, 1)$), we get

$$\rho(e_1, dy, dv) = \delta_{e_1}(dy)\, e^{-\lambda v} \sum_{n\geq 1} \frac{\lambda^{n(\gamma_1+\gamma_0)}}{\Gamma(n(\gamma_1+\gamma_0))} v^{n(\gamma_1+\gamma_0)-1}\, dv$$
$$+ \delta_{e_0}(dy)\, e^{-\lambda v} \sum_{n\geq 0} \frac{\lambda^{(n+1)\gamma_1+n\gamma_0}}{\Gamma((n+1)\gamma_1+n\gamma_0)} v^{(n+1)\gamma_1+n\gamma_0-1}\, dv$$

and

$$\bar\rho(e_1, dy, dv, dw) = \delta_{e_1}(dy) \frac{\lambda^{\gamma_1}}{\Gamma(\gamma_1)} v^{\gamma_1-1}\, e^{-\lambda v}\, \delta_v(dw)\, dv$$
$$+ \delta_{e_1}(dy)\, 1_{\{w>v\}} \frac{1}{\Gamma(\gamma_1)} v^{\gamma_1-1} e^{-\lambda w} \sum_{n\geq 1} \frac{\lambda^{(n+1)\gamma_1+n\gamma_0}}{\Gamma(n(\gamma_1+\gamma_0))} (w-v)^{n(\gamma_1+\gamma_0)-1}\, dvdw$$
$$+ \delta_{e_0}(dy)\, 1_{\{w>v\}} \frac{1}{\Gamma(\gamma_0)} v^{\gamma_0-1} e^{-\lambda w} \sum_{n\geq 0} \frac{\lambda^{(n+1)(\gamma_1+\gamma_0)}}{\Gamma((n+1)\gamma_1+n\gamma_0)} (w-v)^{(n+1)\gamma_1+n\gamma_0-1} dvdw.$$

If $\gamma_1 = \gamma_0 = \lambda$ i.e. $(T_n)_{n\geq 1}$ is a Poisson process with parameter λ, then

$$\rho(e_1, dy, dv) = \delta_{e_1}(dy)\, \frac{1-e^{-2\lambda v}}{2}\, \lambda\, dv + \delta_{e_0}(dy)\, \frac{1+e^{-2\lambda v}}{2}\, \lambda\, dv,$$

$$\bar\rho(e_1, dy, dv, dw) = \delta_{e_1}(dy)\, \lambda\, e^{-\lambda v} \left(\delta_v(dw)\, dv + 1_{\{0<v<w\}} \frac{1-e^{-2\lambda(w-v)}}{2}\, \lambda\, dvdw\right)$$
$$+ \delta_{e_0}(dy)\, 1_{\{0<v<w\}}\, \lambda^2\, e^{-\lambda v}\, \frac{1+e^{-2\lambda(w-v)}}{2}\, dvdw.$$

Formula (5.2) gives $\bar\rho^{x,u}$ as a function of $\rho^{x,u}$. The following lemma gives the reverse.

Lemma 5.1 *Given a Borel function $\varphi : E \times \mathbb{R}_+ \to \mathbb{R}_+$, we have*

$$\int_{E\times\mathbb{R}_+} \varphi(y, v)\, \rho^{x,u}(dy, dv) = \int_{E\times\mathbb{R}_+^2} \int_E \varphi(z, w)\, \beta(y, v; dz)\, \bar\rho^{x,u}(dy, dv, dw). \tag{5.5}$$

Proof Applying (3.1), Chap. 3 and conditioning on (Y_k, T_k) the term of index k in the sum, we get

$$\sum_{k\geq 1}\mathbb{E}_{x,u}(\varphi(Y_k,T_k)\,1_{\{T_k<+\infty\}}) = \int_{E\times\mathbb{R}_+}\varphi(y,v)\,N_0^{x,u}(dy,dv)$$
$$+\sum_{k\geq 1}\mathbb{E}_{x,u}(\varphi(Y_{k+1},T_{k+1})\,1_{\{T_{k+1}<+\infty\}})$$
$$=\int_{E\times\mathbb{R}_+}\varphi(y,v)\,\beta(x,u+v;dy)\,dF_{x,u}(v)$$
$$+\int_{E\times\mathbb{R}_+^2}\int_E\varphi(y_1,v+v_1)\,\beta(y,v_1;dy_1)\,dF_y(v_1)\,\rho^{x,u}(dy,dv).$$

Then the result follows from (5.2) with $\psi(y,v,w)=\int_E\varphi(z,w)\,\beta(y,v;dz)$. ■

In the following lemma, the marginal CSMP distributions are given through the Markov renewal process intensity.

Lemma 5.2 *Let* $(x,u)\in\breve{E}$, $t\geq 0$, *and let* $\psi : E\times[0,t+u]\times[0,t]\to\mathbb{R}$ *be a Borel function.*

If ψ *is nonnegative, then*

$$\mathbb{E}_{x,u}(\psi(Z_t,A_t,t)) = \psi(x,u+t,t)\,\mathbb{P}_{x,u}(T_1>t) + \int_{E\times\mathbb{R}_+}1_{\{v\leq t\}}\,\bar{F}_y(t-v)\,\psi(y,t-v,t)\,\rho^{x,u}(dy,dv). \tag{5.6}$$

If ψ *is bounded, then*

$$\int_{E\times\mathbb{R}_+}1_{\{v\leq t\}}\,\bar{F}_y(t-v)\,|\psi(y,t-v,t)|\,\rho^{x,u}(dy,dv)<+\infty \tag{5.7}$$

and (5.6) holds.

Proof If ψ is nonnegative, then Lemma 5.2 comes from Lemma 2.1, Sect. 2.3. If ψ is bounded, we write $\psi=\psi_+-\psi_-$, where $\psi_+=\max(\psi,0)$ and $\psi_-=\max(-\psi,0)$ are nonnegative functions. Hence $|\psi|=\psi_++\psi_-$, and applying (5.6) to ψ_+ and ψ_- gives (5.7). ■

We will now progressively specify the kernel N and obtain more and more telling intensity decompositions.

5.2 Intensity Decomposition

Given a Borel function $\ell : E\times\mathbb{R}_+\to\mathbb{R}_+$, a Markov kernel β from $E\times\mathbb{R}_+$ to E and a kernel n from E to $E\times\mathbb{R}_+$, we assume that

$$N(x, dy, dv) = \mathbb{P}_x(T_1 > v)\, \ell(x, v)\, \beta(x, v; dy)\, dv + n(x, dy, dv). \tag{5.8}$$

We get

$$N_0^{x,u}(dy, dv) = \mathbb{P}_{x,u}(T_1 > v)\, \ell(x, u+v)\, \beta(x, u+v; dy)\, dv + n_0^{x,u}(dy, dv),$$

where

$$\int_{E\times\mathbb{R}_+} \varphi(y, v)\, n_0^{x,u}(dy, dv) = \frac{1}{\bar{F}_x(u)} \int_{E\times\mathbb{R}_+} 1_{\{v>u\}}\, \varphi(y, v-u)\, n(x, dy, dv).$$

Proposition 5.1 *Let $\psi : E \times \mathbb{R}_+^2 \to \mathbb{R}_+$ be a Borel function, N as in (5.8), and $(x, u) \in \breve{E}$. Then:*

1. *We have $\bar{\rho}^{x,u} = \bar{\rho}_c^{x,u} + \bar{\rho}_D^{x,u}$, where*

$$\int_{E\times\mathbb{R}_+^2} \psi(y, v, w)\, \bar{\rho}_c^{x,u}(dy, dv, dw) = \mathbb{E}_{x,u}\left(\int_{\mathbb{R}_+} \psi(Z_v, A_v, v)\, \ell(Z_v, A_v)\, dv\right)$$

and

$$\int_{E\times\mathbb{R}_+^2} \psi(y, v, w)\, \bar{\rho}_D^{x,u}(dy, dv, dw) = \int_{E\times\mathbb{R}_+} \psi(x, u+v, v)\, n_0^{x,u}(dy, dv)$$
$$+ \int_{(E\times\mathbb{R}_+)^2} \psi(y, v_1, v+v_1)\, n(y, dz, dv_1)\, \rho^{x,u}(dy, dv). \tag{5.9}$$

2. *We have $\rho^{x,u} = \rho_c^{x,u} + \rho_D^{x,u}$, where*

$$\int_{E\times\mathbb{R}_+} \varphi(y, v)\, \rho_c^{x,u}(dy, dv) = \mathbb{E}_{x,u}\left(\int_{E\times\mathbb{R}_+} \varphi(y, v)\, \ell(Z_v, A_v)\, \beta(Z_v, A_v; dy)\, dv\right) \tag{5.10}$$

and $\rho_D^{x,u} = n_0^{x,u} + \rho^{x,u} * n$.

If, moreover, $\mathbb{E}_{x,u}(\sum_{n\geq 1} 1_{\{T_n \leq t\}}) < +\infty$, then

$$\rho_D^{x,u} = n_0^{x,u} + (\rho_c^{x,u} + n_0^{x,u}) * r_1, \quad r_1 = \sum_{k\geq 1} n^{*k}. \tag{5.11}$$

Proof Formula (5.2), Sect. 5.1 gives $\bar{\rho}^{x,u} = \bar{\rho}_c^{x,u} + \bar{\rho}_D^{x,u}$ with $\bar{\rho}_D^{x,u}$ as in (5.9) and

$$\int_{E\times\mathbb{R}_+^2} \psi(y, v, w)\, \bar{\rho}_c^{x,u}(dy, dv, dw) =$$
$$\int_{\mathbb{R}_+} \psi(x, u+v, v)\, \mathbb{P}_{x,u}(T_1 > v)\, \ell(x, u+v)\, dv$$
$$+ \int_{E\times\mathbb{R}_+^2} \psi(y, v_1, v+v_1)\, \bar{F}_y(v_1)\, \ell(y, v_1)\, dv_1\, \rho^{x,u}(dy, dv). \tag{5.12}$$

Then by Lemma 5.2, Sect. 5.1, with $\psi(y, v, w)\,\ell(y, v)$ instead of $\psi(y, v, w)$, we get

$$\int_{E\times\mathbb{R}_+^2} \psi(y, v, w)\,\bar{\rho}_c^{x,u}(dy, dv, dw) = \mathbb{E}_{x,u}\left(\int_{E\times\mathbb{R}_+} \psi(Z_s, A_s, s)\,\ell(Z_s, A_s)\,ds\right),$$

which concludes the proof of the first part.

Define $n_1(x, dy, dv) = \bar{F}_x(v)\,\ell(x, v)\,\beta(x, v; dy)\,dv$. We have $N = n_1 + n$. From (5.4), Sect. 5.1 and $R = I + R * N$ we get

$$\rho^{x,u} = N_0^{x,u} + \rho^{x,u} * N = \rho_c^{x,u} + \rho_D^{x,u},$$

where

$$\rho_c^{x,u}(dy, dv) = \mathbb{P}_{x,u}(T_1 > v)\,\ell(x, u+v)\,\beta(x, u+v; dy)\,dv + \rho^{x,u} * n_1(dy, dv)$$

and $\rho_D^{x,u} = n_0^{x,u} + \rho^{x,u} * n$.

Now given $\varphi : E \times \mathbb{R}_+ \to \mathbb{R}_+$, define $\psi_1(y, v, w) = \int_E \varphi(z, w)\,\beta(y, v; dz)$. We get

$$\int_{E\times\mathbb{R}_+} \varphi(y, v)\,\rho_c^{x,u}(dy, dv) = \int_{\mathbb{R}_+} \left\{\psi_1(x, u+v, v)\,\mathbb{P}_{x,u}(T_1 > v)\,\ell(x, u+v)\right\} dv$$
$$+ \int_{\mathbb{R}_+} \left\{\int_{E\times\mathbb{R}_+} 1_{\{v\le s\}}\,\psi_1(y, s-v, s)\,\bar{F}_y(s-v)\,\ell(y, s-v)\,\rho^{x,u}(dy, dv)\right\} ds.$$

Then Lemma 5.2, Sect. 5.1 gives

$$\int_{E\times\mathbb{R}_+} \varphi(y, v)\,\rho_c^{x,u}(dy, dv) = \mathbb{E}_{x,u}\left(\int_{\mathbb{R}_+} \psi_1(Z_s, A_s, s)\,\ell(Z_s, A_s)\,ds\right),$$

which is (5.10).

Finally, iterating $\rho^{x,u} = \rho_c^{x,u} + n_0^{x,u} + \rho^{x,u} * n$, we get for $k \ge 2$,

$$\rho^{x,u} = \rho_c^{x,u} + n_0^{x,u} + \sum_{j=1}^{k-1} (\rho_c^{x,u} + n_0^{x,u}) * n^{*j} + \rho^{x,u} * n^{*k}.$$

Now $\int_{E\times[0,t]} \rho^{x,u} * n^{*k} \le \int_{E\times[0,t]} \rho^{x,u} * N^{*k} = \sum_{n\ge k+1} \mathbb{E}_{x,u}(1_{\{T_n\le t\}})$, and therefore the assumption $\mathbb{E}_{x,u}(\sum_{n\ge 1} 1_{\{T_n\le t\}}) < +\infty$ gives $\rho_D^{x,u} = n_0^{x,u} + (\rho_c^{x,u} + n_0^{x,u}) * r_1$. ■

In Propositions 5.2 and 5.3 below, measures $\bar{\rho}_c^{x,u}$, $\bar{\rho}_D^{x,u}$, $\rho_c^{x,u}$, $\rho_D^{x,u}$ are interpreted as marked point process intensities, and in Proposition 5.3, formula (5.11) is proved without the assumption $\mathbb{E}_{x,u}(\sum_{n\ge 1} 1_{\{T_n\le t\}}) < +\infty$, resulting in practical conditions to ensure it (see Corollary 5.1 later in this section).

Simple CSMP *P52 def of CSMP simple*

If $(Z_t, A_t)_{t \geq 0}$ is a simple CSMP with parameters ℓ, α, β, we have

$$dF_x(v) = \ell(x, v) \, e^{- \int_0^v \ell(x,w) \, dw} \, 1_{\{v < \alpha(x)\}} \, dv + 1_{\{\alpha(x) < +\infty\}} \, e^{- \int_0^{\alpha(x)} \ell(x,w) \, dw} \, \delta_{\alpha(x)}(dv), \tag{5.13}$$

whence condition (5.8) is satisfied with

$$n(x, dy, dv) = 1_{\{\alpha(x) < +\infty\}} \, e^{- \int_0^{\alpha(x)} \ell(x,w) \, dw} \, \beta(x, \alpha(x); dy) \, \delta_{\alpha(x)}(dv)$$

and

$$n_0^{x,u}(dy, dv) = 1_{\{\alpha(x) < +\infty\}} \, e^{- \int_0^{\alpha(x)-u} \ell(x,u+w) \, dw} \, \beta(x, \alpha(x); dy) \, \delta_{\alpha(x)-u}(dv).$$

Proposition 5.2 and Proposition 5.3 below give an interpretation of the decomposition of respectively $\bar{\rho}^{x,u}$ and $\rho^{x,u}$ given in Proposition 5.1. They allow us to get a decomposition of the intensity of a marked point process related to a parametrized PDMP (Proposition 5.7, Sect. 5.3) which is fundamental to establishing the important results of Chap. 6 (Proposition 6.3, Sect. 6.2 and Theorem 6.1, Sect. 6.2).

Proposition 5.2 *Assume that $(Z_t, A_t)_{t \geq 0}$ is a simple CSMP with parameters ℓ, α, β. Let $\psi : E \times \mathbb{R}_+^2 \to \mathbb{R}_+$ be a Borel function and $(x, u) \in \check{E}$. Then*

$$\bar{\rho}^{x,u} = \bar{\rho}_c^{x,u} + \bar{\rho}_D^{x,u},$$

and the following hold: *first period*

1. $\bar{\rho}_c^{x,u}$ *is the intensity, under $\mathbb{P}_{x,u}$, of the marked point process $((Y_0, A_0 + T_1, T_1), (Y_k, T_{k+1} - T_k, T_{k+1})_{k \geq 1})$ restricted to the jumps that are not caused by Dirac measures, namely,*

$\bar{\rho}^{x,u}$: *jump out of dy*

$\bar{\rho}_c^{x,u}$: *jump outside caused by natural jumps*

by def of $\bar{\rho}^{x,u}$ (5.12)

$$\int_{E \times \mathbb{R}_+^2} \psi(y, v, w) \, \bar{\rho}_c^{x,u}(dy, dv, dw)$$

$$= \int_{E \times \mathbb{R}_+^2} \psi(y, v, w) \, 1_{\{v < \alpha(y)\}} \, \bar{\rho}^{x,u}(dy, dv, dw)$$

$T_1 + u < \alpha(x) \iff T_1 + A_0 < \alpha(x)$

$$= \mathbb{E}_{x,u}(\psi(Y_0, A_0 + T_1, T_1) \, 1_{\{T_1 < \alpha(x)-u\}})$$

$$+ \sum_{k \geq 1} \mathbb{E}_{x,u}(\psi(Y_k, T_{k+1} - T_k, T_{k+1}) \, 1_{\{T_{k+1} < T_k + \alpha(Y_k)\}})$$

Prop 5.1 $$= \int_{\mathbb{R}_+} \mathbb{E}_{x,u}(\psi(Z_v, A_v, v) \, \ell(Z_v, A_v)) \, dv.$$

2. $\bar{\rho}_D^{x,u}$ *is the intensity, under $\mathbb{P}_{x,u}$, of the marked point process $((Y_0, A_0 + T_1, T_1), (Y_k, T_{k+1} - T_k, T_{k+1})_{k \geq 1})$ restricted to the jumps that are caused by Dirac measures, namely,*

$$\begin{aligned}
&\int_{E\times\mathbb{R}_+^2} \psi(y,v,w)\,\bar{\rho}_D^{x,u}(dy,dv,dw) \\
&= \int_{E\times\mathbb{R}_+^2} \psi(y,v,w)\,1_{\{v=\alpha(y)\}}\,\bar{\rho}^{x,u}(dy,dv,dw) \\
&= 1_{\{\alpha(x)<+\infty\}}\,\mathbb{E}_{x,u}(\psi(Y_0, A_0+T_1, T_1)\,1_{\{T_1=\alpha(x)-u\}}) \\
&\quad + \sum_{k\geq 1} \mathbb{E}_{x,u}(\psi(Y_k, T_{k+1}-T_k, T_{k+1})\,1_{\{T_{k+1}=T_k+\alpha(Y_k)<+\infty\}}) \\
&= 1_{\{\alpha(x)<+\infty\}}\,\psi(x,\alpha(x),\alpha(x)-u)\,e^{-\int_u^{\alpha(x)}\ell(x,w)\,dw} \\
&\quad + \int_{E\times\mathbb{R}_+^2} 1_{\{\alpha(y)<+\infty\}}\,\psi(y,\alpha(y),v+\alpha(y))\,e^{-\int_0^{\alpha(y)}\ell(y,w)\,dw}\,\rho^{x,u}(dy,dv).
\end{aligned} \tag{5.14}$$

Proof From (5.2), Sect. 5.1, $1_{\{u+v<\alpha(x)\}}\,dF_{x,u}(v) = \mathbb{P}_{x,u}(T_1 > v)\,\ell(x,u+v)\,dv$ and $1_{\{v_1<\alpha(y)\}}dF_y(v_1) = \bar{F}_y(v_1)\,\ell(y,v_1)\,dv_1$, we deduce that $1_{\{v<\alpha(y)\}}\,\bar{\rho}^{x,u}(dy,dv,dw)$ is the measure $\bar{\rho}_c^{x,u}(dy,dv,dw)$ given by (5.12), Sect. 5.2.

The first part of the second assertion is obtained by writing $\bar{\rho}_D^{x,u} = \bar{\rho}^{x,u} - \bar{\rho}_c^{x,u}$, and the second by conditioning on (Y_k, T_k) the term of index k in the sum. ■

Define

$$\begin{aligned}
N_{B_1,B_2}^c(t) &= 1_{B_1}(Y_0)\,1_{B_2}(Y_1)\,1_{[0,t]}(T_1)\,1_{\{T_1<\alpha(Y_0)-A_0\}} \\
&\quad + \sum_{n\geq 2} 1_{B_1}(Y_{n-1})\,1_{B_2}(Y_n)\,1_{[0,t]}(T_n)\,1_{\{T_n-T_{n-1}<\alpha(Y_{n-1})\}}.
\end{aligned}$$

Conditioning the term of index n on Y_{n-1}, T_{n-1}, T_n, it can be deduced from Proposition 5.2 that

$$\mathbb{E}_{x,u}(N_{B_1,B_2}^c(t)) = \mathbb{E}_{x,u}\left(\int_{B_2\times[0,t]} 1_{B_1}(Z_v)\,\ell(Z_v,A_v)\,\beta(Z_v,A_v;dy)\,dv\right). \tag{5.15}$$

This formula is also a consequence of Proposition 7.3, Sect. 7.1 obtained from the stochastic calculus.

Proposition 5.3 *Assume that $(Z_t, A_t)_{t\geq 0}$ is a simple CSMP with parameters ℓ, α, β. Let $\varphi : E\times\mathbb{R}_+ \to \mathbb{R}_+$ be a Borel function and $(x,u)\in\check{E}$. Then*

$$\rho^{x,u} = \rho_c^{x,u} + \rho_D^{x,u},$$

and the following hold:

1. *$\rho_c^{x,u}$ is the intensity, under $\mathbb{P}_{x,u}$, of the marked point process $(Y_k, T_k)_{k\geq 1}$ restricted to the jumps that are not caused by Dirac measures, namely,*

$$\int_{E\times\mathbb{R}_+} \varphi(y,v)\,\rho_c^{x,u}(dy,dv) = \mathbb{E}_{x,u}(\varphi(Y_1,T_1)\,1_{\{T_1<\alpha(x)-u\}})$$
$$+\sum_{k\geq 1}\mathbb{E}_{x,u}(\varphi(Y_{k+1},T_{k+1})\,1_{\{T_{k+1}<T_k+\alpha(Y_k)\}})$$
$$=\int_{\mathbb{R}_+}\mathbb{E}_{x,u}\left(\int_E \varphi(y,v)\,\ell(Z_v,A_v)\,\beta(Z_v,A_v;dy)\right)dv.$$

2. *$\rho_D^{x,u}$ is the intensity, under $\mathbb{P}_{x,u}$, of the marked point process $(Y_k,T_k)_{k\geq 1}$ restricted to the jumps that are caused by Dirac measures, namely,*

$$\int_{E\times\mathbb{R}_+^2} \varphi(y,v)\,\rho_D^{x,u}(dy,dv) = 1_{\{\alpha(x)<+\infty\}}\,\mathbb{E}_{x,u}(\varphi(Y_1,T_1)\,1_{\{T_1=\alpha(x)-u\}})$$
$$+\sum_{k\geq 1}\mathbb{E}_{x,u}(\varphi(Y_{k+1},T_{k+1})\,1_{\{T_{k+1}=T_k+\alpha(Y_k)<+\infty\}}) \tag{5.16}$$
$$=1_{\{\alpha(x)<+\infty\}}\,e^{-\int_u^{\alpha(x)}\ell(x,w)\,dw}\int_E \varphi(y,\alpha(x)-u)\,\beta(x,\alpha(x);dy)$$
$$+\int_{E\times\mathbb{R}_+^2} 1_{\{\alpha(y)<+\infty\}}\,\varphi(z,v+\alpha(y))\,\beta(y,\alpha(y);dz)\,e^{-\int_0^{\alpha(y)}\ell(y,w)\,dw}\rho^{x,u}(dy,dv).$$

Moreover,

$$\int_{E\times\mathbb{R}_+}\varphi(y,v)\,r_1(y_0,dy,dv) := \sum_{k\geq 1}\int_{E\times\mathbb{R}_+}\varphi(y,v)\,n^{*k}(y_0,dy,dv)$$
$$=\sum_{k\geq 1}\mathbb{E}_{y_0}\left(\varphi(Y_k,T_k)\,1_{\{T_1=\alpha(y_0)<+\infty\}}\,1_{\{T_2-T_1=\alpha(Y_1)<+\infty\}}\cdots 1_{\{T_k-T_{k-1}=\alpha(Y_{k-1})<+\infty\}}\right)$$
$$=\sum_{k\geq 1}\int_{E^k} 1_{\{\alpha(y_0)+\dots+\alpha(y_{k-1})<+\infty\}}\,\varphi(y_k,\alpha(y_0)+\dots+\alpha(y_{k-1}))\,e^{-\int_0^{\alpha(y_0)}\ell(y_0,w)\,dw}\dots$$
$$\dots e^{-\int_0^{\alpha(y_{k-1})}\ell(y_{k-1},w)\,dw}\,\beta(y_0,\alpha(y_0);dy_1)\cdots\beta(y_{k-1},\alpha(y_{k-1});dy_k), \tag{5.17}$$

and

$$\rho_D^{x,u} = n_0^{x,u} + n_0^{x,u} * r_1 + \rho_c^{x,u} * r_1. \tag{5.18}$$

In particular defining $\rho_c(x,dy,dv) = \rho_c^{x,0}(dy,dv)$, $\rho_D(x,dy,dv) = \rho_D^{x,0}(dy,dv)$, we have

$$\rho_D = r_1 + \rho_c * r_1. \tag{5.19}$$

Proof Define $\psi_c(y,v,w) = 1_{\{v<\alpha(y)\}}\int_E \varphi(z,w)\,\beta(y,v;dz)$ and let $\rho_c^{x,u}$ be the intensity, under $\mathbb{P}_{x,u}$, of the marked point process $(Y_k,T_k)_{k\geq 1}$ restricted to the jumps that are not caused by Dirac measures. Proceeding as in the proof of Lemma 5.1, Sect. 5.1 we get

$$\int_{E\times\mathbb{R}_+} \varphi(y,v)\,\rho_c^{x,u}(dy,dv) = \int_{E\times\mathbb{R}_+^2} \psi_c(y,v,w)\,\bar{\rho}_c^{x,u}(dy,dv,dw).$$

Recalling $A_v < \alpha(Z_v)$ a.s., Proposition 5.2 above gives the result for $\rho_c^{x,u}$. We can also get it by taking $B_1 = E$ in (5.15).

By proceeding in the same way, we get the first result for $\rho_D^{x,u}$.

Formula (5.17) is straightforward. Let us now turn to prove (5.18). We write (5.16) as

$$\int_{E\times\mathbb{R}_+} \varphi(y,v)\,\rho_D^{x,u}(dy,dv) = \int_{E\times\mathbb{R}_+} \varphi(y,v)\,n_0^{x,u}(dy,dv) + A + B$$

with

$$A = \sum_{k\geq 1} \mathbb{E}_{x,u}\left(\varphi(Y_{k+1},T_{k+1})\,1_{\{T_1=\alpha(x)-u\}}\cdots 1_{\{T_{k+1}-T_k=\alpha(Y_k)\}}\right)$$

$$= \int_{E\times\mathbb{R}_+} \varphi(y,v)\,n_0^{x,u} * r_1(dy,dv),$$

$$B = \sum_{m\geq 0}\sum_{p\geq 0} \mathbb{E}_{x,u}\left(\varphi(Y_{p+m+2},T_{p+m+2})\,1_{\{T_{m+1}-T_m<\alpha(Y_m)\}}\,1_{\{T_{m+2}-T_{m+1}=\alpha(Y_{m+1})\}}\cdots\right.$$

$$\left.\cdots 1_{\{T_{m+p+2}-T_{m+p+1}=\alpha(Y_{m+p+1})\}}\right).$$

Conditioning on $(Y_n,T_n)_{n\leq m+1}$ the term of index m in the sum, we get

$$B = \sum_{p\geq 0}\int \varphi(y_{p+1}, v+\alpha(y)+\alpha(y_1)+\cdots+\alpha(y_p))\,e^{-\int_0^{\alpha(y)}\ell(y,w)\,dw}\,\beta(y,\alpha(y);dy_1)$$

$$e^{-\int_0^{\alpha(y_1)}\ell(y_1,w)\,dw}\,\beta(y_1,\alpha(y_1),dy_2)\ldots e^{-\int_0^{\alpha(y_p)}\ell(y_p,w)\,dw}\,\beta(y_p,\alpha(y_p),dy_{p+1})$$

$$\rho_c^{x,u}(dy,dv)$$

$$= \int_{E\times\mathbb{R}_+} \varphi(y,v)\,\rho_c^{x,u} * r_1(dy,dv).$$

This concludes the proof. ∎

Example 5.6 Let us go back to Example 3.7, Sect. 3.2, i.e., to Example 2.4, Sect. 2.5 assuming that preventive maintenance is instantaneous and that the item has failure rate λ and repair rate μ and is new at time 0.

The mean number of failures before time t is

$$\rho(e_1,\{e_0\}\times[0,t]) = \rho_c(e_1,\{e_0\}\times[0,t]) = \int_0^t \mathbb{E}_{e_1}(1_{\{Z_v=e_0\}}\,\lambda(A_v))\,dv.$$

The mean number of preventive maintenances before time t is $\rho_D(e_1,\{e_1\}\times[0,t])$. We apply formula (5.19). We have $r_1(e_0,dy,dv)=0$, since $\alpha(e_0)=+\infty$

and

$$r_1(e_1, dy, dv) = \sum_{k \geq 1} e^{-k \int_0^a \lambda(w)\,dw}\, \delta_{e_1}(dy)\, \delta_{ka}(dv).$$

Hence

$$\rho_D(e_1, \{e_1\} \times [0, t]) = \sum_{k=1}^{[t/a]} e^{-k \int_0^a \lambda(w)\,dw} \left(1 + \int_0^{t-ka} \mathbb{E}_{e_1}(1_{\{Z_v = e_0\}}\, \mu(A_v))\, dv \right).$$

Thus the mean numbers of failures and the mean number of preventive maintenances before time t can be obtained as soon as the marginal distributions of the CSMP are known. Chapter 9 gives numerical schemes approximating these marginal distributions. In steady state, the formulas are very simple (Example 8.6, Sect. 8.3).

As already mentioned, formula (5.18) is obtained in Proposition 5.3 without assuming $\mathbb{E}_{x,u}(\sum_{n \geq 1} 1_{\{T_n \leq t\}}) < +\infty$. This assumption, often required, is $\rho^{x,u}(E \times]0, t]) < +\infty$. The following corollary gives a sufficient condition, useful in practical cases, to ensure it.

Corollary 5.1 *Given a simple CSMP* $(Z_t, A_t)_{t \geq 0}$ *with parameters* ℓ, α, β *and* $(x, u) \in \breve{E}$, *let* $E_t^{x,u}$ *be a Borel subset of* E *such that* $\mathbb{P}_{x,u}(\forall s \leq t, Z_s \in E_t^{x,u}) = 1$, *and* $G_t^{x,u} \subset E_t^{x,u} \times \mathbb{R}_+$ *a Borel set such that* $\mathbb{P}_{x,u}(\forall s \leq t, (Z_s, A_s) \in G_t^{x,u}) = 1$.

Assume

$$\sup_{(y,v) \in G_t^{x,u}} \ell(y, v) < +\infty, \tag{5.20}$$

$$\sup_{y_0 \in E_t^{x,u}} \sum_{k \geq 1} \int_{(E_t^{x,u})^k} 1_{\{\alpha(y_0)+\cdots+\alpha(y_{k-1}) \leq t\}}\, e^{-\int_0^{\alpha(y_0)} \ell(y_0, w)\,dw}\, \beta(y_0, \alpha(y_0); dy_1) \ldots$$
$$\ldots e^{-\int_0^{\alpha(y_{k-1})} \ell(y_{k-1}, w)\,dw}\, \beta(y_{k-1}, \alpha(y_{k-1}); dy_k) < +\infty, \tag{5.21}$$

then $\mathbb{E}_{x,u}(\sum_{k \geq 1} 1_{\{T_k \leq t\}}) < +\infty$.

If there exist $A > 0$ *and* $m \geq 1$ *such that*

$$\sup_{y_0 \in E_t^{x,u}} \int_{(E_t^{x,u})^m} 1_{\{\alpha(y_0)+\cdots+\alpha(y_{m-1}) < +\infty\}}\, e^{-A(\alpha(y_1)+\cdots+\alpha(y_m))}\, \beta(y_0, \alpha(y_0); dy_1) \ldots$$
$$\ldots \beta(y_{m-1}, \alpha(y_{m-1}); dy_m) < 1, \tag{5.22}$$

then condition (5.21) holds.

Proof We have $\mathbb{E}_{x,u}(\sum_{k \geq 1} 1_{\{T_k \leq t\}}) = \rho_c^{x,u}(E \times [0, t]) + \rho_D^{x,u}(E \times [0, t])$. From the first assertion of Proposition 5.3 and (5.20) we get $\rho_c^{x,u}(E \times [0, t]) < +\infty$. On the other hand, it is easily seen that for $y_0 \in E_t^{x,u}$, in the computation of $r_1(y_0, E \times$

$$\sum_{0 m + 1 \le k \le (0+1)m} \int_{E_t^{x,u},k} \left\{ \sum \left(\alpha(y_0) + \cdots + \alpha(y_{k-1}) \right) \le t \right\} e^{-\int_0^{\cdot} \ell(y_0,w)\,dw} \cdots$$

$\le m$

5 Intensity of Some Marked Point Processes

$[0, t])$, using (5.17), we may integrate over $(E_t^{x,u})^k$ instead of E^k. Hence (5.21) ensures $r_1(y_0, E \times [0, t]) < +\infty$. The first assertion follows.

Now define a as the first member of (5.22).

From $1_{\{\alpha(y_0)+\cdots+\alpha(y_{k-1})\le t\}} \le 1_{\{\alpha(y_0)+\cdots+\alpha(y_{k-1})<+\infty\}} e^{At} e^{-A(\alpha(y_0)+\cdots+\alpha(y_{k-1}))}$ we get

$$\sum_{k\ge 1} \int_{(E_t^{x,u})^k} 1_{\{\alpha(y_0)+\cdots+\alpha(y_{k-1})\le t\}} e^{-\int_0^{\alpha(y_0)} \ell(y_0,w)\,dw} \beta(y_0, \alpha(y_0); dy_1) \cdots$$

$$\cdots e^{-\int_0^{\alpha(y_{k-1})} \ell(y_{k-1},w)\,dw} \beta(y_{k-1}, \alpha(y_{k-1}); dy_k)$$

$$\le m + e^{At} \sum_{p\ge 1} \sum_{pm+1\le k\le (p+1)m} \int_{(E_t^{x,u})^k} 1_{\{\alpha(y_0)+\cdots+\alpha(y_{k-1})<+\infty\}} e^{-A(\alpha(y_1)+\cdots+\alpha(y_{pm}))}$$

$$\beta(y_0, \alpha(y_0); dy_1) \cdots \beta(y_{k-1}, \alpha(y_{k-1}); dy_k)$$

$$\le m + m\, e^{At} \sum_{p\ge 1} a^p < +\infty.$$

The result follows. ∎

Example 5.7 Return to Example 3.14, Sect. 3.4. Assume that $\bar{\lambda}$ and μ are bounded. Then condition (5.20) is clearly satisfied, and for all $A > 0$,

$$\sup_{y_0\in E} \int 1_{\{\alpha(y_0)+\alpha(y_1)<+\infty\}} e^{-A\alpha(y_1)} \beta(y_0, \alpha(y_0); dy_1)$$

$$= \max(e^{-A(h_{max}-h_{min})/c}, e^{-A(h_{max}-h_{min})/(c_1-c)}) < 1.$$

Indeed,

- for $y_0 = (1_a, x_1, x_2)$ we get $\beta(y_0, \alpha(y_0); \cdot) = \delta_{y_1}$ with $y_1 = (1, h_{min}, x_2)$ and $\alpha(y_1) = (h_{max} - h_{min})/(c_1 - c)$,
- for $y_0 = (1, x_1, x_2)$ we get $\beta(y_0, \alpha(y_0); \cdot) = \delta_{y_1}$ with $y_1 = (1_a, h_{max}, x_2')$ for some x_2' and $\alpha(y_1) = (h_{max} - h_{min})/c$,
- for $y_0 = (0, x_1, x_2)$ we have $\alpha(y_0) = +\infty$.

Example 5.8 Return to Example 3.15, Sect. 3.4 with $a_1 = a_2 = a$, i.e., to Example 2.5, Sect. 2.5 adding a preventive maintenance on a component that reaches age a. Assume that $\bar{\lambda}_i$, $\bar{\lambda}_i'$ and μ_i $(i = 1, 2)$ are bounded. Then condition (5.22) is satisfied with $m = 2$. Indeed, since each $\beta(y, \alpha(y); dz) = q(\phi(y, \alpha(y)); dz) = \delta_{y'}(dz)$ for some y', for all $y_0 \in E$ there exists a sequence $(y_k)_{k\ge 1}$ such that

$$C(y_0, m) = \int 1_{\{\alpha(y_0)+\alpha(z_1)+\cdots+\alpha(z_{m-1})<+\infty\}} e^{-(\alpha(z_1)+\cdots+\alpha(z_m))} \beta(y_0, \alpha(y_0); dz_1) \cdots$$

$$\cdots \beta(z_{m-1}, \alpha(z_{m-1}); dz_m)$$

$$= 1_{\{\alpha(y_0)<+\infty\}} e^{-(\alpha(y_1)+\cdots+\alpha(y_m))}$$

with $e^{-\infty} = 0$.

For $y_0 = (e_4, x_1, x_2)$ we get $\alpha(y_0) = +\infty$, hence $C(y_0, m) = 0$ for all $m \geq 1$.

For $y_0 = (e_3, x_1, x_2)$ (respectively $y_0 = (e_2, x_1, x_2)$) we get for all $k \geq 1$, $y_k = (e_3, 0, \tilde{y}'_k)$ (respectively $y_k = (e_2, \tilde{y}'_k, 0)$) for some $\tilde{y}'_k$, hence $\alpha(y_k) = a$. Therefore, $C(y_0, m) = e^{-am}$ for all $m \geq 1$.

For $y_0 = (e_1, x_1, x_2)$ and $x_1 < x_2$, we have $y_1 = (e_1, a - (x_2 - x_1), 0)$, $\alpha(y_1) = x_2 - x_1$, $y_2 = (e_1, 0, x_2 - x_1)$, $\alpha(y_2) = a - (x_2 - x_1)$; hence $C(y_0, 2) = e^{-a}$. The same is true for $x_2 < x_1$ and $x_1 = x_2$.

5.3 Intensities Related to PDMP

Let $(Y_k, T_k)_{k\geq 0}$ be a zero-delayed Markov renewal process and $(Z_t, A_t)_{t\geq 0}$ its associated CSMP. Let $\phi : \breve{E} \to E$ be a Borel function. Up to and including Proposition 5.5, ϕ is not assumed to be a deterministic Markov flow. Define $X_t = \phi(Z_t, A_t)$.

Then $\rho(x, dy, dv)$ (resp. $\bar{\rho}(x, dy, dv, dw)$) is the intensity of the marked point process $(X_{T_{k+1}}, T_{k+1})_{k\geq 0}$ (resp. $(X_{T_k}, T_{k+1} - T_k, T_{k+1})_{k\geq 0}$) under $\mathbb{P}_x$.

Denote by $\varrho(x, \cdot, \cdot)$ the intensity of the marked point process $(X_{T_{k+1}-}, T_{k+1})_{k\geq 0} = (\phi(Y_k, T_{k+1} - T_k), T_{k+1})_{k\geq 0}$ under $\mathbb{P}_x$:

$$\int_{E\times\mathbb{R}_+} \varphi(y, v)\, \varrho(x, dy, dv) = \sum_{k\geq 0} \mathbb{E}_x(\varphi(\phi(Y_k, T_{k+1} - T_k), T_{k+1}))$$
$$= \int_{E\times\mathbb{R}_+^2} \varphi(\phi(y, v), w)\, \bar{\rho}(x, dy, dv, dw) \quad (5.23)$$

for every Borel function $\varphi : E \times \mathbb{R}_+ \to \mathbb{R}_+$. Therefore, $\varrho(x, dy, dv)$ is the image measure of $\bar{\rho}(x, dy, dv, dw)$ by the mapping $(y, v, w) \to (\phi(y, v), w)$.

Example 5.9 We return to Example 3.4, Sect. 3.1 by starting with the first modeling.

We are first interested in the number of times the stock level is at most a at the time of a delivery that occurs during the time interval $]0, t]$, i.e., $N_a(t) = \sum_{k\geq 0} 1_{\{S_{T_{k+1}-}\leq a,\, T_{k+1}\leq t\}}$ $(a \geq 0)$. For a G1/G/1 queue, it is the number of customers arriving in the time interval $]0, t]$ and waiting for a period of time at most $a \geq 0$ before starting to be served.

Recall that $X_t = (S_t, A_t)$, which is why we will write below (x_1, x_2) instead of x. We get

$$\mathbb{E}_{(x_1,x_2)}(N_a(t)) = \varrho((x_1, x_2), [0, a] \times \mathbb{R}_+ \times]0, t]), \quad (5.24)$$

$\varrho((x_1, x_2), dy_1, dy_2, dv)$ being the intensity of $(S_{T_{k+1}-}, A_{T_{k+1}-}, T_{k+1})_{k\geq 0}$ under $\mathbb{P}_{(x_1,x_2)}$.

Now let us look at the lost production that is the mean number of times the factory must be stopped during a time interval because the stock becomes empty. Denote by $N^{lp}(t)$ the number of times the factory must be stopped during the time interval $]0, t]$ because the stock falls to 0 during this period. We assume $\mu(\{0\}) = 0$. If the stock is empty at time 0, it is empty during the entire time interval $[0, T_1[$, but it does not becomes empty during this period. On the other hand, for $k \geq 1$, since $S_{T_k} > 0$, the stock becomes empty during $[T_k, T_{k+1}[$ iff $S_{T_k} < c(T_{k+1} - T_k)$. We have

$$N^{lp}(t) = 1_{\{S_t=0\}} + \sum_{k\geq 0} 1_{\{S_{T_k} < c(T_{k+1}-T_k),\, T_{k+1}\leq t\}} - 1_{\{S_0=0\}}, \tag{5.25}$$

thus

$$\begin{aligned}\mathbb{E}_{(x_1,x_2)}(N^{lp}(t)) &= \mathbb{P}_{(x_1,x_2)}(S_t = 0) - \mathbb{P}_{(x_1,x_2)}(S_0 = 0) \\ &\quad + \int_{\mathbb{R}_+^4} 1_{\{y_1 < cv,\, w\leq t\}}\, \bar{\rho}((x_1, x_2), dy_1, dy_2, dv, dw),\end{aligned} \tag{5.26}$$

$\bar{\rho}((x_1, x_2), dy_1, dy_2, dv, dw)$ is the intensity of $(S_{T_k}, A_{T_k}, T_{k+1} - T_k, T_{k+1})_{k\geq 0}$ under $\mathbb{P}_{(x_1,x_2)}$. It should be noted that $S_{T_k} < c(T_{k+1} - T_k)$ is not $S_{T_{k+1}-} = 0$, i.e., $S_{T_k} \leq c(T_{k+1} - T_k)$. As a result, $\mathbb{E}_{(x_1,x_2)}(N^{lp}(t))$ cannot be written using ϱ. If $S_{T_k} = c(T_{k+1} - T_k)$, then the stock is not empty during the time period $[T_k, T_{k+1}]$, since $S_{T_{k+1}} > 0$ a.s.

With the second model, we have

$$\mathbb{E}_{(x_1,x_2)}(N^{lp}(t)) = \rho_+((x_1, x_2), \{0\} \times \mathbb{R}_+ \times [0, t]), \tag{5.27}$$

where ρ_+ is the intensity of the Markov renewal process $(S_{T_k^+}, D_{T_k^+}, T_k^+)_{k\geq 1}$ with kernel N_+ defined in (3.6) and (3.7), Sect. 3.1.

Example 5.10 Let us return to Example 3.9, Sect. 3.4. To have easy explicit calculations, assume that $q(0; dy) = q(1; dy) = \mu(dy)$ and that μ is the initial probability distribution of the process $(X_t)_{t\geq 0}$. Then $(Y_n, T_{n+1} - T_n)_{n\geq 0}$ are i.i.d. random variables. As a consequence, $(T_n)_{n\geq 1}$ is a zero-delayed renewal process, the process $(X_t)_{t\geq 0}$ is a regenerative process, and for all n, $(Y_n, T_{n+1} - T_n)$ and T_n are independent. Thus for all n, $X_{T_n} = Y_n$ and T_n are independent, $(X_{T_{n+1}-}, T_{n+1} - T_n) = (\phi(Y_n, T_{n+1} - T_n), T_{n+1} - T_n)$, and the T_n are independent.

Hence the intensity of $(X_{T_n}, T_n)_{n\geq 1}$ is $\rho(dy, dv) = \mu(dy) \sum_{n\geq 1} dF^{*n}(v)$, where dF is the probability distribution of $T_{n+1} - T_n$ $(n \geq 0)$.

Write μ as $\mu = p\,\mu_1 + (1 - p)\,\mu_2$, where μ_1 (respectively μ_2) is a probability distribution on $E_1 =]-\infty, 0[$ (respectively on $E_2 =]1, +\infty[$) and $p = \mu(E_1)$. Define $\tilde{\mu}_1$ (respectively $\tilde{\mu}_2$) as the image of μ_1 (resp. μ_2) by the application $x \to -x$ (resp. $x \to x - 1$), that is, $\int h(x)\, \tilde{\mu}_1(dx) = \int h(-x)\, \mu_1(dx)$, $\int h(x)\, \tilde{\mu}_2(dx) = \int h(x - 1)\, \mu_2(dx)$ for every Borel function $h : E \to \mathbb{R}_+$. We get

$$\begin{aligned}\mathbb{E}(1_{\{0\}}(X_{T_{n+1}-})\, h(T_{n+1} - T_n)) &= \mathbb{E}(1_{]-\infty,0[}(Y_n)\, h(-Y_n)) = \int_{]-\infty,0[} h(-x)\, \mu(dx) \\ &= p \int_{]0,+\infty[} h(x)\, \tilde{\mu}_1(dx).\end{aligned}$$

Similarly, $\mathbb{E}(1_{\{1\}}(X_{T_{n+1}-})\,h(T_{n+1}-T_n)) = (1-p)\int_{]0,+\infty[} h(x)\,\tilde{\mu}_2(dx)$. As a result, the probability distribution of $T_{n+1}-T_n$ $(n \geq 0)$ is $dF = p\tilde{\mu}_1 + (1-p)\tilde{\mu}_2$.

We deduce

$$\int \varphi(y,v)\,\varrho(dy,dv) = \sum_{n\geq 0}\mathbb{E}(\varphi(X_{T_{n+1}-},T_{n+1})) = \sum_{n\geq 0}\mathbb{E}(\varphi(X_{T_{n+1}-},T_n+T_{n+1}-T_n))$$

$$= p\sum_{n\geq 0}\int \varphi(0,u+v)\,dF^{*n}(u)\,\tilde{\mu}_1(dv) + (1-p)\sum_{n\geq 0}\int \varphi(1,u+v)\,dF^{*n}(u)\,\tilde{\mu}_2(dv).$$

Hence

$$\varrho(dy,dv) = p\,\delta_0(dy)\sum_{n\geq 0} dF^{*n} * \tilde{\mu}_1(dv) + (1-p)\,\delta_1(dy)\sum_{n\geq 0} dF^{*n} * \tilde{\mu}_2(dv).$$

Unlike ρ, the measure ϱ is not a product measure, except if $\tilde{\mu}_1 = \tilde{\mu}_2 (= dF)$, that is, μ_1 is the image of some measure (namely dF) by the mapping $x \to -x$ and μ_2 is the image of the same measure by the mapping $x \to 1+x$. In this case, $X_{T_{n+1}-}$ and $T_{n+1}-T_n$ are independent and

$$\varrho(dy,dv) = (p\,\delta_0(dy) + (1-p)\,\delta_1(dy))\sum_{n\geq 1} dF^{*n}.$$

In the following proposition we write N as $N(x,dy,dv) = 1_{\mathbb{R}_+}(v)\,\beta(x,v;dy)\,dF_x(v)$. There is no assumption on the interarrival distributions, but β is specified. For a parametrized PDMP, the results are in Proposition 5.7 and Corollary 5.2, Sect. 5.3.

Proposition 5.4 *Let $(Y_k,T_k)_{k\geq 0}$ be a zero-delayed Markov renewal process with kernel N, $(Z_t,A_t)_{t\geq 0}$ its associated CSMP, and $\phi : E\times\mathbb{R}_+ \to E$ a Borel function. Define $X_t = \phi(Z_t,A_t)$ $(t \geq 0)$. Assume $N(x,dy,dv) = 1_{\mathbb{R}_+}(v)\,Q(\phi(x,v);dy)\,dF_x(v)$. Then*

$$\sum_{k\geq 0}\mathbb{E}_x(\psi(X_{T_{k+1}-},X_{T_{k+1}},T_{k+1})) = \int_{E^2\times\mathbb{R}_+}\psi(y,z,v)\,Q(y;dz)\,\varrho(x,dy,dv)$$

for every Borel function $\psi : E^2\times\mathbb{R}_+ \to \mathbb{R}_+$.

In particular, given two Borel subsets B_1 and B_2 of E, define $N_{B_1,B_2}(t) = \sum_{k\geq 0} 1_{B_1}(X_{T_{k+1}-})\,1_{B_2}(X_{T_{k+1}})\,1_{\{T_{k+1}\leq t\}}$. Then

$$\mathbb{E}_x(N_{B_1,B_2}(t)) = \int_{E\times\mathbb{R}_+} 1_{B_1}(y)\,1_{[0,t]}(v)\,Q(y;B_2)\,\varrho(x,dy,dv).$$

$\psi(x,y,s) = 1_{B_1}(x)1_{B_2}(y) \times 1\{s\leq t\}$

Proof We have $\mathbb{E}_x(\psi(X_{T_{k+1}-},X_{T_{k+1}},T_{k+1}))=\mathbb{E}_x(\psi(\phi(Y_k,T_{k+1}-T_k),Y_{k+1},T_{k+1}))$ $(k\geq 0)$. Since the conditional probability distribution of Y_{k+1} given Y_k, T_k, T_{k+1} is $Q(\phi(Y_k,T_{k+1}-T_k);dy) = Q(X_{T_{k+1}-};dy)$, we get

$$\mathbb{E}_x(\psi(X_{T_{k+1}-}, X_{T_{k+1}}, T_{k+1})) = \mathbb{E}_x\left(\int_E \psi(X_{T_{k+1}-}, y, T_{k+1})\, Q(X_{T_{k+1}-}; dy)\right).$$

The result follows from the definition of ϱ. ■

When $(Z_t, A_t)_{t\geq 0}$ is a simple CSMP and $X_t = \phi(Z_t, A_t)$, Proposition 5.5 below gives a decomposition of ϱ and its interpretation. It is a consequence of Propositions 5.2 and 5.3, Sect. 5.2.

Proposition 5.5 *Let $(Z_t, A_t)_{t\geq 0}$ be a zero-delayed simple CSMP taking values in E with parameters ℓ, α, β. Let $(Y_n, T_n)_{n\geq 0}$ be the associated Markov renewal process and $\phi : E \times \mathbb{R}_+ \to E$ a Borel function. Define $X_t = \phi(Z_t, A_t)$.*

For $x \in E$, define $\varrho_c(x, dy, dv)$ (respectively $\underline{\sigma}(x, dy, dv)$) as the image measure of $\bar{\rho}_c(x, dy, dv, dw)$ (respectively of $\bar{\rho}_D(x, dy, dv, dw)$) by the mapping $(y, v, w) \to (\phi(y, v), w)$, that is,

$$\int_{E\times\mathbb{R}_+} \varphi(y, v)\, \varrho_c(x, dy, dv) = \int_{E\times\mathbb{R}_+^2} \varphi(\phi(y, v), w)\, \bar{\rho}_c(x, dy, dv, dw)$$
$$= \int_{E\times\mathbb{R}_+^2} \varphi(\phi(y, v), w)\, 1_{\{v<\alpha(y)\}}\, \bar{\rho}(x, dy, dv, dw),$$

$$\int_{E\times\mathbb{R}_+} \varphi(y, v)\, \underline{\sigma}(x, dy, dv) = \int_{E\times\mathbb{R}_+^2} \varphi(\phi(y, v), w)\, \bar{\rho}_D(x, dy, dv, dw)$$
$$= \int_{E\times\mathbb{R}_+^2} \varphi(\phi(y, v), w)\, 1_{\{v=\alpha(y)\}}\, \bar{\rho}(x, dy, dv, dw).$$

Then

$$\varrho = \varrho_c + \underline{\sigma},$$

1. *$\varrho_c(x, \cdot, \cdot)$ is the intensity, under $\mathbb{P}_x$, of the marked point process $(X_{T_{k+1}-}, T_{k+1})_{k\geq 0}$ restricted to the jumps that are not caused by Dirac measures, namely,*

$$\int_{E\times\mathbb{R}_+} \varphi(y, v)\, \varrho_c(x, dy, dv) = \sum_{k\geq 0} \mathbb{E}_x(\varphi(X_{T_{k+1}-}, T_{k+1})\, 1_{\{T_{k+1}-T_k<\alpha(Y_k)\}})$$
$$= \int_{\mathbb{R}_+} \mathbb{E}_x(\varphi(X_v, v)\, \ell(Z_v, A_v))\, dv \qquad (5.28)$$

   for every Borel function $\varphi : E \times \mathbb{R}_+ \to \mathbb{R}_+$,

2. *$\underline{\sigma}(x, \cdot, \cdot)$ is the intensity, under $\mathbb{P}_x$, of the marked point process $(X_{T_{k+1}-}, T_{k+1})_{k\geq 0}$ restricted to the jumps caused by Dirac measures, namely,*

$$\int_{E\times\mathbb{R}_+} \varphi(y,v)\,\underline{\sigma}(x,dy,dv) =$$
$$= \sum_{k\geq 0} \mathbb{E}_x(\varphi(X_{T_{k+1}-},T_{k+1})\,1_{\{T_{k+1}=T_k+\alpha(Y_k)<+\infty\}}) \tag{5.29}$$
$$= 1_{\{\alpha(x)<+\infty\}}\varphi(\phi(x,\alpha(x)),\alpha(x))\,e^{-\int_0^{\alpha(x)}\ell(x,w)\,dw}$$
$$+ \int_{E\times\mathbb{R}_+} 1_{\{\alpha(y)<+\infty\}}\varphi(\phi(y,\alpha(y)),v+\alpha(y))\,e^{-\int_0^{\alpha(y)}\ell(y,w)\,dw}\,\rho(x,dy,dv)$$

for every Borel function $\varphi : E\times\mathbb{R}_+\to\mathbb{R}_+$.
Moreover, $\underline{\sigma} = \underline{\sigma}_D + \rho_c * \underline{\sigma}_D$, where

$$\int_{E\times\mathbb{R}_+} \varphi(y,v)\,\underline{\sigma}_D(y_0,dy,dv)$$
$$= \int_{E\times\mathbb{R}_+} \varphi(\phi(y,\alpha(y)),v+\alpha(y))\,e^{-\int_0^{\alpha(y)}\ell(y,w)\,dw}\,(I+r_1)(y_0,dy,dv)$$
$$= \sum_{k\geq 0}\int_{(E\times\mathbb{R}_+)^k} 1_{\{\alpha(y_0)+\cdots+\alpha(y_k)<+\infty\}}\varphi(\phi(y_k,\alpha(y_k)),\alpha(y_0)+\cdots+\alpha(y_k))$$
$$e^{-\int_0^{\alpha(y_0)}\ell(y_0,w)\,dw}\ldots e^{-\int_0^{\alpha(y_k)}\ell(y_k,w)\,dw}\,\beta(y_0,\alpha(y_0);dy_1)\ldots\beta(y_{k-1},\alpha(y_{k-1});dy_k).$$

In formula (5.28), not only the process $(X_t)_{t\geq 0}$ but also the CSMP $(Z_t,A_t)_{t\geq 0}$ appears. This is due to the fact that X is not assumed to be a PDMP in Proposition 5.5. In the following we assume that $(X_t)_{t\geq 0}$ is a PDMP.

Proposition 5.6 *Let $(Y,T)=(Y_n,T_n)_{n\geq 0}$ be a zero-delayed Markov renewal process and $\mathcal{F}_t$ the σ-field generated by this process up to time t. Let $(X_t)_{t\geq 0}$ be a PDMP generated by ϕ and (Y,T). Let A be a Borel subset of E and $s,t\geq 0$.*
We have

$$\mathbb{E}\left(\sum_{n\geq 1} 1_A(Y_n)1_{]t,t+s]}(T_n)\mid\mathcal{F}_t\right) = \rho(X_t, A\times]0,s]).$$

Assume, moreover, that $(X_t)_{t\geq 0}$ is a PDMP with parameters $\phi,\lambda,Q,\alpha,(E_i,q_i)_{i\in I}$. Then

$$\sum_{k\geq 0}\mathbb{P}(X_{T_{k+1}-}\in A,\,T_{k+1}\in]t,t+s],\,T_{k+1}-T_k=\alpha(Y_k)<+\infty\mid\mathcal{F}_t)$$
$$= \underline{\sigma}(X_t, A\times]0,s]). \tag{5.30}$$

Proof Define $(Z_t,A_t)_{t\geq 0}$ as the CSMP associated with (Y,T). Denote by N its kernel. We have $\sum_{n\geq 1}1_A(Y_n)1_{]t,t+s]}(T_n)=\sum_{n\geq 1}1_A(Y_n^t)1_{]0,s]}(T_n^t)$, where $(Y_n^t,T_n^t)_{n\geq 1}$ is the Markov renewal process observed from time t as defined in Theorem 2.1, Sect. 2.4. Given $\mathcal{F}_t$, then $(Y_n^t,T_n^t)_{n\geq 1}$ is a Markov renewal process with initial distribution $N_0^{Z_t,A_t}$ (Corollary 2.2, end of Sect. 2.4). Since $(X_t)_{t\geq 0}$ is a PDMP generated by ϕ and

By def. of PDMP
$N_0^{Z_t, A_t}(\cdot) = N(\phi(Z_t, A_t); \cdot)$
of (3.2)

$(Z_t, A_t)_{t \geq 0}$, we have $N_0^{Z_t, A_t} = N(\phi(Z_t, A_t), \cdot, \cdot) = N(X_t, \cdot, \cdot)$. This concludes the proof of the first formula. The second one is proved with the same arguments. Indeed, from $\phi(Z_t, A_t + T_1') = \phi(X_t, T_1')$ and $\alpha(Z_t) - A_t = \alpha(X_t)$ we deduce that the first member of (5.30) is equal to

$$\mathbb{P}(\phi(X_t, T_1') \in A, 0 < T_1' \leq s, T_1' = \alpha(X_t) < +\infty \mid \mathcal{F}_t)$$
$$+ \sum_{k \geq 1} \mathbb{P}(\phi(Y_k^t, T_{k+1}^t - T_k^t) \in A, 0 < T_{k+1}^t \leq s, T_{k+1}^t - T_k^t = \alpha(Y_k^t) \mid \mathcal{F}_t).$$

Formula (5.30) is then a consequence of Corollary 2.2, end of Sect. 2.4. ∎

In the following, $(X_t)_{t \geq 0}$ is a PDMP generated by ϕ and a zero-delayed Markov renewal process $(Y_n, T_n)_{n \geq 0}$. We do not specify the probability distribution of X_0. We define the measure σ as

$$\sigma(dy, dv) = \int_{\{x : x \in E\}} \underline{\sigma}(x, dy, dv)\, \pi_0(dx),$$

measurable ?

where π_0 is the probability distribution of X_0, that is,

$$\int_{E \times \mathbb{R}_+} \varphi(y, v)\, \sigma(dy, dv) = \sum_{k \geq 0} \mathbb{E}(\varphi(X_{T_{k+1}-}, T_{k+1})\, 1_{\{T_{k+1} = T_k + \alpha(Y_k) < +\infty\}}) \quad (5.31)$$

for every Borel function $\varphi : E \times \mathbb{R}_+ \to \mathbb{R}_+$. We have $\sigma(\Gamma^c \times \mathbb{R}_+) = 0$, where $\Gamma = \{\phi(x, \alpha(x)) : x \in E, \alpha(x) < +\infty\}$.

The partition $(E_i)_{i \in I}$ of the state space leads to a natural decomposition of σ.

Proposition 5.7 *Let $(X_t)_{t \geq 0}$ be a PDMP with parameters $\phi, \lambda, Q, \alpha, (E_i, q_i)_{i \in I}$. Let $(Y_k, T_k)_{k \geq 0}$ be the Markov renewal process associated with $(X_t)_{t \geq 0}$. Define π_s as the probability distribution of X_s, and measures σ_i $(i \in I)$ as*

$$\int \varphi(y, v)\, \sigma_i(dy, dv) = \sum_{k \geq 0} \mathbb{E}(1_{E_i}(Y_k)\, 1_{\{T_{k+1} = T_k + \alpha(Y_k) < +\infty\}}\, \varphi(X_{T_{k+1}-}, T_{k+1}))$$

$$(5.32)$$

for every Borel function $\varphi : E \times \mathbb{R}_+ \to \mathbb{R}_+$, and

$$\Gamma_i^* = \{\phi(y, \alpha(y)) : y \in E_i, \alpha(y) < +\infty, \int_0^{\alpha(y)} \lambda(\phi(y, v))\, dv < +\infty\}.$$

Then

$$\sigma = \sum_{i \in I} \sigma_i, \quad \sigma_i(\Gamma_i^{*c} \times \mathbb{R}_+) = 0, \quad (5.33)$$

and

$$\sum_{k\geq 0}\mathbb{E}(\varphi(X_{T_{k+1}-},T_{k+1}))=\int_{\mathbb{R}_+}\mathbb{E}(\varphi(X_v,v)\,\lambda(X_v))\,dv+\sum_{i\in I}\int_{\Gamma_i^*\times\mathbb{R}_+}\varphi(y,v)\;\sigma_i(dy,dv)$$
$$=\int_{E\times\mathbb{R}_+}\varphi(y,v)\,\lambda(y)\,\pi_v(dy)\,dv+\sum_{i\in I}\int_{\Gamma_i^*\times\mathbb{R}_+}\varphi(y,v)\;\sigma_i(dy,dv)$$

for every Borel function $\varphi: E\times\mathbb{R}_+\to\mathbb{R}_+$. More generally,

$$\sum_{k\geq 0}\mathbb{E}(\psi(X_{T_{k+1}-},X_{T_{k+1}},T_{k+1}))=\mathbb{E}\left(\int_{E\times\mathbb{R}_+}\psi(X_v,z,v)\,\lambda(X_v)\,Q(X_v;dz)\,dv\right)$$
$$+\sum_{i\in I}\int_{E\times\Gamma_i^*\times\mathbb{R}_+}\psi(y,z,v)\,q_i(y;dz)\,\sigma_i(dy,dv)$$
$$=\int_{E^2\times\mathbb{R}_+}\psi(y,z,v)\,\lambda(y)\,Q(y;dz)\,\pi_v(dy)\,dv$$
$$+\sum_{i\in I}\int_{E\times\Gamma_i^*\times\mathbb{R}_+}\psi(y,z,v)\,q_i(y;dz)\,\sigma_i(dy,dv)$$

for every Borel function $\psi: E^2\times\mathbb{R}_+\to\mathbb{R}_+$.

Because of (5.33), σ and σ_i $(i\in I)$ are called boundary measures.

Proof We have:

$$\mathbb{E}(\psi(X_{T_{k+1}-},X_{T_{k+1}},T_{k+1}))=\mathbb{E}(1_{\{T_{k+1}-T_k<\alpha(Y_k)\}}\,\psi(\phi(Y_k,T_{k+1}-T_k),Y_{k+1},T_{k+1}))$$
$$+\sum_{i\in I}\mathbb{E}(1_{E_i}(Y_k)\,1_{\{T_{k+1}-T_k=\alpha(Y_k)\}}\,\psi(\phi(Y_k,T_{k+1}-T_k),Y_{k+1},T_{k+1})).$$

Conditioning with respect to Y_k, T_k, T_{k+1} and applying Proposition 5.2, Sect. 5.2, we get

$$\mathbb{E}(1_{\{T_{k+1}-T_k<\alpha(Y_k)\}}\,\psi(\phi(Y_k,T_{k+1}-T_k),Y_{k+1},T_{k+1}))$$
$$=\mathbb{E}\left(\int_E 1_{\{T_{k+1}-T_k<\alpha(Y_k)\}}\,\psi(\phi(Y_k,T_{k+1}-T_k),z,T_{k+1})\,Q(\phi(Y_k,T_{k+1}-T_k);dz)\right)$$
$$=\int_E\int_{E^2\times\mathbb{R}_+^2}\psi(\phi(y,v),z,w)\,Q(\phi(y,v);dz)\,\bar{\rho}_c(x,dy,dv,dw)\,\pi_0(dx)$$
$$=\int_E\mathbb{E}_x\left(\int_{E\times\mathbb{R}_+}\psi(\phi(Z_v,A_v),z,v)\,Q(\phi(Z_v,A_v);dz)\,\lambda(\phi(Z_v,A_v))\,dv\right)\pi_0(dx)$$
$$=\mathbb{E}\left(\int_{E\times\mathbb{R}_+}\psi(X_v,z,v)\,Q(X_v;dz)\,\lambda(X_v)\,dv\right).$$

Similarly, conditioning with respect to Y_k, T_k, T_{k+1}, we obtain

$$\sum_{i \in I} \mathbb{E}(1_{E_i}(Y_k)\, 1_{\{T_{k+1}-T_k=\alpha(Y_k)\}}\, \psi(\phi(Y_k, T_{k+1}-T_k), Y_{k+1}, T_{k+1}))$$

$$= \sum_{i \in I} \mathbb{E}\left(\int_E 1_{E_i}(Y_k)\, 1_{\{T_{k+1}-T_k=\alpha(Y_k)\}}\, \psi(X_{T_{k+1}-}, z, T_{k+1})\, q_i(X_{T_{k+1}-}; dz)\right)$$

$$= \sum_{i \in I} \int_{E \times \Gamma_i^* \times \mathbb{R}_+} \psi(y, z, v)\, q_i(y; dz)\, \sigma_i(dy, dv).$$

This ends the proof. ∎

It will be seen in Chap. 9 that measures π_v ($v \in \mathbb{R}_+$) and σ can be approximated by numerical schemes. Thus quantities that are expressed through these measures can be computed by these schemes.

Example 5.11 Return to the model described at the end of Example 3.12, Sect. 3.4: on-site jumps have been added when the stock becomes empty. Then the mean number of times the factory must be stopped during the time interval $[0, t]$ because of the lack of commodities is $\sigma(\Gamma_1 \times [0, t])$. Numerical experiments are given Chap. 9. Similarly, by taking the last model in Example 3.13, Sect. 3.4 we have $LP(t) = \sigma(\Gamma_2 \times [0, t])$.

Corollary 5.2 *Let $(X_t)_{t \geq 0}$ be a PDMP with parameters $\phi, \lambda, Q, \alpha, (E_i, q_i)_{i \in I}$, and $(Y_k, T_k)_{k \geq 0}$ the associated Markov renewal process. Given B_1 and B_2, two Borel subsets of E, define $N^c_{B_1,B_2}(t)$ and $N^D_{B_1,B_2}(t)$ as the numbers of jumps up to time t from B_1 to B_2 that are respectively not generated by the Dirac measures (purely random jumps) and that are generated by the Dirac measures, namely*

$$N^c_{B_1,B_2}(t) = \sum_{k \geq 0} 1_{\{T_{k+1} \leq t, T_{k+1}-T_k < \alpha(Y_k)\}}\, 1_{B_1}(X_{T_{k+1}-})\, 1_{B_2}(X_{T_{k+1}}),$$

$$N^D_{B_1,B_2}(t) = \sum_{k \geq 0} 1_{\{T_{k+1} \leq t, T_{k+1}-T_k = \alpha(Y_k)\}}\, 1_{B_1}(X_{T_{k+1}-})\, 1_{B_2}(X_{T_{k+1}}).$$

Then

$$\mathbb{E}(N^c_{B_1,B_2}(t)) = \int_{E \times \mathbb{R}_+} 1_{B_1}(y)\, 1_{[0,t]}(v)\, \lambda(y)\, Q(y; B_2)\, \pi_v(dy)\, dv, \tag{5.34}$$

where π_v is the probability distribution of X_v,

$$\mathbb{E}(N^D_{B_1,B_2}(t)) = \sum_{i \in I} \int_{\Gamma_i^* \times \mathbb{R}_+} 1_{B_1}(y)\, 1_{[0,t]}(v)\, q_i(y; B_2)\, \sigma_i(dy, dv). \tag{5.35}$$

Corollary 5.2 is relevant in applications. As already mentioned, we will see in Chap. 9 how to approximate the probability distributions $(\pi_v)_{v \geq 0}$ and the measures $(\sigma_i)_{i \in I}$ with numerical schemes. In Cocozza-Thivent and Eymard [33] various mean numbers of failures and repairs are computed by applying (5.34). Examples of more general operating costs are given below.

Example 5.12 Return to Example 3.15, Sect. 3.4 with $a_1 = a_2 = a$. We assume that the components are in active redundancy, i.e., the item is up iff at least one component is up. We would like to get the mean cost $C(t)$ generated by the failures during the time interval $[0, t]$. Let us say that this cost is due to the repair cost c of a component, the cost c' for the replacement of a component when it reaches age a, the hourly cost c_h while the item is down, and the cost c_ℓ for the loss of image each time the item breaks down. Hence

$$\begin{aligned}
C(t) &= c\,\mathbb{E}\left(N^c_{E_2,E_1}(t) + N^c_{E_3,E_1}(t) + N^c_{E_4,E_2\cup E_3}(t)\right) \\
&\quad + c'\,\mathbb{E}\left(N^D_{E_1,E_1}(t) + N^D_{E_2,E_2}(t) + N^D_{E_3,E_3}(t)\right) \\
&\quad + c_h\,\mathbb{E}\left(\int_0^t 1_{E_4}(X_s)\,ds\right) + c_\ell\,\mathbb{E}\left(N^c_{E_2,E_4}(t) + N^c_{E_3,E_4}(t)\right) \\
&= c\int_{\mathbb{R}_+\times[0,a[\times[0,t]} \mu_1(x_1)\,\pi_v(\{e_2\}, dx_1dx_2)\,dv \\
&\quad + c\int_{[0,a[\times\mathbb{R}_+\times[0,t]} \mu_2(x_2)\,\pi_v(\{e_3\}, dx_1dx_2)\,dv \\
&\quad + c\int_{\mathbb{R}^2_+\times[0,t]} (\mu_1(x_1)+\mu_2(x_2))\,\pi_v(\{e_4\}, dx_1dx_2)\,dv \\
&\quad + c'\,\sigma(\Gamma\times[0,t]) + c_h\int_0^t \pi_v(E_4)\,dv \\
&\quad + c_\ell\int_{\mathbb{R}_+\times[0,a[\times[0,t]} \bar{\lambda}'_2(x_2)\,\pi_v(\{e_2\}, dx_1dx_2)\,dv \\
&\quad + c_\ell\int_{[0,a[\times\mathbb{R}_+\times[0,t]} \bar{\lambda}'_1(x_1)\,\pi_v(\{e_3\}, dx_1dx_2)\,dv
\end{aligned}$$

with $\Gamma = \{e_1\}\times\{a\}\times[0,a]\cup\{e_1\}\times[0,a[\times\{a\}\cup\{e_2\}\times\mathbb{R}_+\times\{a\}\cup\{(e_3,a)\}\times\mathbb{R}_+$.

Example 5.13 Cost functions that appear in many of the optimization problems discussed in Davis [51] are of the form

$$V(x) = \mathbb{E}_x\left(\int_0^{+\infty} g_1(X_s, s)\,ds + \sum_{k\geq 0} g_2(X_{T_{k+1}-}, T_{k+1})\right).$$

We have

$$V(x) = \int_0^{+\infty} g_1(y,s)\,\pi^x_s(y)\,ds + \int_0^{+\infty} g_2(y,s)\,\underline{\sigma}(x, dy, ds),$$

where π^x_s is the probability distribution of X_s under $\mathbb{P}_x$.

Numerical schemes are particularly well suited to evaluating $V(x)$ for many x, because once the algorithm parameters have been computed, changing x has little impact on the computation time.

Chapter 6
Generalized Kolmogorov Equations

In this chapter we give equations satisfied by the marginal probability distributions of CSMPs, and as a result of PDMPs. They are obtained using the intensities of marked point processes discussed in Chap. 5. As in that chapter, the kernel is progressively specified. The final challenge is to get a formulation that may be used to derive numerical approximation schemes and to prove their convergence, even when there are Dirac measures in the interarrival distributions. This requires Chapman–Kolmogorov equations that are more general than the usual ones, that is, for functions that do not belong to the domain of the extended generator.

To prove the convergence of the numerical schemes introduced in Chap. 9, we need formulas concerning $\mathbb{E}(\psi(Z_t, A_t, t))$ and $\mathbb{E}(g(X_t, t))$ and not only $\mathbb{E}(\varphi(Z_t, A_t))$ and $\mathbb{E}(f(X_t))$. They can all be deduced from results on $\mathbb{E}(f(X_t))$, because a CSMP is a PDMP, and moreover, if $(X_t)_{t\geq 0}$ is a PDMP, then $(X_t, t)_{t\geq 0}$ is also a PDMP. But this is not our choice; we prefer to start working on $\mathbb{E}(\psi(Z_t, A_t, t))$ first so that we do not have to worry about the flow ϕ.

Great use is made of absolutely continuous functions. Readers unfamiliar with this notion should refer to Appendix A, Sect. A.2.

Section 6.1 is about CSMPs, Sect. 6.2 discusses PDMPs. In Sect. 6.3, we assume that the interarrival distributions are absolutely continuous and derive the classical results on the continuity and differentiability of the semigroup. The results of this section are not useful for what follows.

The setting and the notation are as in Chap. 5. Recall that

$$N(x, dy, dv) = 1_{\mathbb{R}_+}(v)\, \beta(x, v; dy)\, dF_x(v), \quad \bar{F}_x(u) = dF_x(]u, +\infty]),$$

$$t^*(x) = \sup\{u \in \mathbb{R}_+ : \bar{F}_x(u) > 0\},$$

$$\breve{E} = \{(x, u) \in E \times \mathbb{R}_+ : u < t^*(x)\}, \quad \hat{E} = \{(x, u) \in E \times \mathbb{R}_+ : u \leq t^*(x)\}.$$

C. Cocozza-Thivent, *Markov Renewal and Piecewise Deterministic Processes*, Probability Theory and Stochastic Modelling 100,
https://doi.org/10.1007/978-3-030-70447-6_6

Except in the inhomogeneous case, (Sect. 6.1) $(Z_t, A_t)_{t\geq 0}$ is a CSMP with kernel N, $(Y_n, T_n)_{n\geq 1}$ is its associated Markov renewal process, and $N_t = \sum_{n\geq 1} 1_{\{T_n \leq t\}}$.

6.1 Kolmogorov Equations for CSMPs

Homogeneous Case

We are going to work a great deal with functions of space and time that are absolutely continuous with respect to their time variable (see Sect. A.2 for the definition of an absolutely continuous function and its main properties).

The following lemma, which is a consequence of Theorem A.2, end of Section A.2, introduces a fundamental notation.

Lemma 6.1 *Given a map* $\psi : E \times \mathbb{R}_+^2 \to \mathbb{R}$ *and* $(x, v_1, v_2) \in E \times \mathbb{R}_+^2$, *define* $\mathcal{X}\psi : E \times \mathbb{R}_+^2 \to \mathbb{R}$ *as*

$$\mathcal{X}\psi(x, v_1, v_2) = \lim_{\Delta \to 0} (\psi(x, v_1 + \Delta, v_2 + \Delta) - \psi(x, v_1, v_2))/\Delta$$

when this limit exists, and as 0 *if not.*

Let $(x, v_1, v_2) \in E \times \mathbb{R}_+^2$ *and* $c > 0$. *Assume that the function* $t \in [0, c] \to \psi(x, v_1 + t, v_2 + t)$ *is absolutely continuous. Then this function is differentiable Lebesgue-almost everywhere and*

$$\forall t \in [0, c], \quad \psi(x, v_1 + t, v_2 + t) - \psi(x, v_1, v_2) = \int_0^t \mathcal{X}\psi(x, v_1 + s, v_2 + s)\, ds.$$

If the function $(v_1, v_2) \to \psi(x, v_1, v_2)$ is of class C^1, then

$$\mathcal{X}\psi(x, v_1, v_2) = \frac{\partial \psi}{\partial v_1}(x, v_1, v_2) + \frac{\partial \psi}{\partial v_2}(x, v_1, v_2).$$

The lemma below describes the trajectory of the process $(Z_t, A_t, t)_{t\geq 0}$ by distinguishing deterministic evolution (Tr) and jumps (S).

Lemma 6.2 *Given* $\psi : \hat{E} \times \mathbb{R}_+ \to \mathbb{R}$ *and* $t > 0$, *we have*

$$\psi(Z_t, A_t, t) - \psi(Z_0, A_0, 0) = Tr + S,$$

where Tr *is the transport part:*

- *for* $t < T_1$, $Tr = \psi(Y_0, A_0 + t, t) - \psi(Y_0, A_0, 0)$,
- *for* $T_1 \leq t$,

$$Tr = \psi(Y_0, A_0 + T_1, T_1) - \psi(Y_0, A_0, 0)$$
$$+ \sum_{k \geq 1} 1_{\{T_{k+1} \leq t\}} (\psi(Y_k, T_{k+1} - T_k, T_{k+1}) - \psi(Y_k, 0, T_k))$$
$$+ \sum_{n \geq 1} 1_{\{T_n \leq t < T_{n+1}\}} (\psi(Y_n, t - T_n, t) - \psi(Y_n, 0, T_n)),$$

No explosion is needed!

where S is the jump part:

$$S = (\psi(Y_1, 0, T_1) - \psi(Y_0, A_0 + T_1, T_1)) 1_{\{T_1 \leq t\}}$$
$$+ \sum_{k \geq 1} 1_{\{T_{k+1} \leq t\}} (\psi(Y_{k+1}, 0, T_{k+1}) - \psi(Y_k, T_{k+1} - T_k, T_{k+1})).$$

Assume that for all $x \in E$ and $a \leq t$, the functions $v \in [0, \min(t, T_1)] \rightarrow \psi(Y_0, A_0 + v, v)$ and $v \in [0, \min(t^*(x), t - a)] \rightarrow \psi(x, v, a + v)$ are absolutely continuous. Then

$$Tr = \int_0^t \mathcal{X}\psi(Z_v, A_v, v) \, dv. \tag{6.1}$$

Proof The proof of the first part is straightforward. The second part is obvious for $t < T_1$. For $T_1 \leq t$, write

$$Tr = \int_0^{T_1} \mathcal{X}\psi(Y_0, A_0 + v, v) \, dv + \sum_{k=1}^{N_t - 1} \int_0^{T_{k+1} - T_k} \mathcal{X}\psi(Y_k, v, T_k + v) \, dv$$
$$+ \int_0^{t - T_{N_t}} \mathcal{X}\psi(Y_{N_t}, v, T_{N_t} + v) \, dv.$$

$\{T_{k+1} \leq t\}$
$= \{N_t \geq k+1\}$
$= \{k \leq N_t - 1\}$

A change of variable gives the result. ∎

In the following proposition, the expression of $\mathbb{E}(S)$ is clarified, using the results of the previous chapter, by the structure of the semi-Markov kernel N being specified. The assumptions are a bit painful to read, because of the areas on which the properties are required. In the first instance, the reader can simply assume that:

- $\mathbb{E}_{x,u}(N_t) < +\infty$,
- for all $(y, v_1, v_2) \in E \times \mathbb{R}_+^2$, the function $v \rightarrow \psi(y, v_1 + v, v_2 + v)$ is absolutely continuous,
- ψ and $\mathcal{X}\psi$ are bounded on $E \times \mathbb{R}_+^2$.

Proposition 6.1 *Let $(x, u) \in \check{E}$, $t \geq 0$ and let $\psi : \hat{E} \times \mathbb{R}_+ \rightarrow \mathbb{R}$ be a Borel function. Define $B_1 = [0, \min(t^*(x) - u, t)]$, $B_2 = \{(y, v, w) \in E \times \mathbb{R}_+^2 : v \leq t^*(y), v \leq w \leq t\}$. Assume:*

- $\mathbb{E}_{x,u}(N_t) < +\infty$,
- *the function $v \in B_1 \rightarrow \psi(x, u + v, v)$ is absolutely continuous,*

- *the function* $v \in B_1 \to |\psi(x, u+v, v)| + |\mathcal{X}\psi(x, u+v, v)|$ *is bounded,*
- *for all* $y \in E$, $a \le t$, *the function* $v \in [0, \min(t^*(y), t-a)] \to \psi(y, v, a+v)$ *is absolutely continuous,*
- *the function* $(y, v, w) \in B_2 \to |\psi(y, v, w)| + |\mathcal{X}\psi(y, v, w)|$ *is bounded.*

Then

1. *In the general case,*

$$\mathbb{E}_{x,u}(\psi(Z_t, A_t, t)) = \psi(x, u, 0) + \int_0^t \mathbb{E}_{x,u}(\mathcal{X}\psi(Z_v, A_v, v))\, dv \tag{6.2}$$
$$+ \int_{\hat{E}\times\mathbb{R}_+} 1_{\{w\le t\}} \int_E (\psi(z, 0, w) - \psi(y, v, w))\, \beta(y, v; dz)\, \bar{\rho}^{x,u}(dy, dv, dw).$$

with $\bar{\rho}^{x,u}$ *defined in (5.1), Sect. 5.1.*

2. *If* N *is given by*

$$N(x, dy, dv) = \mathbb{P}_x(T_1 > v)\, \ell(x, v)\, \beta(x, v; dy)\, dv + n(x, dy, dv),$$

then

$$\mathbb{E}_{x,u}(\psi(Z_t, A_t, t)) = \psi(x, u, 0) + \int_0^t \mathbb{E}_{x,u}(\mathcal{X}\psi(Z_v, A_v, v))\, dv \tag{6.3}$$
$$+ \int_0^t \mathbb{E}_{x,u}\left(\int_E (\psi(y, 0, v) - \psi(Z_v, A_v, v))\, \ell(Z_v, A_v)\, \beta(Z_v, A_v; dy)\right) dv$$
$$+ \int_{\hat{E}\times\mathbb{R}_+} 1_{\{w\le t\}} \int_E (\psi(z, 0, w) - \psi(y, v, w))\, \beta(y, v; dz)\, \bar{\rho}_D^{x,u}(dy, dv, dw)$$

with $\bar{\rho}_D^{x,u}$ *defined in Proposition 5.1, Sect. 5.2.*

3. *If* $(Z_s, A_s)_{s\ge 0}$ *is a simple CSMP with parameters* ℓ, α, β, *then formula (6.3) holds with*

$$\int_{\hat{E}\times\mathbb{R}_+} 1_{\{w\le t\}} \int_E (\psi(z, 0, w) - \psi(y, v, w))\, \beta(y, v; dz)\, \bar{\rho}_D^{x,u}(dy, dv, dw) =$$
$$1_{\{\alpha(x)\le u+t\}} e^{-\int_u^{\alpha(x)} \ell(x,w)\, dw} \int_E (\psi(y, 0, \alpha(x)-u) - \psi(x, \alpha(x), \alpha(x)-u))\beta(x, \alpha(x); dy)$$
$$+ \int_{\widetilde{E}} 1_{\{v+\alpha(y)\le t\}} e^{-\int_0^{\alpha(y)} \ell(y,w)\, dw} \int_E (\psi(z, 0, v+\alpha(y)) - \psi(y, \alpha(y), v+\alpha(y)))$$
$$\beta(y, \alpha(y); dz)\, \rho^{x,u}(dy, dv).$$

with $\rho^{x,u}$ *defined Sect. 5.1.*

Proof Apply Lemma 6.2 and Formula (3.1), Chap. 3. Conditioning respectively on T_1 the first term of $\mathbb{E}(S)$ and on $(Y_k, T_{k+1} - T_k, T_{k+1})$ $(k \ge 1)$ the term of index k in the sum, we get

$$\mathbb{E}_{x,u}(S) = \mathbb{E}_{x,u}\left(1_{\{T_1 \le t\}} \int_E (\psi(z, 0, T_1) - \psi(Y_0, A_0 + T_1, T_1))\, \beta(Y_0, A_0 + T_1; dz)\right)$$

$$+ \sum_{k \ge 1} \mathbb{E}_{x,u}\left(1_{\{T_{k+1} \le t\}} \int_E (\psi(z, 0, T_{k+1}) - \psi(Y_k, T_{k+1} - T_k, T_{k+1}))\, \beta(Y_k, T_{k+1} - T_k; dz)\right)$$

$$= \int_{\hat{E} \times \mathbb{R}_+} 1_{\{w \le t\}} \int_E (\psi(z, 0, w) - \psi(y, v, w))\, \beta(y, v; dz)\, \bar{\rho}^{x,u}(dy, dv, dw).$$

Formula (6.2) follows. The second and the third assertions are respectively due to Proposition 5.1, Sect. 5.2 and Proposition 5.2, Sect. 5.2. ■

Proposition 6.1 can also be proved from formula (5.6), Sect. 5.1 via integrations by parts.

Given a function ψ defined on a subset of a product space $\prod_{i=1}^{k} F_i$, we denote by $\partial_i \psi$ the function defined as the partial derivative of ψ with respect to its ith variable at points where it exists and equal to 0 elsewhere.

The following corollary will be helpful in Chap. 10 to get the generator of a switching process (see the proof of Theorem 10.2, Sect. 10.6). It is a straightforward consequence of Proposition 6.1, since $t = T_{N_t} + A_t$ for all $t \ge 0$. As in Proposition 6.1, the assumptions are hard to read, and to simplify, at first reading it can be assumed that $u = 0$ and:

- $\mathbb{E}_x(N_t) < +\infty$,
- for all $(y, w) \in E \times \mathbb{R}_+$, the function $v \to \psi(y, v, w)$ is absolutely continuous,
- ψ and $\partial_2 \psi$ are bounded on $E \times \mathbb{R}_+^2$.

Corollary 6.1 *Let $(Z_s, A_s)_{s \ge 0}$ be a CSMP with parameters ℓ, α, β and $(x, u) \in \breve{E}$, $t \ge 0$. Define $C_1 = \{v \in \mathbb{R}_+ : v \le t^*(x), 0 \le v - u \le t\}$, $C_2 = \{(y, v, w) \in E \times \mathbb{R}_+^2 : v \le t^*(y), v + w \le t\}$. Given a Borel function $\psi : \hat{E} \times \mathbb{R} \to \mathbb{R}$, assume:*

- $\mathbb{E}_{x,u}(N_t) < +\infty$,
- *the function $v \in [0, \min(t^*(x) - u, t)] \to \psi(x, u + v, -u)$ is absolutely continuous,*
- *the function $v \in C_1 \to |\psi(x, v, -u)| + |\partial_2 \psi(x, v, -u)|$ is bounded,*
- *the function $v \in [0, \min(t^*(y), t - w)] \to \psi(y, v, w)$ is absolutely continuous for all $y \in E$, $w \le t$,*
- *the function $(y, v, w) \in C_2 \to |\psi(y, v, w)| + |\partial_2 \psi(y, v, w)|$ is bounded.*

Then

$$\mathbb{E}_{x,u}(\psi(Z_t, A_t, T_{N_t})) = \psi(x, u, -u) + \int_0^t \mathbb{E}_{x,u}(\partial_2 \psi(Z_v, A_v, T_{N_v}))\, dv$$

$$+ \int_0^t \mathbb{E}_{x,u}\left(\int_E (\psi(y, 0, v) - \psi(Z_v, A_v, T_{N_v}))\, \ell(Z_v, A_v)\, \beta(Z_v, A_v; dy)\right) dv$$

$$+ 1_{\{\alpha(x) \le u+t\}}\, e^{-\int_u^{\alpha(x)} \ell(x,w)\, dw} \int_E (\psi(y, 0, \alpha(x) - u) - \psi(x, \alpha(x), -u))\, \beta(x, \alpha(x); dy)$$

$$+ \int_{\hat{E}} 1_{\{v + \alpha(y) \le t\}}\, e^{-\int_0^{\alpha(y)} \ell(y,w)\, dw} \int_E (\psi(z, 0, v + \alpha(y)) - \psi(y, \alpha(y), v))\, \beta(y, \alpha(y); dz)$$

$$\times\, \rho^{x,u}(dy, dv).$$

Corollary 6.2 *Let $(x,u) \in \breve{E}$, $t \geq 0$, $\varphi : \hat{E} \to \mathbb{R}$ a Borel function, and $(Z_s, A_s)_{s \geq 0}$ a simple CSMP with parameters ℓ, α, β. Assume that $\mathbb{E}_{x,u}(N_t) < +\infty$, the function $v \in [0, \min(t^*(y), t)] \to \varphi(y, v)$ is absolutely continuous for all $y \in E$ and φ and $\partial_2 \varphi$ are bounded on $\hat{E} \cap (E \times [0,t])$. Then*

$$\mathbb{E}_{x,u}(\varphi(Z_t, A_t)) = \varphi(x,u) + \int_0^t \mathbb{E}_{x,u}(\partial_2 \varphi(Z_v, A_v))\, dv$$
$$+ \int_0^t \mathbb{E}_{x,u}\left(\int_E (\varphi(y,0) - \varphi(Z_v, A_v))\, \ell(Z_v, A_v)\, \beta(Z_v, A_v; dy)\right) dv$$
$$+ \int_{\hat{E} \times \mathbb{R}_+} 1_{\{w \leq t\}} \int_E (\varphi(z,0) - \varphi(y,v))\, \beta(y,v;dz)\, \bar{\rho}_D^{x,u}(dy, dv, dw),$$

$\bar{\rho}_D^{x,u}(\Gamma_{CSMP}^c \times \mathbb{R}_+) = 0$ where $\Gamma_{CSMP} = \{(y, \alpha(y)) : y \in E,\ \alpha(y) < t^*(y)\}$, and

$<$ is ok

$$\int_{\hat{E} \times \mathbb{R}_+} 1_{\{w \leq t\}} \int_E (\varphi(z,0) - \varphi(y,v))\, \beta(y,v;dz)\, \bar{\rho}_D^{x,u}(dy,dv,dw) =$$
$$1_{\{\alpha(x) \leq u+t\}} e^{-\int_u^{\alpha(x)} \ell(x,w)\,dw} \int_E (\varphi(y,0) - \varphi(x, \alpha(x)))\, \beta(x, \alpha(x); dy)$$
$$+ \int_{\widetilde{E}} 1_{\{v + \alpha(y) \leq t\}} e^{-\int_0^{\alpha(y)} \ell(y,w)\,dw} \int_E (\varphi(z,0) - \varphi(y, \alpha(y)))\, \beta(y, \alpha(y); dz)\, \rho^{x,u}(dy, dv)$$

with $\rho^{x,u}$ defined Sect. 5.1.

Inhomogeneous Case

This paragraph may be omitted by readers who are not interested in the inhomogenous case.

Let us consider a CSMP $(\widetilde{Z}_t, A_t)_{t \geq 0}$ associated with an inhomogeneous PDMP, (see the end of Sect. 3.1), that is, a zero-delayed CSMP on $\bar{E} = \mathbb{R}_+ \times E$ with kernel

$$\widetilde{N}((s,x), ds_1 dx_1, dv) = 1_{\mathbb{R}_+}(v)\, d\widetilde{F}_{s,x}(v)\, \widetilde{\beta}((s,x), v; dx_1)\, \delta_{s+v}(ds_1), \tag{6.4}$$

where $d\widetilde{F}_{s,x}(v)$ is a probability distribution on $\mathbb{R}_+ \cup \{+\infty\}$ and $\widetilde{\beta}$ a probability kernel from $\bar{E} \times \mathbb{R}_+$ to E.

A CSMP defined as above is said to be inhomogeneous with parameters $\widetilde{\ell}, \widetilde{\alpha}, \widetilde{\beta}$ if

$$1_{\mathbb{R}_+}(v)\, d\widetilde{F}_{s,x}(v) = \widetilde{\ell}(s,x,v)\, e^{-\int_0^v \widetilde{\ell}(s,x,w)\,dw}\, 1_{\{v < \widetilde{\alpha}(s,x)\}}$$
$$+ 1_{\{\widetilde{\alpha}(s,x) < +\infty\}} e^{-\int_0^{\widetilde{\alpha}(s,x)} \widetilde{\ell}(s,x,w)\,dw}\, \delta_{\widetilde{\alpha}(s,x)}(dv).$$

It is a simple CSMP on $\mathbb{R}_+ \times E$ with parameters $\widetilde{\ell}, \widetilde{\alpha}, \widehat{\beta}$, where $\widehat{\beta}$ is defined as $\widehat{\beta}((s,x), v; ds_1, dx_1) = \widetilde{\beta}((s,x), v; dx_1)\, \delta_{s+v}(ds_1)$.

Applying Lemma 3.2, Sect. 3.1, Corollary 6.2, Proposition 5.2, Sect. 5.2 and Formula (3.12), at the end of Sect. 3.1, we get the following proposition. Its assumptions may be relaxed as in Proposition 6.1.

Proposition 6.2 *Let $(\widetilde{Z}_v, A_v)_{v\geq 0}$ be a zero-delayed CSMP with kernel as in (6.4) and $(\widetilde{Y}_k, T_k)_{k\geq 0}$ its associated Markov renewal process. Assume $\widetilde{Y}_0 = (s_0, x_0)$.*

Then $\widetilde{Y}_k = (s_0 + T_k, Y_k)$ for some Y_k $(k \geq 0)$, $\widetilde{Z}_v = (s_0 + T_{N_v}, Z_v)$ with $N_v = \sum_{k\geq 1} 1_{\{T_k\leq v\}}$, and $Z_v = Y_k$ on $\{N_v = k\}$ $(v \geq 0)$.

Let $t \geq 0$. Assume that $(\widetilde{Z}_v, A_v)_{v\geq 0}$ is inhomogeneous with parameters $\widetilde{\ell}, \widetilde{\alpha}, \widetilde{\beta}$ and $\mathbb{E}(N_t) < +\infty$. Let $\varphi : \mathbb{R}_+ \times E \times \mathbb{R}_+$ be a Borel function such that for all $(s, y) \in \mathbb{R}_+ \times E$, the function $v \in [0, t] \to \varphi(s, y, v)$ is absolutely continuous, and φ and $\partial_3\varphi$ are bounded on $[0, s_0 + t] \times E \times [0, t]$. Then

$$\mathbb{E}(\varphi(\widetilde{Z}_t, A_t)) = \varphi(s_0, x_0, 0) + \int_0^t \mathbb{E}(\partial_3\varphi(\widetilde{Z}_v, A_v))\, dv$$
$$+ \int_0^t \mathbb{E}\Big(\int_E (\varphi(s_0 + T_{N_v} + v, y, 0) - \varphi(s_0 + T_{N_v}, Z_v, A_v))\, \widetilde{\ell}(\widetilde{Z}_v, A_v)$$
$$\times \widetilde{\beta}(\widetilde{Z}_v, A_v; dy)\Big)\, dv$$
$$+ \int_{E\times\mathbb{R}_+} 1_{\{v+\widetilde{\alpha}(s_0+v,y)\leq t\}} \int_E (\varphi(s_0 + v + \widetilde{\alpha}(s_0 + v, y), z, 0) - \varphi(s_0 + v, y, \widetilde{\alpha}(s_0 + v, y)))$$
$$\times e^{-\int_0^{\widetilde{\alpha}(s_0+v,y)} \widetilde{\ell}(s_0+v,y,w)\, dw}\, \widetilde{\beta}((s_0 + v), y, \widetilde{\alpha}(s_0 + v, y); dz)\, R((s_0, x_0), dy, dv),$$

where $R(s_0, x_0, dy, dv)$ is the intensity of the marked point process $(Y_n, T_n)_{n\geq 0}$.

6.2 Kolmogorov Equations for PDMPs

Homogeneous Case

In this section, ϕ is a deterministic Markov flow, even if there is no PDMP.

In Kolmogorov's equations for PDMPs, the assumption "$v \to \varphi(x, v)$ (respectively $v \to \psi(x, v_1 + v, v_2 + v)$) is absolutely continuous" will be replaced by "$v \to f(\phi(y, v))$ (respectively $v \to g(\phi(y, v, a + v))$) is absolutely continuous" and the operator $\partial_2\phi$ (respectively $\mathcal{X}$) by the "derivative along the flow" ∂_ϕ (respectively $\partial_{t,\phi}$) defined below.

Definition 6.1 Given $f : E \to \mathbb{R}$, define $\partial_\phi f$ as

$$\partial_\phi f(x) = \lim_{\Delta\to 0} (f(\phi(x, \Delta)) - f(x))/\Delta$$

when this limit exists and $\partial_\phi f(x) = 0$ if not.

Given a Borel function $g : E \times \mathbb{R}_+ \to \mathbb{R}$, define $\partial_{t,\phi} g$ as

$$\partial_{t,\phi} g(x, v) = \lim_{\Delta\to 0} (g(\phi(x, \Delta), v + \Delta) - g(x, v))/\Delta$$

when this limit exists and 0 if not.

Given $f : E \to \mathbb{R}$ and $(x, v) \in E \times \mathbb{R}_+$ such that $\frac{\partial f \circ \phi}{\partial v}(x, v)$ exists, then $\frac{\partial f \circ \phi}{\partial v}(x, v) = \partial_\phi f(\phi(x, v))$. Assume that $E \subset \mathbb{R}^k$, $f : E \to \mathbb{R}$ is differentiable and ϕ is given by the ODE $\frac{\partial \phi}{\partial v}(x, v) = \mathbf{v}(\phi(x, v))$. Then $\partial_\phi f = \nabla f \cdot \mathbf{v}$, where $\nabla f = (\frac{\partial f}{\partial x_1}, \ldots, \frac{\partial f}{\partial x_k})$.

Assume that $E \subset \mathbb{R}^k$ and the functions $g : E \times \mathbb{R}_+ \to \mathbb{R}$ and $v \in \mathbb{R}_+ \to \phi(x, v)$ are differentiable. Then

$$\partial_{t,\phi} g(x, v) = \nabla_x g(x, v) \cdot \frac{\partial \phi}{\partial v}(x, 0) + \frac{\partial g}{\partial v}(x, v)$$

with $x = (x_1, \ldots x_k)$, $\nabla_x g = (\frac{\partial g}{\partial x_1}, \ldots, \frac{\partial g}{\partial x_k})$.
When ϕ is given by the ODE $\frac{\partial \phi}{\partial v}(x, v) = \mathbf{v}(\phi(x, v))$, we get

$$\partial_{t,\phi} g(x, v) = \nabla_x g(x, v) \cdot \mathbf{v}(x) + \frac{\partial g}{\partial v}(x, v).$$

From $\phi(x, u + 1/n) = \phi(\phi(x, u), 1/n)$ we deduce

$$\partial_{t,\phi} g(\phi(x, v_1), v_2) = \mathcal{X}\psi(x, v_1, v_2) \text{ with } \psi(x, v_1, v_2) = g(\phi(x, v_1), v_2). \quad (6.5)$$

The following result is a consequence of Lemma 6.1, Sect. 6.1.

Lemma 6.3 *Let $(x, a) \in E \times \mathbb{R}_+$, $g : E \times \mathbb{R}_+ \to \mathbb{R}$ and $f : E \to \mathbb{R}$ be Borel functions. Assume that the functions $v \in [0, t^*(x)] \to g(\phi(x, v), a + v)$ and $v \in [0, t^*(x)] \to f(\phi(x, v))$ are absolutely continuous. Then these functions are differentiable Lebesgue-almost everywhere and*

$$\forall v \leq t^*(x) : \; g(\phi(x, v), a + v) - g(x, a) = \int_0^v \partial_{t,\phi} g(\phi(x, s), a + s)\, ds,$$

$$\forall v \leq t^*(x) : f(\phi(x, v)) - f(x) = \int_0^v \partial_\phi f(\phi(x, s))\, ds.$$

Kolmogorov's equations for PDMPs can be derived from those for CSMPs, since a PDMP X is of the form $X_t = \phi(Z_t, A_t)$, where $(Z_t, A_t)_{t \geq 0}$ is a zero-delayed CSMP.

Proposition 6.3 *Let $(Y_n, T_n)_{n \geq 0}$ be a zero-delayed Markov renewal process, $(Z_s, A_s)_{s \geq 0}$ its associated CSMP, ϕ a deterministic Markov flow on $\hat{E}$, and $X_s = \phi(Z_s, A_s)$ $(s \geq 0)$.*

Given $x \in E$, $t \geq 0$, $g : E \times \mathbb{R}_+ \to \mathbb{R}$ a Borel function, assume:

- $\mathbb{E}_x(\sum_{n\geq 1} 1_{\{T_n \leq t\}}) < +\infty$,
- *for any* $y \in E$, $a \in [0, t]$ *the function* $v \in [0, min(t^*(y), t-a)] \to g(\phi(y, v), a+v)$ *is absolutely continuous,*
- *the function* $(y, v, w) \to |g(\phi(y, v), w)| + |\partial_{t,\phi} g(\phi(y, v), w)|$ *is bounded on* $\{(y, v, w) \in E \times [0, t]^2 : v \leq t^*(y)\}$.

Then

$$\mathbb{E}_x(g(X_t, t)) = g(x, 0) + \int_0^t \mathbb{E}_x(\partial_{t,\phi} g(X_v, v))\, dv \tag{6.6}$$

$$+ \int_{\hat{E}\times\mathbb{R}_+} 1_{\{w \leq t\}} \int_E (g(z, w) - g(\phi(y, v), w))\, \beta(y, v; dz)\, \bar{\rho}(x, dy, dv, dw)$$

with $\bar{\rho}$ *defined Sect. 5.1.*

If, moreover, $(X_s)_{s\geq 0}$ *is a PDMP with parameters* $\phi, \lambda, Q, \alpha, q$, *we have*

$$\begin{aligned}\mathbb{E}(g(X_t, t)) = \mathbb{E}(g(X_0, 0)) &+ \int_0^t \mathbb{E}(\partial_{t,\phi} g(X_v, v))\, dv \\ &+ \int_0^t \mathbb{E}\left(\lambda(X_v) \int_E (g(y, v) - g(X_v, v))\, Q(X_v; dy)\right) dv \\ &+ \int_{\Gamma^*\times]0,t]} \int_E (g(z, v) - g(y, v))\, q(y; dz)\, \sigma(dy, dv),\end{aligned} \tag{6.7}$$

where $\Gamma^* = \{\phi(y, \alpha(y)) : y \in E, \alpha(y) < +\infty, \int_0^{\alpha(y)} \lambda(\phi(y, v))\, dv < +\infty\}$ *and* σ *is as in (5.31), Sect. 5.3.*

Proof Writing (6.2), Sect. 6.1 with $\psi(y, v, w) = g(\phi(y, v), w)$ and using (6.5), we obtain (6.6). Then (6.7) is given by (6.3), Sect. 6.1 where $\bar{\rho}_D$ is as in (5.14), Sect. 5.2 with $u = 0$. ■

Proceeding in a like manner, we get the following result under the same assumptions as in the first part of Proposition 6.3:

$$\mathbb{E}_x(g(X_t, A_t)) = g(x, 0) + \int_0^t \mathbb{E}_x(\partial_{t,\phi} g(X_v, A_v))\, dv$$

$$+ \int_{E\times\mathbb{R}_+\times]0,t]} \int_E (g(z, 0) - g(\phi(y, v), v)\, \beta(y, v; dz)\, \bar{\rho}(x, dy, dv, dw).$$

When $(X_s)_{s\geq 0}$ is a PDMP with parameters $\phi, \lambda, Q, \alpha, q$, $\mathbb{E}(g(X_t, A_t))$ is obtained by replacing $(g(y, v) - g(X_v, v))\, Q(X_v; dy)$ by $(g(y, 0) - g(X_v, v))Q(X_v; dy)$ in (6.7) and $(g(z, v) - g(y, v))\, q(y; dz)$ by $(g(z, 0) - g(y, v))\, q(y, dz)$.

Corollary 6.3 *Let* $(X_s)_{s\geq 0}$ *be a PDMP with parameters* $\phi, \lambda, Q, \alpha, q$, *and denote* $(Y_n, T_n)_{n\geq 0}$ *its associated Markov renewal process. Given* $t \geq 0$, *a Borel function* $f : E \to \mathbb{R}$, *assume:*

- $\mathbb{E}(\sum_{n\geq 1} 1_{\{T_n\leq t\}}) < +\infty$,
- *for all* $y \in E$, *the function* $v \in [0, \min(t^*(y), t)] \to f(\phi(y, v))$ *is absolutely continuous,*
- *the functions* $f \circ \phi$ *and* $(\partial_\phi f) \circ \phi$ *are bounded on* $\{(y, v) \in E \times [0, t] : v \leq t^*(y)\}$.

Then

$$\begin{aligned}\mathbb{E}(f(X_t)) &= \mathbb{E}(f(X_0)) + \int_0^t \mathbb{E}(\partial_\phi f(X_v)\,dv) \\ &+ \int_0^t \mathbb{E}\left(\lambda(X_v)\int_E (f(y) - f(X_v))\,Q(X_v; dy)\right) dv \\ &+ \int_{\Gamma^*\times]0,t]}\int_E (f(y) - f(x))\,q(x; dy)\,\sigma(dx, dv), \qquad (6.8)\end{aligned}$$

where Γ^* *and* σ *are as in Proposition 6.3.*

Actually, for a PDMP with parameters $\phi, \lambda, Q, \alpha, (E_i, q_i)_{i\in I}$, the condition "the function $v \in [0, \min(t^*(y), t)] \to f(\phi(y, v))$ is absolutely continuous for all $y \in E$" is an excessive restriction. Indeed, functions f with discontinuities on the boundary Γ^* may be considered, and their prolongations by left limits on it (basically by $\lim_{v\uparrow\alpha(y)} f(\phi(y, v))$) must appear in (6.8). It is the same for the similar condition on g in Proposition 6.3.

To get a better understanding, let us go back to Example 3.8, Sect. 3.2 with $E = \mathbb{R}$, $n = 1$ and $\alpha_1 = 0$. The condition "the function $v \in [0, \min(t^*(y), t)] \to f(\phi(y, v))$ is absolutely continuous for all $y \in E$" is equivalent to "the function f is absolutely continuous on $\mathbb{R}$." When we take $\tilde{E} = \{a\}\times]-\infty, 0[\cup\{b\} \times [0, +\infty[$ for the state space, the condition becomes "the function $v \to f(a, v)$ is absolutely continuous on $]-\infty, 0]$ and the function $v \to f(b, v)$ is absolutely continuous on $[0, +\infty[$." This is a weaker condition, which gives rise to stating separately the conditions on $]-\infty, 0]$ and on $[0, +\infty[$, keeping in mind that f is absolutely continuous on $]-\infty, 0]$ if it is on $]-\infty, 0[$ and if in addition f is left-continuous at point 0. This is the reason for the introduction of functions g_i in Theorem 6.1 below.

Theorem 6.1 *Let* $(X_s)_{s\geq 0}$ *be a PDMP with parameters* $\phi, \lambda, Q, \alpha, (E_i, q_i)_{i\in I}$, *and* $(Y_n, T_n)_{n\geq 0}$ *its associated Markov renewal process. Given* $t \geq 0$, $g : E \times \mathbb{R}_+ \to \mathbb{R}$ *a Borel function, assume:*

- $\mathbb{E}(\sum_{n\geq 1} 1_{\{T_n\leq t\}}) < +\infty$,
- $(y, v, w) \to g(\phi(y, v), w)$ *is bounded on* $\{(y, v, w) \in E \times [0, t]^2, v \leq t^*(y)\}$,
- *for all* $i \in I$, *there exists a function* $g_i : (E_i \cup \Gamma_i^*) \times \mathbb{R}_+ \to \mathbb{R}$ *such that* $g_i = g$ *on* $E_i \times \mathbb{R}_+$, *the function* $v \in [0, \min(t^*(y), t)] \to g_i(\phi(y, v), a + v)$ *is absolutely continuous for all* $y \in E_i$, $a \in [0, t]$, *and* $\sup\{|\partial_{t,\phi} g_i(\phi(y, v), w)| : i \in I, (y, v, w) \in E_i \times [0, t]^2, v \leq t^*(y)\} < +\infty$.

Then

$$\mathbb{E}(g(X_t,t)) = \mathbb{E}(g(X_0,0)) + \int_0^t \mathbb{E}(\partial_{t,\phi} g(X_v,v))\,dv$$
$$+ \int_0^t \mathbb{E}\left(\lambda(X_v)\int_E (g(y,v) - g(X_v,v))\,Q(X_v;dy)\right)dv$$
$$+ \sum_{i\in I}\int_{\Gamma_i^*\times]0,t]}\int_E (g(y,v) - g_i(x,v))\,q_i(x;dy)\,\sigma_i(dx,dv), \quad (6.9)$$

and

$$\mathbb{E}(g(X_t,A_t)) = \mathbb{E}(g(X_0,0)) + \int_0^t \mathbb{E}(\partial_{t,\phi} g(X_v,A_v))\,dv$$
$$+ \int_0^t \mathbb{E}\left(\lambda(X_v)\int_E (g(y,0) - g(X_v,v))\,Q(X_v;dy)\right)dv$$
$$+ \sum_{i\in I}\int_{\Gamma_i^*\times]0,t]}\int_E (g(y,0) - g_i(x,v))\,q_i(x;dy)\,\sigma_i(dx,dv),$$

where $\Gamma_i^* = \{\phi(y,\alpha(y)) : y \in E_i, \alpha(y) < +\infty, \int_0^{\alpha(y)} \lambda(\phi(y,w))\,dw < +\infty\}$ *and* $\int \varphi(y,v)\,\sigma_i(dy,dv) = \sum_{k\geq 0}\mathbb{E}(1_{E_i}(Y_k)\,1_{\{T_{k+1}=T_k+\alpha(Y_k)<+\infty\}}\,\varphi(X_{T_{k+1}-},T_{k+1}))$ *for every Borel function* $\varphi : E\times\mathbb{R}_+ \to \mathbb{R}_+$.

A different way to introduce the functions g_i is the following: assume that for all $(y,a) \in E\times\mathbb{R}_+$ the function $v \in [0,t^*(y)] \cap [0,\alpha(y)[\to g(\phi(y,v),a+v)$ is absolutely continuous and for all $i \in I$ there exists a Borel function $g_i : \Gamma_i^*\times\mathbb{R}_+ \to \mathbb{R}$ such that $g_i(\phi(y,\alpha(y)),a+v) = \lim_{v\uparrow\alpha(y)} g(\phi(y,v),a+v)$ for all $(y,a) \in E_i\times\mathbb{R}_+$.

Proof Write $g(X_t,t) - g(X_0,t) = Tr + S$ as in Lemma 6.2, Sect. 6.1 with Tr for the transport part governed by the flow and S for the jump part, while introducing the functions g_i. We get

$$Tr = \sum_{i\in I}\sum_{k\geq 0} 1_{\{T_{k+1}\leq t\}}\,1_{E_i}(Y_k)\,(g_i(\phi(Y_k,T_{k+1}-T_k),T_{k+1}) - g_i(Y_k,T_k))$$
$$+ \sum_{i\in I}\sum_{n\geq 0} 1_{\{T_n\leq t<T_{n+1}\}}\,1_{E_i}(Y_n)\,(g_i(\phi(Y_n,t-T_n),t) - g_i(Y_n,T_n)),$$

$$S = \sum_{i\in I}\sum_{k\geq 0} 1_{\{T_{k+1}\leq t\}}\,1_{E_i}(Y_k)(g(Y_{k+1},T_{k+1}) - g_i(\phi(Y_k,T_{k+1}-T_k),T_{k+1})).$$

For $a \in \mathbb{R}_+$, $y \in E_i$, $v < t^*(y)$ we have $\phi(y,v) \in E_i$ and $g(\phi(y,v),a+v) = g_i(\phi(y,v),a+v)$. Hence $v \in [0,t^*(y)\wedge t[\to g(\phi(y,v),a+v)$ is absolutely continuous and $\partial_{t,\phi} g(\phi(y,v),a+v) = \partial_{t,\phi} g_i(\phi(y,v),a+v)$. Therefore, we have for $T_{k+1} \leq t$, $Y_k \in E_i$,

$$g_i(\phi(Y_k, T_{k+1} - T_k), T_{k+1}) - g_i(Y_k, T_k) = \int_0^{T_{k+1}-T_k} \partial_{t,\phi} g_i(\phi(Y_k, v), T_k + v)\, dv$$
$$= \int_0^{T_{k+1}-T_k} \partial_{t,\phi} g(X_{T_k+v}, T_k + v) = \int_{T_k}^{T_{k+1}} \partial_{t,\phi} g(X_w, w)\, dw,$$

and hence $Tr = \int_0^t \partial_{t,\phi} g(X_w, w)\, dw$.

On the other hand, Proposition 5.7, Sect. 5.3 gives

$$\mathbb{E}(S) = \sum_{k\geq 0} \mathbb{E}(1_{\{T_{k+1}\leq t\}}\, g(X_{T_{k+1}}, T_{k+1})) - \sum_{i\in I}\sum_{k\geq 0} \mathbb{E}(1_{\{T_{k+1}\leq t\}}\, 1_{E_i}(Y_k)\, g_i(X_{T_{k+1}-}, T_{k+1}))$$
$$= \mathbb{E}\Big(\int_{E\times\mathbb{R}_+} 1_{\{v\leq t\}}\, g(z, v)\, \lambda(X_v)\, Q(X_v; dz)\, dv\Big)$$
$$+ \sum_{i\in I} \int_{E\times\Gamma_i^*\times\mathbb{R}_+} g(z, v)\, q_i(y; dz)\, \sigma_i(dy, dv)$$
$$- \sum_{k\geq 0} \mathbb{E}(1_{\{T_{k+1}\leq t\}}\, 1_{\{T_{k+1}-T_k<\alpha(Y_k)\}}\, g(X_{T_{k+1}-}, T_{k+1}))$$
$$- \sum_{i\in I} \int_{\Gamma_i^*\times\mathbb{R}_+} 1_{\{v\leq t\}} g_i(y, v)\, \sigma_i(dy, dv).$$

From (5.28), Sect. 5.3 we get

$$\sum_{k\geq 0} \mathbb{E}(1_{\{T_{k+1}\leq t\}}\, 1_{\{T_{k+1}-T_k<\alpha(Y_k)\}}\, g(X_{T_{k+1}-}, T_{k+1})) = \mathbb{E}\Big(\int_{\mathbb{R}_+} 1_{\{v\leq t\}}\, g(X_v, v)\, \lambda(X_v)\, dv\Big).$$

This ends the proof of the first result. The second result is proved in a similar way. ∎

Denote by π_t the probability distribution of X_t and assume that there exists $T > 0$ such that $g(x, t) = 0$ for all $(x, t) \in E \times [T, +\infty[$. Letting t tend to infinity in (6.9), we get

$$0 = \int_E g(x, 0)\, \pi_0(dx) + \int_{E\times\mathbb{R}_+} \partial_{t,\phi} g(x, s)\, \pi_s(dx)\, ds$$
$$+ \int_{E\times\mathbb{R}_+} \lambda(x) \Big(\int_E g(y, s)\, Q(x; dy) - g(x, s)\Big)\, \pi_s(dx)\, ds$$
$$+ \sum_{i\in I} \int_{\Gamma_i^*\times\mathbb{R}_+} \Big(\int_E g(y, s)\, q_i(x; dy) - g_i(x, s)\Big)\, \sigma_i(dx, ds). \qquad (6.10)$$

Formula (6.10) is the basis for the derivation of the numerical schemes given in Chap. 9. Theorem C.1, end of Sect. C.1, states that under suitable conditions, which are continuity conditions of the PDMP parameters on each E_i, formula (6.10) characterizes measures $\pi_s(dx)\, ds$ and boundary measures $(\sigma_i(dx, ds))_{i\in I}$.

If for some $i \in I$, a part of Γ_i^* is in the topological interior of the state space, to get the uniqueness result we have to consider functions g that may not be continuous on $\Gamma_i^* \times \mathbb{R}_+$. As a consequence, a partition $(E_i)_{i\in I}$ and the functions g_i must be introduced even if the partition is not required a priori as in Example 3.8, Sect. 3.2. This need is hidden in Assumptions (H3), Sect. 9.4.

Indeed, if there is a "boundary" in the topological interior of the state space, as in Examples 3.8, Sect. 3.2, 3.10, 3.12, and 3.14, Sect. 3.4, then the function α is discontinuous on this boundary.

Remark 6.1 We saw in Examples 3.12 and 3.13, Sect. 3.4 that a boundary can be artificially added to get more information. Assume that Γ_{i_0} is such a boundary and g satisfies the assumptions required in Theorem 6.1 for the model where the boundary is not added. When $q_{i_0}(x; dy) = \delta_x(dy)$ for all $x \in \Gamma_{i_0}$ as in Example 3.13, Sect. 3.4, then the term with index i_0 in formula (6.9) vanishes.

Inhomogeneous Case

This paragraph is intended for readers interested in inhomogeneous PDMPs. It is not necessary for understanding the rest of the book.

Throughout this section, the terminology and the notation are those of the subsection inhomogeneous PDMPs at the end of Sect. 3.1.

Let $\phi : \mathbb{R}_+ \times E \times \mathbb{R}_+ \to E$ be an inhomogeneous Markov flow, let $\widetilde{\lambda} : \mathbb{R}_+ \times \mathbb{E} \to \mathbb{R}_+$ and $\widetilde{\alpha} : \mathbb{R}_+ \times E \to]0, +\infty]$ be Borel functions such that

$$\widetilde{\alpha}(s+v, \phi(s,x,s+v)) = \widetilde{\alpha}(s,x) - v, \quad (s,x) \in \mathbb{R}_+ \times E,\ v < \widetilde{\alpha}(s,x).$$

Let $(E_i)_{i\in I}$ be a countable partition of E. Define

$$\Gamma_i = \{\phi(s,x,s+\widetilde{\alpha}(s,x)) : (s,x) \in \mathbb{R}_+ \times E_i, \widetilde{\alpha}(s,x) < +\infty\},$$

$$\Gamma_i^* = \{\phi(s,x,s+\widetilde{\alpha}(s,x)) \in \Gamma_i : \int_0^{\widetilde{\alpha}(s,x)} \widetilde{\lambda}(s+v, \phi(s,x,s+v))\, dv < +\infty\}.$$

Assume:

- $\forall i \in I, \forall (s,x) \in \mathbb{R}_+ \times E_i, \forall v < \widetilde{\alpha}(s,x) : \phi(s,x,s+v) \in E_i$,
- $\forall i \in I, E_i \cap \Gamma_i = \emptyset$.

Let $\widetilde{Q}$ be a probability kernel from $\mathbb{R}_+ \times E$ to E and $\widetilde{q}_i$ $(i \in I)$ probability kernels from $\mathbb{R}_+ \times \Gamma_i^*$ to E respectively.

Definition 6.2 An inhomogeneous PDMP on E with parameters $\phi, \widetilde{\lambda}, \widetilde{Q}, \widetilde{\alpha}$, $(E_i, \widetilde{q}_i)_{i\in I}$ is an inhomogeneous PDMP generated by ϕ and $\bar{N}$ defined as

$$\bar{N}((s,x), dx_1, dv) = \widetilde{\ell}(s,x,v)\, e^{-\int_0^v \widetilde{\ell}(s,x,w)\,dw}\, 1_{\{v<\widetilde{\alpha}(s,x)\}}\, \widetilde{\beta}((s,x), v; dx_1)\, dv$$
$$+\, 1_{\{\widetilde{\alpha}(s,x)<+\infty\}}\, e^{-\int_0^{\widetilde{\alpha}(s,x)} \widetilde{\ell}(s,x,w)\,dw}\, \widetilde{\beta}((s,x), v; dx_1)\, \delta_{\widetilde{\alpha}(s,x)}(dv)$$

with

$$\widetilde{\ell}(s,x,v) = \widetilde{\lambda}(s+v, \phi(s,x,s+v)),$$

$$\widetilde{\beta}((s,x),v;dx_1) = \begin{cases} \widetilde{Q}((s+v,\phi(s,x,s+v));dx_1) & v < \widetilde{\alpha}(s,x), \\ \widetilde{q}_i((s+\widetilde{\alpha}(s,x),\phi(s,x,s+\widetilde{\alpha}(s,x)));dx_1), & v = \widetilde{\alpha}(s,x), x \in E_i. \end{cases}$$

Define $\widetilde{t}_1^*(s,x) = \sup\{v \in \mathbb{R}_+ : \int_0^v \widetilde{\lambda}(s+w,\phi(s,x,s+w))\,dw < +\infty\}$ and $\widetilde{t}^*(s,x) = \min(\widetilde{\alpha}(s,x), \widetilde{t}_1^*(s,x))$.

The terminology inhomogeneous PDMP is justified, since condition (3.11), Sect. 3.1 is satisfied and $(t, X_t)_{t\geq 0}$ is a PDMP (Proposition 3.3, Sect. 3.1). Its kernel is $\widetilde{N}$ given by (3.10), Sect. 3.1.

Theorem 6.1, Sect. 6.2 applied to the PDMP $(t, X_t)_{t\geq 0}$ gives the following proposition. A result with weaker assumptions on f can be obtained by introducing functions f_i.

Proposition 6.4 *Let $(X_t)_{t\geq 0}$ be an inhomogeneous PDMP on E with parameters $\phi, \widetilde{\lambda}, \widetilde{Q}, \widetilde{\alpha}, (E_i, \widetilde{q}_i)_{i\in I}$. Assume $\mathbb{E}(\sum_{k\geq 1} 1_{\{T_k\leq t\}}) < +\infty$, where $(T_k)_{k\geq 0}$ is as in Lemma 3.2, end of Sect. 3.1.*

Let $f : \bar{E} = \mathbb{R}_+ \times E \to \mathbb{R}$ a Borel function. Define $\widetilde{\partial}_\phi f$ as

$$\widetilde{\partial}_\phi f(s,x) = \lim_{\Delta\to 0} (f(s+\Delta, \phi(s,x,s+\Delta)) - f(s,x))/\Delta$$

when this limit exists and 0 if not. Assume that the function $v \in [0, t^*(s,x)] \to f(s+v, \phi(s,x,s+v))$ is absolutely continuous for all $(s,x) \in \bar{E}$ and that $(s,x,v) \to |f(s+v,\phi(s,x,s+v)| + |\widetilde{\partial}_\phi f(s+v,\phi(s,x,s+v)|$ is bounded on $\{(s,x,v) \in \bar{E} \times \mathbb{R}_+, v \leq \widetilde{t}^*(s,x), s+v \leq t\}$ for all $t > 0$.

Then for all $t \geq 0$,

$$\begin{aligned} \mathbb{E}(f(t,X_t)) &= f(0,x_0) + \int_0^t \mathbb{E}(\widetilde{\partial}_\phi f(v,X_v))\,dv \\ &+ \int_0^t \mathbb{E}_{0,x_0}\left(\widetilde{\lambda}(v,X_v)\int_E (f(v,y) - f(v,X_v))\,\widetilde{Q}((v,X_v);dy)\right)dv \\ &+ \sum_{i\in I}\int_{\widetilde{\Gamma}_i^*\times]0,t]} 1_{\{v\leq t\}}\int_E (f(v,z) - f(v,y))\,\widetilde{q}_i((v,y);dz)\,\widetilde{\sigma}_i(dy,dv) \end{aligned}$$

with

$$\int g(y,v)\,\widetilde{\sigma}_i(dy,dv) = \mathbb{E}\left(\sum_{k\geq 0} 1_{E_i}(X_{T_k})\,g(X_{T_{k+1}-}, T_{k+1})\,1_{\{T_{k+1}-T_k=\widetilde{\alpha}(T_k,Y_k)\}}\right)$$

for any Borel function $g : \mathbb{R}_+ \times E \to \mathbb{R}_+$.

6.3 Some Semigroup Properties in the Absence of Dirac Measures

The case in which the interarrival distributions have densities with respect to the Lebesgue measure is handled in the books of M. H. A Davis and M. Jacobsen. The scope of this section is to take advantage of the approach using CSMPs to derive some semigroup properties while gradually hardening the conditions imposed on the test function. See the books of Davis ([51], Sect. 27) and Jacobsen ([66], Sect. 7) for other results.

CSMP Case

Throughout this section, $(Z_t, A_t)_{t\geq 0}$ is a CSMP without boundary, with parameters ℓ and β, i.e., its kernel N is

$$N(x, dy, dv) = e^{-\int_0^v \ell(x,s)\,ds}\,\ell(x, v)\,\beta(x, v; dy)\,dv.$$

The function ℓ is assumed to be bounded on $E \times [0, a]$ for all $a \in \mathbb{R}_+$, and hence $\mathbb{E}_{x,u}(N_t) < +\infty$ for all $(x, u, t) \in \breve{E} \times \mathbb{R}_+$.

Let $(P_t)_{t\geq 0}$ be the semigroup of the Markov process $(Z_t, A_t)_{t\geq 0}$ defined as

$$P_t\varphi(x, u) = \mathbb{E}_{x,u}(\varphi(Z_t, A_t))$$

for every bounded Borel function $\varphi : E \times \mathbb{R}_+ \to \mathbb{R}$.

Formula (2.21), end of Sect. 2.5, leads to

$$\begin{aligned} P_t\varphi(x, u) &= \varphi(x, u + t)\,e^{-\int_0^t \ell(x,u+s)\,ds} \\ &+ \int_{E\times[0,t]} P_{t-v}\varphi(y, 0)\,e^{-\int_0^v \ell(x,s+u)\,ds}\,\ell(x, u + v)\,\beta(x, u + v; dy)\,dv. \end{aligned} \tag{6.11}$$

This is the integral backward Kolmogorov equation for $(P_t)_{t\geq 0}$.

Given a Borel function $\varphi : E \times \mathbb{R}_+ \to \mathbb{R}$ that is bounded on $E \times [0, a]$ for all $a > 0$ and such that the function $u \in \mathbb{R}_+ \to \varphi(x, u)$ is absolutely continuous for all $x \in E$, define

$$L\varphi(x, u) = \partial_2\varphi(x, u) + \int_E (\varphi(y, 0) - \varphi(x, u))\,\ell(x, u)\,\beta(x, u; dy).$$

Proposition 6.5 *Let $\varphi : E \times \mathbb{R}_+ \to \mathbb{R}$ be a Borel function that is bounded on $E \times [0, a]$ for all $a \in R_+$.*

(A) Assume that the function $v \in \mathbb{R}_+ \to \varphi(x, v)$ is continuous for all $x \in E$. Then:

1. *for all $(x, u) \in \breve{E}$, the function $t \to P_t\varphi(x, u)$ is continuous,*

2. *for all* $(x, t) \in E \times \mathbb{R}_+$, *the function* $u \in [0, t^*(x)[\to P_t\varphi(x, u)$ *is continuous.*

(B) Assume that the function $v \in \mathbb{R}_+ \to \varphi(x, v)$ *is absolutely continuous for all* $x \in E$ *and* $\partial_2\varphi$ *is bounded on* $E \times [0, a]$ *for all* $a \in \mathbb{R}_+$. *Then:*

1. *for all* $(x, u, t) \in \breve{E} \times \mathbb{R}_+$,

$$P_t\varphi(x, u) = \varphi(x, u) + \int_0^t P_s L\varphi(x, u)\, ds. \tag{6.12}$$

Thus for all $(x, u) \in \breve{E}$, *the function* $t \to P_t\varphi(x, u)$ *is absolutely continuous, and for Lebesgue-almost every* t,

$$\frac{\partial}{\partial t} P_t\varphi(x, u) = P_t L\varphi(x, u), \tag{6.13}$$

2. *for all* $(x, t) \in E \times \mathbb{R}_+$, *the function* $u \in [0, t^*(x)[\to P_t\varphi(x, u)$ *is absolutely continuous and for Lebesgue-almost every* u *such that* $(x, u) \in \breve{E}$,

$$P_t L\varphi(x, u) = L P_t\varphi(x, u). \tag{6.14}$$

Proof (A) The continuity of $t \to P_t\varphi(x, u)$ is an application of (6.11) and Lebesgue's dominated convergence theorem. In view of this result and of the formula

$$P_t\varphi(x, u) = \varphi(x, u + t)\, e^{-\int_u^{u+t} \ell(x,s)\, ds}$$
$$+ e^{\int_0^u \ell(x,s)\, ds} \int_u^{u+t} \ell(x, v)\, e^{-\int_0^v \ell(x,s)\, ds} \int_E P_{t+u-v}\varphi(y, 0)\, \beta(x, v; dy)\, dv, \tag{6.15}$$

which is a rewriting of (6.11), Lebesgue's dominated convergence theorem gives the continuity of $u \to P_t\varphi(x, u)$.

(B) Corollary 6.2, Sect. 6.1 yields (6.12), and the first assertion follows.

We deduce from (6.15) that

$$e^{-\int_0^u \ell(x,w)\, dw}\, P_t\varphi(x, u) = P_{t+u}\varphi(x, 0) - \int_{E\times]0,u]} P_{t+u-v}\varphi(y, 0)\, N(x, dy, dv), \tag{6.16}$$

and from (6.12) that

$$P_{t+u-v}\varphi(y, 0) = P_t\varphi(y, 0) + \int_0^{u-v} P_{t+w} L\varphi(y, 0)\, dw.$$

Thus

$$e^{-\int_0^u \ell(x,w)\, dw}\, P_t\varphi(x, u) = P_{t+u}\varphi(x, 0) - \int_0^u \int_E P_t\varphi(y, 0)\, N(x, dy, dv)$$
$$+ \int_0^u \int_{E\times\mathbb{R}_+} 1_{\{0<v\le s\}}\, P_{t+s-v}\varphi(y, 0)\, N(x, dy, dv)\, ds.$$

We deduce that $u \to P_t\varphi(x,u)$ is absolutely continuous, and for Lebesgue-almost every u,

$$
\begin{aligned}
&-\ell(x,u)\, e^{-\int_0^u \ell(x,w)\,dw}\, P_t\varphi(x,u) + e^{-\int_0^u \ell(x,w)\,dw}\, \partial_2 P_t\varphi(x,u)\\
&= P_{t+u}L\varphi(x,0) - \ell(x,u)\, e^{-\int_0^u \ell(x,w)\,dw} \int_E P_t\varphi(y,0)\,\beta(x,u;dy)\\
&\quad - \int_{E\times\mathbb{R}_+} 1_{\{0<v\le u\}}\, P_{t+u-v}L\varphi(y,0)\, N(x,dy,dv).
\end{aligned}
$$

Using again (6.16), we get

$$
\begin{aligned}
&e^{-\int_0^u \ell(x,w)\,dw}\left(\partial_2 P_t\varphi(x,u) + \int_E (P_t\varphi(y,0) - P_t\varphi(x,u))\,\ell(x,u)\,\beta(x,u;dy)\right)\\
&= P_{t+u}L\varphi(x,0) - \int_{E\times\mathbb{R}_+} 1_{\{0<v\le u\}}\, P_{t+u-v}L\varphi(y,0)\, N(x,dy,dv)\\
&= e^{-\int_0^u \ell(x,w)\,dw}\, P_t L\varphi(x,u),
\end{aligned}
$$

and the last result follows. ∎

With stronger assumptions, we get the differentiability of the functions $t \to P_t\varphi(x,u)$ and $u \to P_t\varphi(x,u)$ and the Feller property for the semigroup $(P_t)_{t\ge 0}$.

Proposition 6.6 *Assume that for all $x \in E$, the function $u \in [0, t^*(x)[\to \ell(x,u) \int_E h(y)\,\beta(x,u;dy)$ is continuous for every bounded and continuous function $h : E \to \mathbb{R}$ and that ℓ is bounded.*

Then $(P_t)_{t\ge 0}$ is a Feller semigroup, that is, for every continuous bounded function $\varphi : E \times \mathbb{R}_+ \to \mathbb{R}$ and all $t \in \mathbb{R}_+$, the function $P_t\varphi : \breve{E} \to \mathbb{R}$ is continuous.

Let $\varphi : E \times \mathbb{R}_+ \to \mathbb{R}$ be a bounded Borel function such that for all $x \in E$, the function $v \to \varphi(x,v)$ is continuously differentiable and $\partial_2\varphi$ is bounded. Then:

- *for all $(x,u) \in \breve{E}$, the function $t \in \mathbb{R}_+ \to P_t\varphi(x,u)$ is continuously differentiable, and formula (6.13) holds for all $(x,u,t) \in \breve{E} \times \mathbb{R}_+$,*
- *for all $(x,t) \in E \times \mathbb{R}_+$, the function $u \in [0, t^*(x)[\to P_t\varphi(x,u)$ is continuously differentiable, and formula (6.14) holds for all $(x,u,t) \in \breve{E} \times \mathbb{R}_+$.*

Proof The proof of the Feller property is given in Davis [51] (Theorem 27.6 page 77) for a PDMP. Following M. H. A. Davis, let us outline it in the special case of a CSMP.

Given a bounded Borel function $\varphi : E \times \mathbb{R}_+ \to \mathbb{R}$, define the operator S on bounded Borel functions $\psi : E \times \mathbb{R}_+^2 \to \mathbb{R}$ as

$$
S\psi(x,u,t) = \mathbb{E}_{x,u}(\varphi(Z_t, A_t)\, 1_{\{t<T_1\}} + \psi(Y_1, 0, t - T_1)\, 1_{\{T_1\le t\}}).
$$

We have

$$S\psi(x,u,t) = \varphi(x,u+t)\, e^{-\int_0^t \ell(x,u+s)\,ds}$$
$$+ \int_0^t \left(\int_E \psi(z,0,t-v)\, \beta(x,u+v;dz) \right) \ell(x,u+v)\, e^{-\int_0^v \ell(x,u+s)\,ds}\, dv.$$

From Lebesgue's dominated convergence theorem, we get that if ψ is continuous and bounded, then so is $S\psi$, and therefore, $S^n\psi$. By induction it can be proved that

$$S^n\psi(x,u,t) = \mathbb{E}_{x,u}(\varphi(Z_t,A_t)\, 1_{\{t<T_n\}} + \psi(Y_n,0,t-T_n)\, 1_{\{T_n\le t\}}).$$

Thus

$$\begin{aligned}|P_t\varphi(x,u) - S^n\psi(x,u,t)| &\le \mathbb{E}_{x,u}\left((|\varphi(Z_t,A_t)| + |\psi(Y_n,0,t-T_n)|)\, 1_{\{T_n\le t\}} \right)\\ &\le (||\varphi||_\infty + ||\psi||_\infty)\, \mathbb{P}_{x,u}(T_n \le t).\end{aligned}$$

Since $(x,u) \to S^n\psi(x,u,t)$ is continuous, to prove the continuity of $(x,u) \to P_t\varphi(x,u)$ it is sufficient to check that $\mathbb{P}_{x,u}(T_n \le t)$ converges to 0 uniformly in (x,u) as n tends to infinity. This is given by Proposition 2.6, end of Sect. 2.3, since ℓ is bounded.

Now we turn to proving the second part of Proposition 6.6.

For all $y \in E$, the function $v \to L\varphi(y,v)$ is continuous, and therefore, from Proposition 6.5 assertion A.1, we get that $s \to P_s L\varphi(x,u)$ is continuous. Therefore, formula (6.13) for all $(x,u,t) \in \tilde{E} \times \mathbb{R}_+$ comes from (6.12).

On the other hand, the end of Proposition 6.5 yields

$$P_t\varphi(x,u) = P_t\varphi(x,0) + \int_0^u \partial_2 P_t\varphi(x,v)\,dv = P_t\varphi(x,0)$$
$$+\int_0^u \left(P_t L\varphi(x,v) - \ell(x,v) \int_E (P_t\varphi(y,0) - P_t\varphi(x,v))\, \beta(x,v;dy) \right) dv, \quad (6.17)$$

and thus the function $u \to P_t\varphi(x,u)$ is continuous for all $x \in E$ and consequently, the assumptions and the Feller property of the semigroup ensure that $v \to \ell(x,v) \int_E (P_t\varphi(y,0) - P_t\varphi(x,v))\, \beta(x,v;dy)$ is continuous. They also ensure that $u \to L\varphi(x,u)$ is continuous and bounded on $[0,a]$ for all a. So, by Proposition 6.5 Part A, the function $v \to P_t L\varphi(x,v)$ is continuous. Consequently, we deduce from (6.17) that the function $u \to P_t\varphi(x,u)$ is differentiable and

$$\frac{\partial P_t\varphi}{\partial u}(x,u) = P_t L\varphi(x,u) - \ell(x,u) \int_E (P_t\varphi(y,0) - P_t\varphi(x,u))\, \beta(x,u;dy).$$

This ends the proof. ∎

Example 6.1 In the case of a Markov jump process, we have

$$N(x, dy, dv) = \lambda(x)\, e^{-\lambda(x)v} Q(x; dy)\, dv.$$

Assume that $\sup_{x \in E} \lambda(x) < +\infty$ and let $f : E \to \mathbb{R}$ be a bounded Borel function. We get

$$Lf(x) = \int_E (f(y) - f(x))\, \lambda(x)\, Q(x; dy), \quad \frac{d}{dt} P_t f = P_t L f = L P_t f.$$

These are the classical Chapman–Kolmogorov equations.

Application to PDMPs

Throughout this section, $(X_t)_{t \geq 0}$ is a PDMP on E, without boundary and with parameters ϕ, λ, Q. The function λ is assumed to be bounded on $E \times [0, a]$ for all $a \in \mathbb{R}_+$.

Denote by $(\mathcal{P}_t)_{t \geq 0}$ its semigroup: $\mathcal{P}_t f(x) = \mathbb{E}_x(f(X_t))$ for every bounded Borel function $f : E \to \mathbb{R}$.

Given a bounded Borel function $f : E \to \mathbb{R}$ such that for all $x \in E$, the function $v \in \mathbb{R}_+ \to f(\phi(x, v))$ is absolutely continuous, define

$$\mathcal{L} f(x) = \partial_\phi f(x) + \int_E (f(y) - f(x))\, \lambda(x)\, Q(x; dy).$$

Let $(Z_t, A_t)_{t \geq 0}$ be the CSMP associated to $(X_t)_{t \geq 0}$ and $(P_t)_{t \geq 0}$ its semigroup. Let $f : E \to \mathbb{R}_+$ be a Borel function and define $\varphi = f \circ \phi$. From $\mathbb{E}_{x,u}(f(\phi(Z_t, A_t))) = \mathbb{E}_{\phi(x,u)}(f(\phi(Z_t, A_t)))$ (Theorem 3.1, Sect. 3.1) and $\partial_2 \varphi(x, u) = \partial_\phi f(\phi(x, u))$, we get

$$P_t \varphi(x, u) = \mathcal{P}_t f(\phi(x, u)), \quad L\varphi(x, u) = \mathcal{L} f(\phi(x, u)),$$

$$P_t L \varphi(x, u) = \mathcal{P}_t \mathcal{L} f(\phi(x, u)), \quad L P_t \varphi(x, u) = \mathcal{L} \mathcal{P}_t f(\phi(x, u)).$$

Therefore, the following result is a consequence of Proposition 6.5.

Proposition 6.7 *Let $f : E \to \mathbb{R}$ be a Borel function such that $f \circ \phi$ is bounded on $E \times [0, a]$ for all $a > 0$.*

(A) Assume that for all $x \in E$, the function $v \in \mathbb{R}_+ \to f \circ \phi(x, v)$ is continuous. Then:

1. *for all $x \in E$, the function $t \in \mathbb{R}_+ \to \mathcal{P}_t f(x)$ is continuous,*
2. *for all $(x, t) \in E \times \mathbb{R}_+$ the function $u \in [0, t^*(x)[\to \mathcal{P}_t f(\phi(x, u))$ is continuous.*

(B) Assume that for all $x \in E$, the function $v \in \mathbb{R}_+ \to f \circ \phi\,(x, v)$ is absolutely continuous and the function $(\partial_\phi f) \circ \phi$ is bounded on $E \times [0, a]$ for all $a > 0$. Then:

1. *for all $(x, t) \in E \times \mathbb{R}_+$,*

$$\mathcal{P}_t f(x) = f(x) + \int_0^t \mathcal{P}_s \mathcal{L} f(x)\, ds,$$

and thus for all $x \in E$, the function $t \in \mathbb{R}_+ \to \mathcal{P}_t f(x)$ is absolutely continuous, and for Lebesgue-almost every t:

$$\frac{\partial}{\partial t}\mathcal{P}_t f(x) = \mathcal{P}_t \mathcal{L} f(x), \tag{6.18}$$

2. *for all $(x,t) \in E \times \mathbb{R}_+$, the function $u \in [0, t^*(x)[\to \mathcal{P}_t f(\phi(x,u))$ is absolutely continuous, and for Lebesgue-almost every $u \in [0, t^*(x)[$,*

$$\mathcal{P}_t \mathcal{L} f(\phi(x,u)) = \mathcal{L}\mathcal{P}_t f(\phi(x,u)). \tag{6.19}$$

As a consequence of Proposition 6.6 we get the following result.

Proposition 6.8 *Assume that the function $x \in E \to \lambda(x) \int_E h(y)\, Q(x; dy)$ is continuous for every bounded and continuous function $h : E \to \mathbb{R}$ and that λ is bounded. Then $(\mathcal{P}_t)_{t\geq 0}$ is a Feller semigroup.*

Let f be a bounded Borel function such that for all $x \in E$, the function $v \in \mathbb{R}_+ \to f \circ \phi(x,v)$ is continuously differentiable and the function $\partial_\phi f$ is bounded.

Then:

1. *for all $x \in E$, the function $t \in \mathbb{R}_+ \to \mathcal{P}_t f(x)$ is continuously differentiable,*
2. *for all $(x,t) \in E \times \mathbb{R}_+$, the function $u \in [0, t^*(x)[\to \mathcal{P}_t f(\phi(x,u))$ is continuously differentiable,*
3. $\frac{\partial}{\partial t}\mathcal{P}_t f = \mathcal{P}_t \mathcal{L} f = \mathcal{L}\mathcal{P}_t f.$

In Propositions 6.7 and 6.8 we do not seek to refine the assumptions but to show how results for CSMPs immediately give results for PDMPs. Namely, the assumption that $v \in \mathbb{R}_+ \to f \circ \phi(x,v)$ is absolutely continuous can be replaced by the assumption that $v \in [0, t^*(x)] \to f \circ \phi(x,v)$ is absolutely continuous.

Chapter 7
A Martingale Approach

Contrary to the usual practice, martingales are not at the heart of our study of PDMPs, but a book devoted to PDMPs cannot fail to mention those related to them. This chapter contains the classic results on these martingales. It is based on the results mentioned in Appendix A (Sect. A.4) about the compensator of a marked point process and the associated stochastic calculus. The reader who is not specifically interested in martingales can skip this chapter without prejudicing the reading of the following ones.

As usual, we begin by looking at CSMPs before moving on to PDMPs. The last part of the chapter provides information on the infinitesimal generator and the extended generator of a Markov process; this educational part is not specific to PDMPs.

We exhibit (local) martingales, generally with respect to the filtration $(\mathcal{F}_t)_{t\geq 0}$ generated by the CSMP. This filtration is also generated by the marked point process (Y, T) (see the end of Sect. 2.4) The results also hold for the completed filtration, and we shall not mention this again. Since on-site jumps are allowed, these filtrations are larger than those generated by a PDMP.

Notation:
$B_t \equiv C_t$ means that the process $t \to B_t - C_t$ is a martingale,
$B_t \equiv_{loc} C_t$ means that the process $t \to B_t - C_t$ is a local martingale.
If nothing is mentioned, the filtration is the filtration $(\mathcal{F}_t)_{t\geq 0}$ generated by the CSMP.

7.1 Martingales Related to Markov Renewal Processes

As an immediate consequence of Proposition A.9, Sect. A.4 we get the following proposition.

C. Cocozza-Thivent, *Markov Renewal and Piecewise Deterministic Processes*, Probability Theory and Stochastic Modelling 100,
https://doi.org/10.1007/978-3-030-70447-6_7

Proposition 7.1 *Let $(Y_n, T_n)_{n\geq 1}$ be a Markov renewal process on E with kernel N and initial distribution $N_0^{x,u}$. Let $(Z_t, A_t)_{t\geq 0}$ be its associated CSMP.*

Then the compensator of the marked point process $p = \sum_{n\geq 1} \delta_{Y_n, T_n}$ with respect to the filtration $(\mathcal{F}_t)_{t\geq 0}$ is $\tilde{p}$ defined as

$$\int_{E\times\mathbb{R}_+} h(y, v, \omega)\, 1_{]0,T_1]}(v)\, \tilde{p}(dy, dv)$$
$$= \int_{E\times\mathbb{R}_+} h(y, v, \omega)\, 1_{\{0<v\leq T_1\}} \frac{1}{1 - N_0^{x,u}(E\times[0, v[)} N_0^{x,u}(dy, dv),$$

and for all $k \geq 1$,

$$\int_{E\times\mathbb{R}_+} h(y, v, \omega)\, 1_{]T_k,T_{k+1}]}(v)\, \tilde{q}(dy, dv)$$
$$= \int_{E\times\mathbb{R}_+} h(y, v + T_k, \omega)\, 1_{\{0<v\leq T_{k+1}-T_k\}} \frac{1}{1 - N(Y_k, E\times[0, v[)} N(Y_k, dy, dv).$$

When $(Z_t, A_t)_{t\geq 0}$ is a simple CSMP with parameters ℓ, α, β, then

$$\tilde{p}(dy, dv) = \ell(Z_v, A_v)\, \beta(Z_v, A_v; dy)\, dv$$
$$+\ 1_{\{\alpha(x)-u=T_1<+\infty\}}\, \beta(x, \alpha(x); dy)\, \delta_{\alpha(x)-u}(dv)$$
$$+ \sum_{k\geq 1} 1_{\{T_k+\alpha(Y_k)=T_{k+1}<+\infty\}}\, \beta(Y_k, \alpha(Y_k); dy)\, \delta_{T_k+\alpha(Y_k)}(dv).$$

The following propositions give examples of how stochastic calculus can be applied.

Recall that $\rho^{x,u}$, defined Sect. 5.1, is the intensity of the marked point process $(Y_k, T_k)_{k\geq 1}$ under $\mathbb{P}_{x,u}$.

Proposition 7.2 *Let $(Y_n, T_n)_{n\geq 1}$ be a Markov renewal process on E associated with a simple CSMP $(Z_t, A_t)_{t\geq 0}$ with parameters ℓ, α, β and initial distribution $N_0^{x,u}$.*

Given B_1 and B_2 two Borel subsets of E, denote by $N_{B_1,B_2}(t)$ the number of jumps from B_1 to B_2 up to time t, that is,

$$N_{B_1,B_2}(t) = 1_{B_1}(x)\, 1_{B_2}(Y_1)\, 1_{]0,t]}(T_1) + \sum_{n\geq 2} 1_{B_1}(Y_{n-1})\, 1_{B_2}(Y_n)\, 1_{]0,t]}(T_n).$$

Then

$$N_{B_1,B_2}(t) \equiv_{loc} \int_0^t 1_{B_1}(Z_v)\, 1_{B_2}(y)\, \ell(Z_v, A_v)\, \beta(Z_v, A_v; B_2)\, dv + \tag{7.1}$$
$$1_{\{0<\alpha(x)-u=T_1\leq t\}} 1_{B_1}(x)\, \beta(x, \alpha(x); B_2) + \sum_{k\geq 1} 1_{\{T_k+\alpha(Y_k)=T_{k+1}\leq t\}} 1_{B_1}(Y_k) \beta(Y_k, \alpha(Y_k); B_2)$$

and

$$\mathbb{E}_{x,u}(N_{B_1,B_2}(t)) = \mathbb{E}_{x,u}\left(\int_0^t 1_{B_1}(Z_v)\,\ell(Z_v, A_v)\,\beta(Z_v, A_v; B_2)\,dv\right)$$
$$+\ 1_{\{0<\alpha(x)-u\le t\}}\, e^{-\int_u^{\alpha(x)} \ell(x,w)\,dw}\, 1_{B_1}(x)\,\beta(x, \alpha(x); B_2)$$
$$+ \int_{E\times\mathbb{R}_+} 1_{\{v+\alpha(y)\le t\}}\, e^{-\int_0^{\alpha(y)} \ell(y,w)\,dw}\, 1_{B_1}(y)\,\beta(y, \alpha(y); B_2)\,\rho^{x,u}(dy, dv).$$

Proof We have

$$N_{B_1,B_2}(t) = \int_{E\times\mathbb{R}_+} h(z, s, \omega)\, 1_{\{s\le t\}}\, p(dz, ds)$$

with $h(z, s, \omega) = 1_{B_1}(Z_{s-}(\omega))\, 1_{B_2}(z)$. The function h is left-continuous, hence predictable. Moreover, $h \in L^1_{loc}(p)$ with $\sigma_n = T_n$ for the localization. Applying Proposition A.10, Sect. A.4, we get the first assertion.

Denote by M_t the difference between the two members of (7.1). Writing $\mathbb{E}_{x,u}(M_{t\wedge T_n}) = 0$ and letting n tend to infinity, we get the second assertion. ∎

The following proposition provides the interpretation expected for the first term of the second member of (7.1).

Proposition 7.3 *Let $(Y_n, T_n)_{n\ge1}$ be a Markov renewal process on E associated with a simple CSMP $(Z_t, A_t)_{t\ge0}$ with parameters ℓ, α, β and initial distribution $N_0^{x,u}$.*

Given B_1 and B_2 two Borel subsets of E, define $N^c_{B_1,B_2}(t)$ as the number of jumps from B_1 to B_2 up to time t that are not caused by the Dirac measures (that is, the "purely random" jumps):

$$N^c_{B_1,B_2}(t) = 1_{B_1}(x)\, 1_{B_2}(Y_1)\, 1_{[0,t]}(T_1)\, 1_{\{T_1<\alpha(x)-u\}}$$
$$+ \sum_{n\ge2} 1_{B_1}(Y_{n-1})\, 1_{B_2}(Y_n)\, 1_{[0,t]}(T_n)\, 1_{\{T_n-T_{n-1}<\alpha(Y_{n-1})\}}.$$

Then

$$N^c_{B_1,B_2}(t) \equiv_{loc} \int_0^t 1_{B_1}(Z_s)\,\ell(Z_s, A_s)\,\beta(Z_s, A_s; B_2)\,ds$$

and

$$\mathbb{E}_{x,u}(N^c_{B_1,B_2}(t)) = \mathbb{E}_{x,u}\left(\int_0^t 1_{B_1}(Z_s)\,\ell(Z_s, A_s)\,\beta(Z_s, A_s; B_2)\,ds\right).$$

Proof We proceed as in the proof of Proposition 7.2. We apply Proposition A.10, Sect. A.4 with $h(z, s, \omega) = 1_{B_1}(Z_{s-}(\omega))\, 1_{B_2}(z)\, 1_{\{A_{s-}(\omega)\ne\alpha(Z_{s-}(\omega))\}}$ and then we notice that for Lebesgue-almost every s, $Z_{s-}(\omega) = Z_s(\omega)$, $A_{s-}(\omega) = A_s(\omega)$ and $A_{s-}(\omega) \ne \alpha(Z_{s-}(\omega))$. We conclude with the same arguments as in the proof of Proposition 7.2. ∎

Define p_c as the marked point process generated by the jumps that are not caused by the Dirac measures:

$$p_c(dy, dv) = 1_{\{T_1 < \alpha(x) - u\}} \, \delta_{Y_1, T_1}(dy, dv) + \sum_{n \geq 2} 1_{\{T_n - T_{n-1} < \alpha(Y_{n-1})\}} \, \delta_{Y_n, T_n}(dy, dv).$$

The first part of Proposition 7.3 states that $\tilde{p}_c(dy, dv) = \ell(Z_v, A_v)\, \beta(Z_v, A_v; dy)\, dv$ is the compensator of p_c. The second part is (5.15), Sect. 5.2.

7.2 Martingales Related to Simple CSMPs

In this section, let $(Z_t, A_t)_{t \geq 0}$ be a simple CSMP with parameters ℓ, α, β. Recall that

$$t^*(x) = \alpha(x) \wedge t^*_\ell(x), \quad \text{where } t^*_\ell(x) = \sup\{v > 0 : \int_0^v \ell(x, w)\, dw < +\infty\},$$

$$\hat{E} = \{(x, u) \in E \times \mathbb{R}_+ : u \leq t^*(x)\}.$$

Proposition 7.4 *Let $(Z_t, A_t)_{t \geq 0}$ be a simple CSMP with parameters ℓ, α, β, with $(Y_n, T_n)_{n \geq 1}$ its associated Markov renewal process and $p = \sum_{n \geq 1} \delta_{Y_n, T_n}$.*

Let $\psi : \hat{E} \times \mathbb{R}_+ \to \mathbb{R}$ be a Borel function. Define

$$\tilde{\mathcal{B}}\psi(x, v, \omega) = \psi(x, 0, v) - \psi(Z_{v-}(\omega), A_{v-}(\omega), v).$$

Assume that the function $v \in [0, T_1] \cap \mathbb{R}_+ \to \psi(Z_0, A_0 + v, v)$ is absolutely continuous, the function $v \in [0, t^*(x)] \cap \mathbb{R}_+ \to \psi(x, v, a + v)$ is absolutely continuous for all $(x, a) \in E \times \mathbb{R}_+$, and $(x, v, \omega) \to \tilde{\mathcal{B}}\psi(x, v, \omega)\, 1_{]0,t]}(v) \in L^1_{loc}(p)$ for all $t \in \mathbb{R}_+$.

Then

$$\psi(Z_t, A_t, t) - \psi(Z_0, A_0, 0) \equiv_{loc} \int_0^t \mathcal{X}\psi(Z_v, A_v, v)\, dv$$
$$+ \int_{E \times]0,t]} (\psi(y, 0, v) - \psi(Z_v, A_v, v))\, \ell(Z_v, A_v)\, \beta(Z_v, A_v; dy)\, dv$$
$$+ \sum_{v: 0 < v \leq t} 1_{\{A_{v-} = \alpha(Z_{v-})\}} \int_E (\psi(y, 0, v) - \psi(Z_{v-}, A_{v-}, v))\, \beta(Z_{v-}, A_{v-}; dy).$$

If $\tilde{\mathcal{B}}\psi\, 1_{[0,t]} \in L^1(p)$, then $\equiv_{loc}$ is replaced by $\equiv$.

Proof Applying Lemma 6.2, Sect. 6.1 we obtain

$$\psi(Z_t, A_t, t) - \psi(0, Z_0, A_0) = \int_0^t \mathcal{X}\psi(Z_v, A_v, v)\, dv + S_t,$$

where $S_t = \int_{E\times]0,t]} \tilde{\mathcal{B}}\psi(y, v, \omega)\, p(dy, dv)$. Since $\tilde{\mathcal{B}}\psi$ is left-continuous, hence predictable, we get from Proposition A.10, Sect. A.4 and Proposition 7.1 that

$$S_t \equiv_{loc} \int_{E\times]0,t]} (\psi(y, 0, v) - \psi(Z_{v_-}, A_{v_-}, v))\, \tilde{p}(dy, dv).$$

The result follows from the two following remarks: $Z_{v_-} = Z_v, A_{v_-} = A_v$ for Lebesgue-almost every v, and $A_{v-} = \alpha(Z_{v-})$ means either $v = T_1 = \alpha(Z_0) - A_0$ or there exists $k \geq 1$ such that $v = T_k + \alpha(Y_k)$. ■

Corollary 7.1 *Let $(Z_t, A_t)_{t\geq 0}$ be a simple CSMP with parameters ℓ, α, β, and $(Y_n, T_n)_{n\geq 1}$ its associated Markov renewal process. Recall $p = \sum_{n\geq 1} \delta_{Y_n, T_n}$.*

Let $\varphi : \hat{E} \to \mathbb{R}$ be a Borel function. Define

$$\mathcal{B}\varphi(x, v, \omega) = \varphi(x, 0) - \varphi(Z_{v-}(\omega), A_{v-}(\omega)).$$

Assume that the function $v \in [0, T_1] \cap \mathbb{R}_+ \to \varphi(Z_0, A_0 + v)$ is absolutely continuous, the function $v \in [0, t^*(x)] \cap \mathbb{R}_+ \to \varphi(x, v)$ is absolutely continuous for all $x \in E$, and $(x, v, \omega) \to \mathcal{B}\varphi(x, v, \omega)\, 1_{]0,t]}(v) \in L^1_{loc}(p)$ for any $t \in \mathbb{R}_+$.

Then

$$\begin{aligned}\varphi(Z_t, A_t) - \varphi(Z_0, A_0) &\equiv_{loc} \int_0^t L\varphi(Z_v, A_v)\, dv \\ &+ \sum_{v:0<v\leq t} 1_{\{A_{v-}=\alpha(Z_{v-})\}} \int_E (\varphi(y, 0) - \varphi(Z_{v-}, A_{v-}))\, \beta(Z_{v-}, A_{v-}; dy),\end{aligned}$$

with

$$L\varphi(x, u) = \partial_2\varphi(x, u) + \int_E (\varphi(y, 0) - \varphi(x, u))\, \ell(x, u)\, \beta(x, u; dy).$$

If $\mathcal{B}\varphi\, 1_{]0,t]} \in L^1(p)$, then $\equiv_{loc}$ is replaced by $\equiv$.

7.3 Martingales Related to Parametrized PDMPs

In this section we consider only PDMPs with parameters $\phi, \lambda, Q, \alpha, q$. Since we do not introduce the functions g_i $(i \in I)$, unlike what is done in Theorem 6.1, Sect. 6.2, introducing the partition $(E_i)_{i\in I}$ is not useful most of the time. However, when needed, as in Example 3.10, Sect. 3.4, you just have to replace $q(X_{T_{k+1}-}; dy)$ by $\sum_{i\in I} 1_{E_i}(X_{T_k})\, q_i(X_{T_{k+1}-}; dy)$ in formulas (7.2) and (7.3) below.

Applying Proposition 7.4 with $A_0 = 0$ and $\psi(x, v, w) = g(\phi(x, v), w)$, we get the following result.

Proposition 7.5 *Let $(X_t)_{t\geq 0}$ be a PDMP on E with parameters $\phi, \lambda, Q, \alpha, q$. Let $(Y_n, T_n)_{n\geq 0}$ be an associated Markov renewal process and $p = \sum_{n\geq 1} \delta_{Y_n, T_n}$. Define*

$$t^*(x) = \alpha(x) \wedge t_1^*(x) \text{ where } t_1^*(x) = \sup\{v > 0 : \int_0^v \lambda(\phi(x, w))\, dw < +\infty\}.$$

Let $g : E \times \mathbb{R}_+ \to \mathbb{R}$ be a Borel function. Assume that the function $v \in [0, t^*(x)] \cap \mathbb{R}_+ \to g(\phi(x, v), a + v)$ is absolutely continuous for all $(x, a) \in E \times \mathbb{R}_+$ and the function $(x, v, \omega) \to (g(x, v) - g(X_{v-}(\omega), v))\, 1_{]0,t]}(v) \in L^1_{loc}(p)$ for all $t \in \mathbb{R}_+$. Then

$$g(X_t, t) - g(X_0, 0) \equiv_{loc}$$
$$\int_0^t \partial_{t,\phi}(X_v, v)\, dv + \int_{E\times]0,t]} (g(y, v) - g(X_v, v))\, \lambda(X_v)\, Q(X_v; dy)\, dv$$
$$+ \sum_{k\geq 0} 1_{\{T_{k+1} = T_k + \alpha(Y_k) \leq t\}} \int_E (g(y, T_{k+1}) - g(X_{T_{k+1}-}, T_{k+1}))\, q(X_{T_{k+1}-}; dy). \quad (7.2)$$

If $(x, v, \omega) \to (g(x, v) - g(X_{v-}(\omega), v))\, 1_{]0,t]}(v) \in L^1(p)$ for all $t \in \mathbb{R}_+$, then $\equiv_{loc}$ is replaced by $\equiv$.

Recall that the notations $\equiv_{loc}$ and $\equiv$ are related to the filtration $(\mathcal{F}_t)_{t\geq 0}$ generated by the Markov renewal process. It is bigger than the filtration $(\mathcal{G}_t)_{t\geq 0}$ generated by the PDMP.

Remark 7.1 Denote by $(\mathcal{G}_t)_{t\geq 0}$ the filtration generated by $(X_t)_{t\geq 0}$. Consider a PDMP with closed boundary Γ and parameters $\phi, \lambda, Q, \Gamma, q$. The event $\{X_{v-} \in \Gamma\}$ means that there exists $k \geq 0$ such that $v = T_{k+1} = T_k + \alpha(Y_k)$. Therefore, the last term in (7.2) is

$$\sum_{k\geq 0} 1_{\{T_{k+1} = T_k + \alpha(Y_k) \leq t\}} \int_E (g(y, T_{k+1}) - g(X_{T_{k+1}-}, T_{k+1}))\, q(X_{T_{k+1}-}; dy)$$

$$= \sum_{v : v \leq t} 1_{\{X_{v-} \in \Gamma\}} \int_E (g(y, v) - g(X_{v-}, v)\, q(X_{v-}; dy).$$

Hence all terms in (7.2) are $\mathcal{G}_t$ measurable, and the notations $\equiv_{loc}$ and $\equiv$ can be considered to be related to the filtration $(\mathcal{G}_t)_{t\geq 0}$.

Corollary 7.2 *Let $(X_t)_{t\geq 0}$ be a PDMP on E with parameters $\phi, \lambda, Q, \alpha, q$ and $(Y_n, T_n)_{n\geq 0}$, p, t^* as in Proposition 7.5.*

Let $f : E \to \mathbb{R}$ be a Borel function such that the function $v \in [0, t^*(x)] \cap \mathbb{R}_+ \to f(\phi(x, v))$ is absolutely continuous for all $x \in E$ and $(x, v, \omega) \to (f(x) - f(X_{v-}(\omega)))\, 1_{]0,t]}(v) \in L^1_{loc}(p)$ for all $t \in \mathbb{R}_+$.

Define

$$\mathcal{L}f(x) = \partial_\phi f(x) + \int_E (f(y) - f(x))\, \lambda(x)\, Q(x; dy).$$

Then

$$f(X_t) - f(X_0) \equiv_{loc} \int_0^t \mathcal{L}f(X_v)\, dv + \sum_{k\geq 0} 1_{\{T_{k+1}=T_k+\alpha(Y_k)\leq t\}} \int_E (f(y) - f(X_{T_{k+1}-}))\, q(X_{T_{k+1}-}; dy). \tag{7.3}$$

If $(x, v, \omega) \to (f(x) - f(X_{v-}(\omega)))\, 1_{]0,t]}(v) \in L^1(p)$, then $\equiv_{loc}$ is replaced by $\equiv$.

If $(X_t)_{t\geq 0}$ is a PDMP with closed boundary Γ and parameters $\phi, \lambda, Q, \Gamma, q$, then

$$f(X_t) - f(X_0) \equiv_{loc} \int_0^t \mathcal{L}f(X_v)\, dv + \sum_{v:v\leq t} 1_{\{X_{v-}\in\Gamma\}} \int_E (f(y) - f(X_{v-}))\, q(X_{v-}; dy),$$

and the notations $\equiv_{loc}$ and $\equiv$ can be considered to be related to the filtration $(\mathcal{G}_t)_{t\geq 0}$ generated by the PDMP.

We choose to write $B_t \equiv C_t$ (respectively $B_t \equiv_{loc} C_t$), since writing explicitly the martingale (respectively local martingale) $M_t = B_t - C_t$ does not seem useful to us. That is not the choice of other authors. M. H. A. Davis considers PDMPs with closed boundary Γ and parameters $\phi, \lambda, Q, \Gamma, q$. He calls the formula

$$f(X_t) - f(X_0) = \int_0^t \mathcal{L}f(X_s)\, ds + \sum_{s:s\leq t} 1_{\{X_{s-}\in\Gamma\}} \int_E (f(y) - f(X_{s-}))\, q(X_{s-}; dy) + \int_{E\times\mathbb{R}_+} (f(y) - f(X_{s-}(\omega))\, 1_{\{s\leq t\}}(p(dy, ds) - \tilde{p}(dy, ds))$$

the PDP differential formula ([51] Theorem 31.3 page 83). J. Jacobsen, who studies PDMPs without boundary, calls it Itô's formula ([66] Sect. 7.6).

7.4 Generators

Define $B(E)$ as the set of bounded Borel functions $f : E \to \mathbb{R}$ endowed with the norm $||f||_\infty = \sup_{x\in E} |f(x)|$. It is a Banach space.

Consider a Markov process $(X_t)_{t\geq 0}$ taking values in E, with semigroup $(P_t)_{t\geq 0}$: $P_t f(x) = \mathbb{E}_x(f(X_t))$ for any $x \in E$, $f \in B(E)$.

Infinitesimal Generator

In this subsection (and only in this one), the limits are taken with respect to the norm $||\cdot||_\infty$, and the integrals are Riemann integrals in $B(E)$.

Definition 7.1 Given $B_c(E) \subset B(E)$, the semigroup $(P_t)_{t\geq 0}$ is strongly continuous on B_c if

$$||P_t f - f||_\infty \underset{t\to 0}{\longrightarrow} 0 \tag{7.4}$$

for all $f \in B_c$.

The largest subset of $B(E)$ on which the semigroup $(P_t)_{t\geq 0}$ is strongly continuous is the set $B_0(E)$ of functions $f \in B(E)$ for which (7.4) holds. It is a closed subset of $B(E)$, hence a Banach space.

Definition 7.2 Define the infinitesimal generator $\widehat{\mathcal{U}}$ of the semigroup $(P_t)_{t\geq 0}$ as follows:

- its domain $\mathcal{D}(\widehat{\mathcal{U}})$ is the set of functions $f \in B_0(E)$ such that $\lim_{t\to 0}(P_t f - f)/t$ exists for the norm $||\cdot||_\infty$,
- for $f \in \mathcal{D}(\widehat{\mathcal{U}})$, define $\widehat{\mathcal{U}} f$ as

$$\widehat{\mathcal{U}} f = \lim_{t\to 0} \frac{1}{t}(P_t f - f).$$

Proposition 7.6 *(Ethier and Kurtz [57] Proposition 1.5 page 9)*
Let $f \in \mathcal{D}(\widehat{\mathcal{U}})$. Then $P_t f \in \mathcal{D}(\widehat{\mathcal{U}})$ for all $t \in \mathbb{R}_+$, $t \to P_t f$ is differentiable for the norm $||\cdot||_\infty$, and

$$\frac{d}{dt}P_t f = P_t \widehat{\mathcal{U}} f = \widehat{\mathcal{U}} P_t f,$$

$$P_t f = f + \int_0^t P_s \hat{\mathcal{U}} f \, ds. \tag{7.5}$$

Generally, for a PDMP with boundary, the set of bounded k-times differentiable functions ($k \in \mathbb{N}^*$) with compact support is not a subset of $B_0(E)$, and Proposition 7.6 can be applied only to a small set of functions with which it is difficult to work. To understand this, consider a CSMP with parameters ℓ, α, β. Let $\varphi : E \times \mathbb{R}_+ \to \mathbb{R}$ be a k-times differentiable function with compact support. Let $x \in E$ be such that $\alpha(x) < +\infty$. From

$$||P_t\varphi - \varphi||_\infty \geq \lim_{\substack{u\to\alpha(x)\\ u<\alpha(x)}} |\mathbb{E}_{x,u}(\varphi(Z_t, A_t) - \varphi(x,u))|,$$

$$\mathbb{P}_{x,u}(t < T_1) \leq 1_{\{t<\alpha(x)-u\}} \underset{\substack{u\to\alpha(x)\\ u<\alpha(x)}}{\longrightarrow} 0,$$

we get

$$||P_t\varphi - \varphi||_\infty \geq \lim_{\substack{u\to\alpha(x)\\ u<\alpha(x)}} \left| \int_{E\times\mathbb{R}_+} \mathbb{E}_y(\varphi(Z_{t-v}, A_{t-v}) - \varphi(x,u))\, 1_{\{v\leq t\}}\, N_0^{x,u}(dy, dv) \right|$$
$$= \left| \int_E \mathbb{E}_y(\varphi(Z_{t-}, A_{t-}) - \varphi(x, \alpha(x)))\, \beta(x, \alpha(x); dy) \right|.$$

Hence

$$\liminf_{t\to 0} ||P_t\varphi - \varphi||_\infty \geq \left| \int_E (\varphi(y,0) - \varphi(x,\alpha(x)))\, \beta(x,\alpha(x); dy) \right|.$$

Therefore functions $\varphi \in B_0(E)$ must satisfy

$$\int (\varphi(y,0) - \varphi(x,, \alpha(x))\, \beta(x, \alpha(x); dy) = 0$$

for all $x \in E$ such that $\alpha(x) < +\infty$.

Dynkin's formula

From formula (7.5), we get the following Dynkin's formula, which holds for all $f \in \mathcal{D}(\widehat{\mathcal{U}})$, $t > 0$, $x \in E$:

$$\mathbb{E}_x(f(X_t)) = f(x) + \mathbb{E}_x\left(\int_0^t \widehat{\mathcal{U}} f(X_s)\, ds\right). \tag{7.6}$$

The difficulty is to check that a function f belongs to $\mathcal{D}(\widehat{\mathcal{U}})$. However, Propositions 7.7 and 7.8 below, which are derived from Corollary 7.1, Sect. 7.2 and Corollary 7.2, Sect. 7.3, provide Dynkin-type formulas.

Proposition 7.7 *Let $(Z_t, A_t)_{t\geq 0}$ be a simple CSMP on E with parameters ℓ, α, β, $(Y_n, T_n)_{n\geq 1}$ its associated Markov renewal process, and $p = \sum_{n\geq 1} \delta_{Y_n, T_n}$.*

Let $x \in E$ and u satisfying $\mathbb{P}_x(T_1 > u) > 0$, $\varphi : E \times \mathbb{R}_+ \to \mathbb{R}$ a Borel function such that:

1. *for all $y \in E$, the function $v \in [0, \alpha(y)] \cap \mathbb{R}_+ \to \varphi(y, v)$ is absolutely continuous,*
2. *for all $y \in E$ such that $\alpha(y) < +\infty$, we have*

$$\varphi(y, \alpha(y)) = \int_E \varphi(z, 0)\, \beta(y, \alpha(y); dz)$$

,

3. *under $\mathbb{P}_{x,u}$, $(z, s, \omega) \to (\varphi(z, 0) - \varphi(Z_{s-}(\omega), A_{s-}(\omega)))\, 1_{]0,t]}(s) \in L^1(p)$ for all $t \in \mathbb{R}_+$.*

Define

$$L\varphi(x, v) = \partial_2\varphi(x, v) + \int_E (\varphi(y, 0) - \varphi(x, v))\, \ell(x, v)\, \beta(x, v; dy). \tag{7.7}$$

Then for all $t \geq 0$,

$$\mathbb{E}_{x,u}(\varphi(Z_t, A_t)) = \varphi(x, u) + \mathbb{E}_{x,u}\left(\int_0^t L\varphi(Z_v, A_v)\, dv\right).$$

Proposition 7.8 *Let $(X_t)_{t\geq 0}$ be a PDMP on E with parameters $\phi, \lambda, Q, \alpha, q$, $(Y_n, T_n)_{n\geq 1}$ an associated Markov renewal process, $p = \sum_{n\geq 1} \delta_{Y_n,T_n}$. Define*

$$\Gamma^* = \{\phi(y, \alpha(y)) : y \in E, \alpha(y) < +\infty, \int_0^{\alpha(y)} \lambda(\phi(y, v))\, dv < +\infty\}.$$

Let $x \in E$, $f : E \to \mathbb{R}$ a Borel function such that:

1. *for all $y \in E$, the function $v \in [0, \alpha(y)] \cap \mathbb{R}_+ \to f(\phi(y, v))$ is absolutely continuous,*
2. *for all $y \in \Gamma^*$, $f(y) = \int_E f(z)\, q(y; dz)$,*
3. *under $\mathbb{P}_x$, $(y, s, \omega) \to (f(y) - f(X_{s-}(\omega)))\, 1_{]0,t]}(s) \in L^1(p)$ for all $t \in \mathbb{R}_+$.*

Define

$$\mathcal{L}f(y) = \partial_\phi f(y) + \int_E (f(z) - f(y))\, \lambda(x)\, Q(y; dz). \tag{7.8}$$

Then for all $t \geq 0$,

$$\mathbb{E}_x(f(X_t)) = f(x) + \mathbb{E}_x\left(\int_0^t \mathcal{L}f(X_v)\, dv\right).$$

Actually, Proposition 7.7 (respectively Proposition 7.8), with Condition 3 replaced by conditions $\mathbb{E}_{x,u}(\sum_{n\geq 1} 1_{\{T_n\leq t\}}) < +\infty$ (respectively $\mathbb{E}_x(\sum_{n\geq 1} 1_{\{T_n\leq t\}}) < +\infty$) for all $t \geq 0$ and φ (respectively f) bounded, can also be derived from Corollary 6.2, Sect. 6.1 (respectively Corollary 6.3, Sect. 6.2). As a matter of fact, their assumptions are stronger than above, but those are the ones that can be checked in practice.

In fact, formula (7.6) for all $x \in E$ and $t \geq 0$ is equivalent to a martingale property.

Proposition 7.9 *Let $(X_t)_{t\geq 0}$ be a Markov process on E with semigroup $(P_t)_{t\geq 0}$. Let L be an operator on $\mathcal{D}(L) \subset B(E)$, and let $f \in \mathcal{D}(L)$ be such that $\int_0^t Lf(X_s)\, ds$ exists and is $\mathbb{P}_x$-integrable for all $(x, t) \in E \times \mathbb{R}_+$. Then the following properties are equivalent:*

1. *for all $(x, t) \in E \times \mathbb{R}_+$, $\mathbb{E}_x(f(X_t)) = f(x) + \mathbb{E}_x\left(\int_0^t Lf(X_s)\, ds\right)$,*
2. *for all $x \in E$, $t \to M_t = f(X_t) - f(X_0) - \int_0^t Lf(X_s)\, ds$ is a $\mathbb{P}_x$-martingale for the filtration generated by the Markov process.*

Proof By writing $\mathbb{E}_x(M_t) = \mathbb{E}_x(M_0)$, we get assertion 1 from assertion 2.

Now assume that assertion 2 holds. Let $(\mathcal{G}_t)_{t\geq 0}$ be the filtration generated by the Markov process. We get

$$\begin{aligned}\mathbb{E}(M_{t+u} - M_t \mid \mathcal{G}_t) &= \mathbb{E}(f(X_{t+u}) \mid \mathcal{G}_t) - f(X_t) - \int_0^u \mathbb{E}(Lf(X_{t+s}) \mid \mathcal{G}_t)\, ds \\ &= \mathbb{E}_{X_t}(f(X_u)) - f(X_t) - \mathbb{E}_{X_t}\left(\int_0^u Lf(X_s)\, ds\right) = 0.\end{aligned}$$

This concludes the proof. ■

Extended Generator

Definition 7.3 Let $(X_t)_{t\geq 0}$ be a Markov process taking values in E. Define $\mathcal{D}(\mathcal{U})$ as the set of Borel functions $f : E \to \mathbb{R}$ that satisfy the following property: there exists a Borel function $\psi : E \to \mathbb{R}$ such that for all $x \in E$, $\mathbb{P}_x$-almost surely the function $s \to \psi(X_s)$ is Lebesgue integrable on $[0, t]$ for all $t \geq 0$ and

$$t \to f(X_t) - f(X_0) - \int_0^t \psi(X_s)\, ds$$

is a local martingale with respect to the filtration generated by the Markov process. Define $\mathcal{U} f = \psi$. The operator $(\mathcal{D}(\mathcal{U}), \mathcal{U})$ is the extended generator of the Markov process.

From (7.6) and Proposition 7.9 we deduce that if $(\mathcal{D}(\widehat{\mathcal{U}}), \widehat{\mathcal{U}})$ is the infinitesimal generator and $(\mathcal{D}(\mathcal{U}), \mathcal{U})$ is the extended generator, then $\mathcal{D}(\widehat{\mathcal{U}}) \subset \mathcal{D}(\mathcal{U})$ and $\mathcal{U} = \widehat{\mathcal{U}}$ on $\mathcal{D}(\widehat{\mathcal{U}})$.

From Corollary 7.1, Sect 7.2 and Corollary 7.2, Sect. 7.3, we get the following two propositions.

Proposition 7.10 *Let $(Z_t, A_t)_{t\geq 0}$ be a simple CSMP on E with parameters ℓ, α, β, $(Y_n, T_n)_{n\geq 1}$ its associated Markov renewal process, and $p = \sum_{n\geq 1} \delta_{Y_n, T_n}$.*

Let $(\mathcal{D}(\mathcal{U}), \mathcal{U})$ be the extended generator of the CSMP. Define $\mathcal{D}$ as the set of Borel functions $\varphi : E \times \mathbb{R}_+ \to \mathbb{R}$ such that:

1. *for all $x \in E$, the function $v \in [0, \alpha(x)] \cap \mathbb{R}_+ \to \varphi(x, v)$ is absolutely continuous,*
2. *for all $x \in E$ such that $\alpha(x) < +\infty$, we have*

$$\varphi(x, \alpha(x)) = \int_E \varphi(y, 0)\, \beta(x, \alpha(x); dy)$$

,

3. *for all $(x, u, t) \in E \times \mathbb{R}_+^2$ such that $\mathbb{P}_x(T_1 > u) > 0$, we have $(z, s, \omega) \to (\varphi(z, 0) - \varphi(Z_{s-}(\omega), A_{s-}(\omega)))1_{]0,t]}(s) \in L^1_{loc}(p)$ under $\mathbb{P}_{x,u}$.*

Then $\mathcal{D} \subset \mathcal{D}(\mathcal{U})$, and on $\mathcal{D}$ we have $\mathcal{U} = L$, where L is as in (7.7).

Proposition 7.11 *Let $(X_t)_{t\geq 0}$ be a PDMP on E with parameters $\phi, \lambda, Q, \alpha, q$, $(Y_n, T_n)_{n\geq 1}$ an associated Markov renewal process, $p = \sum_{n\geq 1} \delta_{Y_n,T_n}$.*

Let $(\mathcal{D}(\mathcal{U}), \mathcal{U})$ be the extended generator of the PDMP. Define $\mathcal{D}_0$ as the set of Borel functions $f : E \to \mathbb{R}$ such that:

1. *for all $x \in E$, the function $v \in [0, \alpha(x)] \cap \mathbb{R}_+ \to f(\phi(x, v))$ is absolutely continuous,*
2. *for all $x \in \Gamma^*$, $f(x) = \int_E f(y) q(x; dy)$,*
3. *for all $(x, t) \in E \times \mathbb{R}_+$, we have $(z, s, \omega) \to (f(z) - f(X_{s-}(\omega))\, 1_{]0,t]}(s) \in L^1_{loc}(p)$ under $\mathbb{P}_x$.*

Then $\mathcal{D}_0 \subset \mathcal{D}(\mathcal{U})$, and on $\mathcal{D}_0$ we have $\mathcal{U} = \mathcal{L}$, where $\mathcal{L}$ is as in (7.8).

In [51], M. H. A. Davis describes exactly $\mathcal{D}(U)$ for a PDMP with a boundary Γ that is a subset of the topological state space (or without boundary) when the flow is governed by an ODE and on-site jumps are not allowed (Theorem (26.14) page 69). The difference with $\mathcal{D}_0$ is that assertion 1 is replaced by the following: the function $v \in [0, t^*(x)[\to f(\phi(x, v))$ is absolutely continuous for all $x \in E$, and for all $x \in \Gamma$, $f(x) = \lim_{t\downarrow 0} f(\phi(x, -t))$. This is the same idea as in the introduction of functions g_i in Theorem 6.1, Sect. 6.2.

Chapter 8
Stability

The purpose of this chapter is to study the relationships between the stability of different processes, especially between the driving Markov chain and the CSMP and consequently between this driving chain and the PDMP. A large proportion of this work is inspired by Jacod [67]. Steady-sate intensities are also derived and applications are given. No special assumption about the structure of the interarrival distributions is required.

The asymptotic behavior of PDMPs as a result of its semiregenerative property is recalled in Sect. 8.6 in the nonarithmetic case.

Definition 8.1 Given a Markov chain $Y = (Y_n)_{n \geq 0}$ taking values in E, a nonnegative σ-finite measure ν on E is said to be invariant (or stationary) for Y if we have

$$\int_E \mathbb{E}(f(Y_n) \mid Y_0 = y)\, \nu(dy) = \int_E f(y)\, \nu(dy) \tag{8.1}$$

for every Borel function $f : E \to \mathbb{R}_+$ and $n \geq 1$.

Given a Markov process $(X_t)_{t \geq 0}$ taking values in E, a nonnegative σ-finite measure m on E is said to be invariant (or stationary) for $(X_t)_{t \geq 0}$ if we have

$$\int_E \mathbb{E}(f(X_t) \mid X_0 = x)\, m(dx) = \int_E f(x)\, m(dx)$$

for every Borel function $f : E \to \mathbb{R}_+$ and $t \geq 0$.

If ν (respectively m) is invariant, then $c\nu$ (respectively cm) is invariant for all $c > 0$. Therefore, an invariant measure can be unique only up to a multiplicative constant.

C. Cocozza-Thivent, *Markov Renewal and Piecewise Deterministic Processes*, Probability Theory and Stochastic Modelling 100,
https://doi.org/10.1007/978-3-030-70447-6_8

When Y is a Markov chain with probability kernel P, (8.1) holds for all $n \geq 1$ if and only if it holds for $n = 1$, that is,

$$\nu P = \nu \tag{8.2}$$

with the notation of (2.2), Sect. 2.1.

The formula that replaces (8.2) for a Markov process is $\int_E \mathcal{L}f(y)\, m(dy) = 0$ for all f in the domain $\mathcal{D}(\mathcal{L})$ of the (infinitesimal or extended) generator. In the case of a PDMP, there is no general expression for it. Even if we limit ourselves to parametrized PDMPs, the presence of Dirac measures in the kernel makes the generator domain difficult to use, and a specific work, carried out in Davis [51], Sect. 34, is necessary to get a more general formula than $\int_E \mathcal{L}f(y)\, m(dy) = 0$ for all $f \in \mathcal{D}(\mathcal{L})$. We give more details below, at the end of Sect. 8.4. We have chosen another approach that allows us to make no assumption about the structure of interarrival distributions.

All the measures introduced in this chapter are assumed to be σ-finite even if it is not mentioned.

Let $(Z_t, A_t)_{t\geq 0}$ be a CSMP with a proper kernel N, that is, $N(x, E \times \mathbb{R}_+) = 1$ for all $x \in E$. Hence the measure $N_0^{x,u}$ defined in (2.7), Sect. 2.3, is a probability distribution on $\mathbb{R}_+$. Recall that $X_0 = x$, $A_0 = u$ means that the probability distribution of (Y_1, T_1) is $N_0^{x,u}$. We defined $\mathbb{E}_{x,u}(\cdot\,) = \mathbb{E}(\cdot \mid Z_0 = x, A_0 = u)$ and $\mathbb{E}_x = \mathbb{E}_{x,0}$. If $(Z_t, A_t)_{t\geq 0}$ is a CSMP associated with a PDMP $(X_t)_{t\geq 0}$, we have also $\mathbb{E}_x(\cdot\,) = \mathbb{E}(\cdot \mid X_0 = x)$.

A measure m_0 is invariant for $(Z_t, A_t)_{t\geq 0}$ if

$$\int_{E\times\mathbb{R}_+} E_{x,u}(\varphi(Z_t, A_t))\, m_0(dx, du) = \int_{E\times\mathbb{R}_+} \varphi(x, u)\, m_0(dx, du)$$

for all $t > 0$ and every Borel function $\varphi : E \times \mathbb{R}_+ \to \mathbb{R}_+$.

In Chap. 2 we defined $\bar{Z}_t$ and W_t as follows:

$$\bar{Z}_t = Y_{n+1}, \quad W_t = T_{n+1} - t \quad \text{on } T_n \leq t < T_{n+1}.$$

Throughout this chapter, m_0 is a nonnegative measure on $E \times \mathbb{R}_+$. Measures m and m_1 are defined as

$$m(dx, du, dy, dv) = N_0^{x,u}(dy, dv)\, m_0(dx, du), \tag{8.3}$$

$$m_1(dy, dv) = \int_{\{(x,u)\in E\times\mathbb{R}_+\}} m(dx, du, dy, dv). \tag{8.4}$$

The following result explains the role played by measures m and m_1. It is an immediate consequence of Corollary 2.2, end of Sect. 2.4.

Proposition 8.1 *Assume that m_0 is an invariant measure for $(Z_t, A_t)_{t\geq 0}$. Then*

$$\int_{E\times\mathbb{R}_+} \mathbb{E}_{x,u}(f(Z_t, A_t, \bar{Z}_t, W_t))\, m_0(dx, du)$$

$$\int_{(E\times\mathbb{R}_+)^2} f(x, u, y, v)\, N_0^{x,u}(dy, dv)\, m_0(dx, du)$$

for every Borel function $f : (E \times \mathbb{R}_+)^2 \to \mathbb{R}_+$. In particular,

$$\int_{E\times\mathbb{R}_+} \mathbb{E}_{x,u}(\varphi(\bar{Z}_t, W_t))\, m_0(dx, du) = \int_{E\times\mathbb{R}_+} \varphi(y, v)\, m_1(dy, dv)$$

for every Borel function $f : E \times \mathbb{R}_+ \to \mathbb{R}_+$.

Hence if $(Z_t, A_t)_{t\geq 0}$ is in a steady state, it is not only the probability distribution of (Z_t, A_t) that does not depend on t, but also the probability distribution of $(Z_t, A_t, \bar{Z}_t, W_t)$.

8.1 Connections Between Invariant Measures of a CSMP and those of Its Driving Chain

Invariant Measure for the CSMP From That of Its Driving Chain

Let $(Z_t, A_t)_{t\geq 0}$ be a CSMP with kernel N and $(Y_n, T_n)_n$ its associated Markov renewal process. Let ν be an invariant measure for the driving Markov chain $(Y_n)_n$. Taking into account equation (2.22), end of Sect. 2.5, Proposition 2.12 and Theorem 2.3, Sect. 2.7, it is expected that the measure $m_0(dx, du) = \bar{F}_x(u)\, \nu(dx)\, du$ is invariant for $(Z_t, A_t)_{t\geq 0}$. We are going to prove this under weaker assumptions than those in Theorem 2.3.

When $m_0(dx, du) = \bar{F}_x(u)\, \nu(dx)\, du$, the measures m and m_1 defined respectively by (8.3) and (8.4) are

$$\int_{(E\times\mathbb{R}_+)^2} \psi(x, u, y, v)\, m(dx, du, dy, dv) =$$

$$\int_{(E\times\mathbb{R}_+)^2} \psi(x, u, y, v - u)\, 1_{\{v>u\}}\, N(x, dy, dv)\, \nu(dx)\, du,$$

$$\int_{E\times\mathbb{R}_+} \varphi(y, v)\, m_1(dy, dv) = \int_{(E\times\mathbb{R}_+)^2} \varphi(y, v)\, 1_{\{u>v\}}\, N(x, dy, du)\, \nu(dx)\, dv \tag{8.5}$$

for all nonnegative Borel functions ψ and φ.

Recall that $R = \sum_{n\geq 0} N^{*n}$.

Lemma 8.1 *Assume that ν is an invariant measure for the driving chain Y, and define $m_0(dx, du) = \bar{F}_x(u)\,\nu(dx)\,du$ and m_1 as in (8.5).*

Let $h : E \to \mathbb{R}_+$ be a Borel function such that $\int_E h(x)\,\nu(dx) < +\infty$. Then for all $p > 0$:

$$\int_{E\times\mathbb{R}_+} h(x)\,e^{-ps}\,m_1 * R\,(dx, ds) = \frac{1}{p}\int_E h(x)\,\nu(dx) - \frac{1}{p}\lim_{n\to+\infty}\int_E \mathbb{E}_x\left(h(Y_n)e^{-pT_n}\right)\,\nu(dx).$$

Proof For $p \in \mathbb{R}_+^*$ and $n \geq 0$, we get from (8.5) that

$$\int_{E\times\mathbb{R}_+} h(x)\,e^{-ps}\,m_1 * N^{*n}(dx, ds) = \frac{1}{p}\int_{(E\times\mathbb{R}_+)^2\times E} (1 - e^{-pv})\,h(z)\,e^{-pw} N^{*n}(y, dz, dw)\,N(x, dy, dv)\,\nu(dx). \quad (8.6)$$

Since ν is an invariant measure for the driving chain Y, we have

$$\int_{(E\times\mathbb{R}_+)^2\times E} h(z)\,e^{-pw}\,N^{*n}(y, dz, dw)\;N(x, dy, dv)\,\nu(dx) = \int_{E\times\mathbb{R}_+\times E} h(z)\,e^{-pw}\,N^{*n}(y, dz, dw)\,\nu(dy) \leq \int_E h(y)\,\nu(dy) < +\infty,$$

and (8.6) turns into

$$\int_{E\times\mathbb{R}_+} h(x)\,e^{-ps}\quad m_1 * N^{*n}(dx, ds) = \frac{1}{p}\int_{E^2\times\mathbb{R}_+} h(z)\,e^{-pw}\,N^{*n}(y, dz, dw)\,\nu(dy) - \frac{1}{p}\int_{E^2\times\mathbb{R}_+} h(z)\,e^{-pw}\,N^{*(n+1)}(y, dz, dw)\,\nu(dy).$$

The result follows. ∎

Proposition 8.2 *Assume that ν is an invariant measure for the driving chain Y, define the measure m_0 as*

$$m_0(dx, du) = \bar{F}_x(u)\,\nu(dx)\,du$$

and the measure m_1 as in (8.5). Assume that the following condition holds:
(AC) There exists a nondecreasing sequence of Borel sets $E_k \subset E$ such that $\nu(E_k) < +\infty$ for any k, $\cup_k E_k = E$, and

$$\forall\, p \in \mathbb{R}_+^*,\quad \lim_{n\to+\infty}\int_E \mathbb{E}_x(1_{E_k}(Y_n)\,e^{-pT_n})\,\nu(dx) = 0.$$

Then

$$m_1 * R\,(dx, ds) = \nu(dx)\, 1_{\mathbb{R}_+}(s)\, ds.$$

Proof Given a bounded Borel function $h : E \to \mathbb{R}_+$, define $h_k = h\, 1_{E_k}$. Applying Lemma 8.1 to h_k and then letting k tend to infinity, for all $p \in \mathbb{R}_+^*$ we get $\int_{E\times\mathbb{R}_+} h(x)\, e^{-ps}\, m_1 * R(dx, ds) = \int_{E\times\mathbb{R}_+} h(x)\, e^{-ps}\, \nu(dx)\, ds$. This gives the result by usual techniques: the Stone–Weierstrass theorem, then the approximation of $1_{[a,b]}$ ($0 \le a < b < +\infty$) by an increasing sequence of continuous functions with compact support and finally the monotone class theorem. ∎

If $\nu(E) < +\infty$, condition (AC) holds, since it is assumed that $\lim_{n\to+\infty} T_n = +\infty$. Proposition 2.6, end of Sect. 2.3, gives a condition for $\lim_{n\to+\infty} T_n = +\infty$ to hold. It gives also a condition for (AC) to hold whether $\nu(E)$ is finite or not. Indeed, let (E_k) be such that $\nu(E_k) < +\infty$ and $\cup_k E_k = E$; then on applying Proposition 2.6 with $h_1 = 1_{E_k}$ and $h_2(u) = e^{-pu}$, we get the following proposition.

Proposition 8.3 *Assume that there exists an r.v. V with values in $\bar{\mathbb{R}}_+$ such that*

$$\forall (x, y) \in E,\ \forall t > 0, \quad \mathbb{P}_x(T_1 > t \mid Y_1 = y) \ge \mathbb{P}(V > t),$$

and $\mathbb{P}(V > 0) \neq 0$. Then condition (AC) holds.

The following lemma is a rewrite of (2.21), Sect. 2.5 using (2.20).

Lemma 8.2 *Let $\varphi : E \times \mathbb{R}_+ \to \mathbb{R}_+$ be a Borel function, m_0 a nonnegative measure on $E \times \mathbb{R}_+$, and m_1 as in (8.4). Then*

$$\int_{E\times\mathbb{R}_+} \mathbb{E}_{x,u}(\varphi(Z_t, A_t))\, m_0(dx, du) \tag{8.7}$$

$$= \int_{E\times\mathbb{R}_+} \varphi(x, u+t)\, \frac{\bar{F}_x(u+t)}{\bar{F}_x(u)}\, m_0(dx, du) + m_1 * R * g(t).$$

Theorem 8.1 *Assume that ν is an invariant measure for the driving chain Y, and define m_0 as $m_0(dx, du) = \bar{F}_x(u)\, \nu(dx)\, du$. Assume that (AC) holds. Then m_0 is invariant for the CSMP $(Z_t, A_t)_{t\ge 0}$.*

Proof Lemma 8.2 and Proposition 8.2 yield

$$\int_{E\times\mathbb{R}_+} \mathbb{E}_{x,u}(\varphi(Z_t, A_t))\, m_0(dx, du) = \int_{E\times\mathbb{R}_+} \varphi(x, u)\, \bar{F}_x(u)\, \nu(dx)\, du,$$

and therefore m_0 is an invariant measure for $(Z_t, A_t)_{t\ge 0}$. ∎

Example 8.1 Let us return to the time elapsed since the last jump in a renewal process (Example 3.2, Sect. 3.1). Theorem 8.1 shows that $m_0(dv) = \bar{F}(v)\, dv$ is invariant for $(A_t)_{t\ge 0}$.

Assume $c = \int v\, dF(v) = \int \bar{F}(v)\, dv < +\infty$. From Proposition 8.1, we deduce that in steady state, i.e., when the probability distribution of A_0 is $\bar{F}(v)\, dv/c$, we have $\mathbb{E}(\varphi(A_t, W_t)) = \int \varphi(u, v-u)\, 1_{\{v>u\}}\, dF(v)\, du/c$. A well-known and surprising consequence is that the probability distribution of the length $A_t + W_t$ of the interval that contains t is not $dF(v)$ but $v\, dF(v)/c$. When dF is an exponential distribution, its mean length is $2\mathbb{E}(dF)$.

More generally, for a Markov renewal process such that $c = \int \mathbb{E}_y(T_1)\, \nu(dy) < +\infty$, in steady state, i.e., when $\bar{F}_x(u)\, \nu(dx)\, du/c$ is the probability distribution of (Z_0, A_0), the probability distribution of the length $A_t + W_t$ of the interval that contains t is $\int_{\{x \in E\}} v\, dF_x(v)\, \nu(dx)/c$.

Example 8.2 Let us assume that $(Z_t, A_t)_{t\geq 0}$ is associated with an alternating Markov renewal process (see Example 2.2, Sect. 2.3). Define $\bar{F}_i(u) = dF_i(]u, +\infty])$. The Markov chain Y has one and only one invariant measure up to a multiplicative constant; it is $\nu(e_1) = \nu(e_0) = 1$. Hence

$$m_0(dx, du) = \delta_{e_1}(dx)\, \bar{F}_1(u)\, du + \delta_{e_0}(dx)\, \bar{F}_0(u)\, du \tag{8.8}$$

is an invariant measure for the CSMP. We have $m_0(E \times \mathbb{R}_+) < +\infty$ if and only if $\mathbb{E}(dF_i) = \int_0^{+\infty} \bar{F}_i(u)\, du < +\infty$ for $i \in \{0, 1\}$.

Example 8.3 We return to Example 2.4, Sect. 2.5. Define $p = dF_1([a, +\infty])$.

When preventive maintenance is instantaneous, the invariant probability distribution ν of the Markov chain Y is given in Example 2.7, Sect. 2.7. We have $\nu(e_1) = 1$, $\nu(e_0) = 1 - p$. Hence, without other assumption, we deduce that

$$m_0(dx, du) = \delta_{e_1}(dx)\, 1_{\{u<a\}}\, \bar{F}_1(u)\, du + (1-p)\, \delta_{e_0}(dx)\, \bar{F}_0(u)\, du$$

is invariant for the CSMP. When $p < 1$, we have $m_0(E \times \mathbb{R}_+) < +\infty$ if and only if $\mathbb{E}(dF_0) = \int_0^{+\infty} \bar{F}_0(u)\, du < +\infty$.

When preventive maintenance is not instantaneous, the probability kernel of the Markov chain Y is $P(e_1, e_2) = p$, $P(e_1, e_0) = 1 - p$, $P(e_2, e_1) = P(e_0, e_1) = 1$, $P(e_2, e_0) = P(e_0, e_2) = P(e_i, e_i) = 0\ (i = 0, 1)$. Its invariant measure, up to a multiplicative constant, is $\nu(e_1) = 1$, $\nu(e_2) = p$, $\nu(e_0) = 1 - p$; hence the measure

$$\begin{aligned} m_0(dx, du) &= \delta_{e_1}(dx)\, 1_{\{u<a\}}\, \bar{F}_1(u)\, du + p\, \delta_{e_2}(dx)\, \bar{F}_2(u)\, du \\ &\quad + (1-p)\, \delta_{e_0}(dx) \bar{F}_0(u)\, du \end{aligned}$$

is invariant for the CSMP. Assuming $0 < p < 1$, we have $m_0(E \times \mathbb{R}_+) < +\infty$ if and only if $\mathbb{E}(dF_i) = \int_0^{+\infty} \bar{F}_i(u)\, du < +\infty$ for $i \in \{1, 2\}$.

The uniqueness of the invariant measure (up to a multiplicative constant) will be a consequence of results of the following Sections; e.g., the CSMP of Example 8.3 above is Harris-recurrent (Theorem 8.3, Sect. 8.2). See also Example 8.4, end of Sect. 8.2, which goes back to Example 8.2, and Example 8.8, Sect. 8.4.

Invariant Measure for the Chain From That of the CSMP

Proposition 8.4 *Assume that m_0 is an invariant measure for $(Z_t, A_t)_{t\geq 0}$ and define m_1 as in (8.4), Sect. 8.1. Then there exists a nonnegative measure ν on E such that*

$$m_1 * R\,(dx, du) = \nu(dx)\, 1_{\mathbb{R}_+}(u)\, du.$$

Proof Given a Borel set $A \subset E$, define $m_A(B) = m_1 * R\,(A \times B)$. The measure m_A is translation invariant. Define (Y_n^t, T_n^t) as in Theorem 2.1, Sect. 2.4. From Corollary 2.1, Sect. 2.3, we get for all $s, t \in \mathbb{R}_+$,

$$\begin{aligned} N_0^{x,u} * R(A\times]t, t+s]) &= E_{x,u}\left(\sum_{n\geq 1} 1_A(Y_n)\, 1_{]t,t+s]}(T_n)\right) \\ &= \mathbb{E}_{x,u}\left(\sum_{n\geq 1} 1_A(Y_n^t)\, 1_{]0,s]}(T_n^t)\right). \end{aligned}$$

According to Corollary 2.2, end of Sect. 2.4,

$$\mathbb{E}_{x,u}\left(\sum_{n\geq 1} 1_A(Y_n^t)\, 1_{]0,s]}(T_n^t)\right) = \mathbb{E}_{x,u}(\varphi(Z_t, A_t))$$

for some Borel function φ. Hence

$$\begin{aligned} m_1 * R\,(A\times]t, t+s]) &= \int_{E\times\mathbb{R}_+} \mathbb{E}_{x,u}(\varphi(Z_t, A_t))\, m_0(dx, du) \\ &= \int_{E\times\mathbb{R}_+} \mathbb{E}_{x,u}(\varphi(Z_0, A_0))\, m_0(dx, du) \\ &= m_1 * R\,(A\times]0, s]). \end{aligned}$$

The measure m_A being translation invariant, there exists c_A such that $m_1 * R\,(A \times B) = c_A \int_B ds$. From $c_A = m_1 * R\,(A \times [0, 1])$ we conclude that $A \to c_A$ is a measure ν. This ends the proof. ■

Theorem 8.2 *Let $(Z_t, A_t)_{t\geq 0}$ be a CSMP and $(Y, T) = (Y_n, T_n)_n$ its associated Markov renewal process. Assume that m_0 is an invariant measure for $(Z_t, A_t)_{t\geq 0}$ and define m_1 as in (8.4). Then there exists a measure ν such that*

$$m_0(dx, du) = \bar{F}_x(u)\, \nu(dx)\, du, \quad m_1 * R(dx, du) = \nu(dx)\, 1_{\mathbb{R}_+}(u)\, du,$$

$$\int_E \mathbb{E}_x(f(Y_n))\, \nu(dx) = \int_E f(x)\, \nu(dx)$$

for every $n \in \mathbb{N}^*$ *and Borel function* $f : E \to \mathbb{R}_+$.

Proof Proposition 8.4 gives $m_1 * R(dx, du) = \nu(dx)\, 1_{\mathbb{R}_+}(u)\, du$ for some measure ν. From (8.7) and the invariance of m_0 for the CSMP we get

$$\int_{E\times\mathbb{R}_+} \varphi(x,u)\, m_0(dx,du) = \int_{E\times\mathbb{R}_+} \mathbb{E}_{x,u}(\varphi(Z_t, A_t))\, m_0(dx,du) =$$
$$\int_{E\times\mathbb{R}_+} \varphi(x,u+t)\, \frac{\bar{F}_x(u+t)}{\bar{F}_x(u)}\, m_0(dx,du) + \int_{E\times\mathbb{R}_+} 1_{\{0\le u\le t\}} \varphi(x,u)\, \bar{F}_x(u)\, \nu(dx)\, du. \tag{8.9}$$

We take $\varphi = 1_{A\times[0,M]}$ and $t > M$, we get $m_0(A\times[0,M]) = \int_{A\times[0,M]} \bar{F}_x(u)\nu(dx)\,du$; hence $m_0(dx,du) = \bar{F}_x(u)\,\nu(dx)\,du$.

It remains to be proved that $\nu P = \nu$ where $P(x,dy) = \int_{\{v\in\mathbb{R}_+\}} N(x,dy,dv)$ is the probability kernel of the driving Markov chain Y. From $m_0(dx,du) = \bar{F}_x(u)\,\nu(dx)\,du$ and (8.5), we have

$$\int_{E\times\mathbb{R}_+} \varphi(y,v)\, m_1(dy,dv) = \int_{(E\times\mathbb{R}_+)^2} \varphi(y,w)\, 1_{\{w<v\}}\, N(x,dy,dv)\, \nu(dx)\, dw.$$

We also have $R = I + R * N$, and thus $m_1 * R = m_1 + m_1 * R * N$. Hence for every Borel function $\varphi : E\times\mathbb{R}_+ \to \mathbb{R}_+$, we get

$$\begin{aligned}
\int_{E\times\mathbb{R}_+} \varphi(y,v)\,\nu(dy)\,dv &= \int_{E\times\mathbb{R}_+} \varphi(y,v)\, m_1 * R(dy,dv)\\
&= \int_{(E\times\mathbb{R}_+)^2} \varphi(y,w)\, 1_{\{w<v\}}\, N(x,dy,dv)\,\nu(dx)\,dw\\
&\quad + \int_{(E\times\mathbb{R}_+)^2} \varphi(y,s+v)\, N(x,dy,dv)\,\nu(dx)\,ds\\
&= \int_{(E\times\mathbb{R}_+)^2} \varphi(y,w)\, N(x,dy,dv)\,\nu(dx)\,dw = \int_{E^2\times\mathbb{R}_+} \varphi(y,w)\, P(x,dy)\,\nu(dx)\,dw.
\end{aligned}$$

Taking $\varphi(y,v) = f(y)\, 1_{[0,1]}(v)$ concludes the proof. ∎

Let P be the probability kernel of the Markov chain Y. Theorem 8.2 leads to the following result: if equation $\nu P = \nu$ has one and only one solution $\nu \neq 0$ (up to a multiplicative constant), then the CSMP has only one invariant measure (up to a multiplicative constant). This uniqueness can be obtained in a computational way as in Examples 8.1 to 8.3.

This can also be the case because the Markov chain Y is Harris-recurrent (see, e.g., Jamison and Orey [72]). We will see in the following section that then the CSMP is also Harris-recurrent.

8.2 CSMP Recurrence from That of Its Driving Chain

Definition 8.2 Let $\mu \neq 0$ be a nonnegative σ-finite measure on E. A Markov process $(X_t)_{t\geq 0}$ (respectively a Markov chain $(X_n)_{n\geq 1}$) taking values in E is said to be μ-recurrent if for every Borel set $A \subset E$ such that $\mu(A) > 0$, we have $\int_0^{+\infty} 1_A(X_t)\,dt = +\infty$ (respectively $\sum_{n\geq 1} 1_A(X_n) = +\infty$) $\mathbb{P}_x$-almost surely for all $x \in E$.

A Markov chain is said to be Harris-recurrent if it is μ-recurrent for some measure $\mu \neq 0$.

This section is a transcription for the process $(Z_t, A_t, \bar{Z}_t)_{t\geq 0}$ of some of the results in Jacod [67] for the process $(\bar{Z}_t, W_t)_{t\geq 0}$. Therefore, only outlines of the proofs are given. For details, the reader is invited to refer to [67].

Recall that in this chapter, N is assumed to be proper. Denote by P the probability kernel of the driving chain Y and by $dF_{x,y}$ the probability distribution of $T_{n+1} - T_n$ given $Y_n = x, Y_{n+1} = y$. We have $N(x, dy, dv) = dF_{x,y}(dv)\, P(x, dy)$.

Assuming that the driving chain Y is Harris-recurrent, it has a unique (up to a multiplicative constant) invariant measure ν and is ν-recurrent. Define

$$\bar{m}(dx, du, dy) = \int_{\{v:v>u\}} N(x, dy, dv)\,\nu(dx)\,du = P(x, dy)\, dF_{x,y}(]u, +\infty[)\,\nu(dx)\,du,$$

$$m_0(dx, du) = \int_{\{y\in E\}} \bar{m}(dx, du, dy).$$

We have

$$m_0(dx, du) = \bar{F}_x(u)\,\nu(dx)\,du. \tag{8.10}$$

As mentioned Sect. 2.3, in contrast to what is assumed in the remainder of this book, in the following theorem we do not ask for T_n to tend to infinity as $n \to +\infty$. It is a consequence of the recurrence of the driving chain.

Theorem 8.3 *Assume that the driving chain Y is ν-recurrent. Then $\lim_{n\to\infty} T_n = +\infty$, and the process $(Z_t, A_t, \bar{Z}_t)$ is $\bar{m}$-recurrent. Hence the CSMP $(Z_t, A_t)_{t\geq 0}$ is m_0-recurrent.*

Proof The proof follows that given by J. Jacod for the process $(\bar{Z}_t, W_t)_{t\geq 0}$ ([67] Lemma 1 page 94). Let $A \subset E \times \mathbb{R}_+ \times E$ be a Borel set such that $\bar{m}(A) > 0$.

Given $\alpha > 0$, define $b^\alpha_{x,y} = \sup\{v : dF_{x,y}(]v, +\infty[) > \alpha\}$. There exists $\alpha > 0$ such that $\int_{E^2} (\int_0^{b^\alpha_{x,y}} 1_A(x, u, y)\,du)\,\nu(dx)\,P(x, dy) > 0$. Hence, there exists $\varepsilon > 0$ such that $\int_{H_\varepsilon} \nu(dx)\,P(x, dy) > 0$, where $H_\varepsilon = \{(x, y) : \int_0^{b^\alpha_{x,y}} 1_A(x, u, y)\,du \geq \varepsilon\}$.

If $(x, y) \in H_\varepsilon$, then

$$\int_0^{b^\alpha_{x,y}-\varepsilon/2} 1_A(x, u, y)\,du \geq \frac{\varepsilon}{2}, \tag{8.11}$$

and thus $b^{\alpha}_{x,y} - \varepsilon/2 > 0$.

From the ν-recurrence of Y we deduce the $\tilde{\nu}$-recurrence of the Markov chain $\widetilde{Y} = (Y_{n-1}, Y_n)$, where $\tilde{\nu}(dx, dy) = \nu(dx)\, P(x, dy)$. Define $(\tau_k)_{k\geq 1}$ as the sequence of times when the chain $\widetilde{Y}$ enters H_ε. Since $\tilde{\nu}(H_\varepsilon) > 0$, we get $\sum_k 1_{\{\tau_k<+\infty\}} = +\infty$ a.s.

Let $\mathcal{F}_{\tau_k}$ be the σ-algebra generated by the Markov chain $\widetilde{Y}$ up to time τ_k and

$$\Omega_n = \left\{\tau_n < +\infty,\, T_{\tau_n} - T_{\tau_n-1} > b^{\alpha}_{Y_{\tau_n-1}, Y_{\tau_n}} - \frac{\varepsilon}{2}\right\}.$$

From $\mathbb{P}(\Omega_n \,|\, \mathcal{F}_{\tau_n}) \geq 1_{\{\tau_n<+\infty\}}\, dF_{Y_{\tau_n-1}, Y_{\tau_n}}(]b^{\alpha}_{Y_{\tau_n-1}, Y_{\tau_n}} - \frac{\varepsilon}{2}, +\infty[) \geq \alpha\, 1_{\{\tau_n<+\infty\}}$ we deduce $\sum_{n\geq 1} \mathbb{P}(\Omega_n \,|\, \mathcal{F}_{\tau_n}) = +\infty$. Hence by the generalized Borel–Cantelli lemma we get $\mathbb{P}(\limsup_k \Omega_k) = 1$.

Therefore, for an infinite number of n, which is a subsequence of $(\tau_k)_{k\geq 1}$, we have $(Y_{n-1}, Y_n) \in H_\varepsilon$, $T_n - T_{n-1} \geq b^{\alpha}_{Y_{n-1}, Y_n} - \varepsilon/2$, and from (8.11) we get

$$T_n - T_{n-1} \geq \int_0^{T_n - T_{n-1}} 1_A(Y_{n-1}, u, Y_n)\, du \geq \int_0^{b^{\alpha}_{Y_{n-1}, Y_n} - \varepsilon/2} 1_A(Y_{n-1}, u, Y_n)\, du \geq \frac{\varepsilon}{2}.$$

Hence $\lim_n T_n = +\infty$ and

$$\int_0^{+\infty} 1_A(Z_t, A_t, \bar{Z}_t)\, dt \geq \sum_{n\geq 1} \int_{T_{n-1}}^{T_n} 1_A(Y_{n-1}, t - T_{n-1}, Y_n)\, dt = +\infty\ a.s.,$$

and thus the process $(Z_t, A_t, \bar{Z}_t)_{t\geq 0}$ is Harris-recurrent.

Finally, given a Borel set $B \subset E \times \mathbb{R}_+$ such that $m_0(B) > 0$, take $A = B \times E$. From $\bar{m}(A) > 0$ we deduce $\int_0^{+\infty} 1_B(Z_t, A_t) = \int_0^{+\infty} 1_A(Z_t, A_t, \bar{Z}_t)\, dt = +\infty\ a.s.$ Hence $(Z_t, A_t)_{t\geq 0}$ is m_0-recurrent. ■

In Theorem 8.4 below, condition (AC) of Theorem 8.1 is replaced by the Harris-recurrence of the driving chain Y. Its proof is based on the following lemma.

Lemma 8.3 (Azéma–Duflo–Revuz *[6]* Propositions 2.1 page 26 and 2.3 page 27) *Let $(X_t)_{t\geq 0}$ be a Markov process taking values in a locally compact space with a countable base and μ a σ-finite measure. Define the measure V_1 as $V^1 f(x) = \int_{\mathbb{R}_+} e^{-t}\, \mathbb{E}_x(f(X_t))\, dt$ and the measure μV^1 as $\mu V^1(dy) = \int_{\{x\in E\}} \mu(dx)\, V^1(x, dy)$. Assume that $(X_t)_{t\geq 0}$ is μ-recurrent and $\int f\, d(\mu V^1) \leq \int f\, d\mu$ for every Borel function $f : E \to \mathbb{R}_+$. Then $\mu V^1 = \mu$, and thus μ is an invariant measure.*

We are now ready to establish the announced theorem.

Theorem 8.4 *The Polish space E is assumed to be locally compact. Assume that the driving chain Y is Harris-recurrent with invariant measure ν. Define the measure m_0 as*

$$m_0(dx, du) = \bar{F}_x(u)\, \nu(dx)\, du.$$

Then the CSMP is m_0-recurrent and m_0 is its invariant measure (up to a multiplicative constant).

Proof In view of Theorem 8.3 and Lemma 8.3, we have only to check that $m_0 V^1 \leq m_0$, where $\int \varphi \, d(m_0 V^1) := \int_{E\times\mathbb{R}_+^2} e^{-t}\, \mathbb{E}_{x,u}(\varphi(Z_t, A_t))\, m_0(dx, du)\, dt$. Given a Borel function $\varphi : E \times \mathbb{R}_+ \to \mathbb{R}_+$ such that $\int \varphi \, dm_0 < +\infty$, define g as $g(x, u) = \varphi(x, u)\, \bar{F}_x(u)$. From (8.7), Sect. 8.1 and Lemma 8.1, Sect. 8.1 we get

$$\begin{aligned}\int \varphi \, d(m_0 V^1) &= \int_{E\times\mathbb{R}_+^2} e^{-t}\, g(x, u+t)\, \nu(dx)\, du\, dt + \int_0^{+\infty} e^{-t}\, m_1 * R * g(t)\, dt \\ &\leq \int_{E\times\mathbb{R}_+} (1 - e^{-u})\, g(x, u)\, \nu(dx)\, du + \int_{E\times\mathbb{R}_+} e^{-u}\, g(x, u)\, \nu(dx)\, du \\ &= \int \varphi \, dm_0.\end{aligned}$$

If $\int \varphi \, dm_0 = +\infty$, the inequality is obvious; hence the proof is complete. ■

Example 8.4 Let us go back to Example 8.2, Sect. 8.1. The Markov chain Y is Harris recurrent; hence the measure m_0 defined in (8.8), Sect. 8.1 is the only invariant measure (up to a multiplicative constant) for the CSMP. Assume $c_i = \mathbb{E}(dF_i) = \int_0^{+\infty} u\, dF_i(u) < +\infty$ for $i \in \{0, 1\}$; the unique steady state is $(\delta_{e_1}(dx)\, \bar{F}_1(u)\, du + \delta_{e_0}(dx)\, \bar{F}_0(u)\, du)/(c_1 + c_0)$. For instance, when it is a model for an item subject to failure and the failure state is e_0, in steady state the probability that the item is down for a period more than a is $\int_a^{+\infty} \bar{F}_0(u)\, du/(c_0 + c_1) = \mathbb{E}((U_0 - a)_+)/(\mathbb{E}(U_1) + \mathbb{E}(U_0))$, where U_i is an r.v. with probability distribution dF_i $(i = 0, 1)$.

8.3 Intensities of Stable CSMPs

Let us begin by recalling the results of the previous sections, which will be of use here.

Let $(Z_t, A_t)_{t\geq 0}$ be a CSMP, $(Y, T) = (Y_k, T_k)_k$ its associated Markov renewal process, and ν an invariant measure for the Markov chain Y. We assume that condition (AC) Sect. 8.1 holds or that Y is Harris-recurrent. Then $m_0(dx, du) = \bar{F}_x(u)\, \nu(dx)\, du$ is invariant for $(Z_t, A_t)_{t\geq 0}$ and m_1 defined in (8.4), Sect. 8.1 and specified in (8.5), satisfies $m_1 * R(dx, du) = \nu(dx)\, 1_{\mathbb{R}_+}(u)\, du$.

We return to the notation of Chap. 5. The function φ (respectively ψ) is a Borel function from $E \times \mathbb{R}_+$ (respectively $E \times \mathbb{R}_+^2$) to $\mathbb{R}_+$.

We define what we call a stationary intensity as follows: if $U^{x,u}$ is the intensity of some marked point process $(\zeta_k, S_k)_k$ under $\mathbb{P}_{x,u}$, i.e., $\int \varphi(y, v)\, U^{x,u}(dy, dv) = \sum_k \mathbb{E}_{x,u}(\varphi(\zeta_k, S_k))$, then the corresponding steady-state intensity is defined as

$$\int \varphi(y, v)\, {}^{s}U(dy, dv) = \int \varphi(y, v)\, U^{x,u}(dy, dv)\, m_0(dx, du).$$

Since m_0 is not assumed to be a probability distribution, when $m_0(E \times \mathbb{R}_+) = \int_E \mathbb{E}_x(T_1)\,\nu(dx) < +\infty$ we introduce the steady-state intensity ${}^e U$ (where e stands for "equilibrium"), which is the intensity when the probability distribution of (Z_0, A_0) is $m_0/m_0(E \times \mathbb{R}_+)$.

Recall that $\int_{E\times\mathbb{R}_+} \varphi(y, v)\, \rho^{x,u}(dy, dv) = \sum_{k\geq 1} \mathbb{E}_{x,u}(\varphi(Y_k, T_k)\, 1_{\{T_k<+\infty\}})$. From $\rho^{x,u} = N_0^{x,u} * R$ (formula (5.4), Sect. 5.1) we deduce

$$\begin{aligned}\int_{E\times\mathbb{R}_+} \varphi(y, v)\, {}^s\rho(dy, dv) &= \int_{E\times\mathbb{R}_+} \varphi(y, v)\, \rho^{x,u}(dy, dv)\, m_0(dx, du)\\ &= \int_{E\times\mathbb{R}_+} \varphi(y, v)\, m_1 * R(dy, dv)\\ &= \int_{E\times\mathbb{R}_+} \varphi(y, v)\, \nu(dy)\, dv. \end{aligned} \tag{8.12}$$

In particular, if $\int_E \mathbb{E}_y(T_1)\, \nu(dy) < +\infty$, then the normalized steady-state intensity ρ is

$${}^e\rho(A\times]t, t+h]) = \frac{h\,\nu(A)}{\int_E \mathbb{E}_y(T_1)\,\nu(dy)}.$$

We now move on to $\bar{\rho}$. Recall that

$$\begin{aligned}\int_{E\times\mathbb{R}_+^2} \psi(y, v, w)\, \bar{\rho}^{x,u}(dy, dv, dw) &= \mathbb{E}_{x,u}(\psi(x, u + T_1, T_1)\, 1_{\{T_1<+\infty\}})\\ &+ \sum_{k\geq 1} \mathbb{E}_{x,u}(\psi(Y_k, T_{k+1} - T_k, T_{k+1}) 1_{\{T_{k+1}<+\infty\}}).\end{aligned}$$

From (5.2), Sect. 5.1 we deduce

$$\begin{aligned}\int_{E\times\mathbb{R}_+^2} \psi(y, v, w)\, {}^s\bar{\rho}(dy, dv, dw) &= \int_{E^2\times\mathbb{R}_+^3} \psi(y, v, w)\, \bar{\rho}^{x,u}(dy, dv, dw)\, m_0(dx, du)\\ &= \int_{E\times\mathbb{R}_+^2} \psi(x, v, v-u)\, 1_{\{v>u\}}\, dF_x(v)\, \nu(dx)\, du\\ &+ \int_{E\times\mathbb{R}_+^2} \psi(y, v, w)\, 1_{\{w>v\}}\, dF_y(v)\, \nu(dy)\, dw\\ &= \int_{E\times\mathbb{R}_+^2} \psi(y, v, w)\, dF_y(v)\, \nu(dy)\, dw. \end{aligned} \tag{8.13}$$

Example 8.5 Let us go back to the alternating renewal process that models an item subject to failure. We take the notation of Example 2.3, Sect. 2.5. In Example 5.5, Sect. 5.1 we had to specify dF_i $(i = 0, 1)$ to be able to assess the intensities and thus the mean numbers of failures and repairs. In steady state, the formulas are much simpler. The only assumption about dF_i $(i = 0, 1)$ is $\mathbb{E}(dF_i) := \int_{\mathbb{R}_+} x\, dF_i(x) < +\infty$.

The only invariant measure (up to a multiplicative constant) of the driving chain is $\nu(e_0) = \nu(e_1) = 1$. The mean number of failures (respectively completed repairs) in any time interval of length h is therefore

$$^e\rho(\{e_0\} \times (t, t+h)) = {}^e\rho(\{e_1\} \times (t, t+h)) = \frac{h}{\mathbb{E}(dF_0) + \mathbb{E}(dF_1)},$$

while the mean number of repairs completed in any time interval of length h whose duration is less than a is

$$^e\bar{\rho}(\{e_0\} \times [0, a] \times (t, t+h)) = \frac{dF_0([0, a])\, h}{\mathbb{E}(dF_0) + \mathbb{E}(dF_1)}.$$

Example 8.6 We consider the example of a component with an instantaneous preventive maintenance (Example 2.4, Sect. 2.5). Define $p = e^{-\int_0^a \lambda(v)dv}$. We have seen in Example 2.7, Sect. 2.7 that $\nu(e_1) = 1$, $\nu(e_0) = 1 - p$. We deduce (Example 5.4, Sect. 5.1) that in steady state, the mean number of failures in the time interval $(t, t+h)$ is

$$^e\rho(\{e_0\} \times (t, t+h)) = {}^e\bar{\rho}(\{e_1\} \times [0, a[\times (t, t+h)) = \frac{(1-p)\, h}{\mathbb{E}(dF_1) + (1-p)\, \mathbb{E}(dF_0)},$$

and the mean number of preventive maintenances in the same time interval is

$$^e\bar{\rho}(\{e_1\} \times \{a\} \times (t, t+h)) = \frac{p\, h}{\mathbb{E}(dF_1) + (1-p)\, \mathbb{E}(dF_0)}.$$

Assume now that $(Z_t, A_t)_{t\geq 0}$ is a CSMP with parameters ℓ, α, β. From the first part of Proposition 5.3, Sect. 5.2 and the invariance of m_0 for $(Z_t, A_t)_{t\geq 0}$ we get

$$\int_{E\times\mathbb{R}_+} \varphi(y, v)\, {}^s\rho_c(dy, dv) =$$
$$\int_{E\times\mathbb{R}_+} 1_{\{u<\alpha(x)\}}\, \ell(x, u)\, e^{-\int_0^u \ell(x,w)\, dw} \left(\int_{E\times\mathbb{R}_+} \varphi(y, v)\, \beta(x, u; dy)\, dv \right) \nu(dx)\, du\,.$$

The second part of Proposition 5.3 and (8.12) yield

$$\int_{E\times\mathbb{R}_+} \varphi(y, v)\, {}^s\rho_D(dy, dv) =$$
$$\int_E 1_{\{\alpha(x)<+\infty\}} e^{-\int_0^{\alpha(x)} \ell(x,w)\, dw} \left(\int_{E\times\mathbb{R}_+} 1_{\{u<\alpha(x)\}}\, \varphi(y, \alpha(x) - u) \beta(x, \alpha(x); dy) du \right) \nu(dx)$$
$$+ \int_E 1_{\{\alpha(y)<+\infty\}} e^{-\int_0^{\alpha(y)} \ell(y,w)\, dw} \left(\int_{E\times\mathbb{R}_+} \varphi(z, v + \alpha(y))\, \beta(y, \alpha(y); dz)\, dv \right) \nu(dy);$$

hence

$$\int_{E\times\mathbb{R}_+} \varphi(y,v)\,{}^s\rho_D(dy,dv) =$$
$$\int_E 1_{\{\alpha(x)<+\infty\}} e^{-\int_0^{\alpha(x)} \ell(x,w)\,dw} \left(\int_{E\times\mathbb{R}_+} \varphi(y,v)\,\beta(x,\alpha(x);dy)\,dv\right)\nu(dx).$$

At first glance, ${}^s\rho_c + {}^s\rho_D = {}^s\rho$ is not obvious. Actually,

$$\int_{E\times\mathbb{R}_+} \varphi(y,v)\,{}^s\rho_c(dy,dv) + \int_{E\times\mathbb{R}_+} \varphi(y,v)\,{}^s\rho_D(dy,dv) = \int_{E^2} f(y)\,P(x,dy)\,\nu(dx),$$

where P is the probability kernel of the Markov chain Y and $f(y) = \int_{\mathbb{R}_+} \varphi(y,v)\,dv$. Writing $\nu P = \nu$, we get ${}^s\rho_c + {}^s\rho_D = {}^s\rho$.

Example 8.7 Return to Example 8.6, assuming that the component has failure rate λ and repair rate μ. From Example 5.6, Sect. 5.2 we deduce that in steady state, the mean number of failures in the time interval $(t, t+h)$ is ${}^e\rho_c(\{e_0\}\times[0,t])$ and the mean number of preventive maintenances in the time interval $(t, t+h)$ is ${}^e\rho_D(\{e_1\}\times[0,t])$. We get ${}^s\rho_c(\{e_0\}\times[0,t]) = h\,\nu(e_0)\int_{\mathbb{R}_+}\lambda(u)\,\bar{F}_0(u)\,du = h\nu(e_0) = h\,(1-p)$ and ${}^s\rho_D(\{e_1\}\times[0,t]) = h\,e^{-\int_0^a \lambda(w)\,dw}\,\nu(e_1) = h\,p$. We find, fortunately, the results of Example 8.6.

In the same way, Proposition 5.2, Sect. 5.2 and (8.13) give

$$\int_{E\times\mathbb{R}_+^2} \psi(y,v,w)\,{}^s\bar{\rho}_c(dy,dv,dw) =$$
$$\int_{E\times\mathbb{R}_+^2} \psi(y,v,w)\,1_{\{v<\alpha(y)\}}\,\ell(y,v)\,e^{-\int_0^v \ell(y,s)\,ds}\,dv\,\nu(dy)\,dw, \tag{8.14}$$

$$\int_{E\times\mathbb{R}_+^2} \psi(y,v,w)\,{}^s\bar{\rho}_D(dy,dv,dw) = \int_{E\times\mathbb{R}_+} \psi(y,\alpha(y),w)\,e^{-\int_0^{\alpha(y)} \ell(y,s)\,ds}\,\nu(dy)\,dw. \tag{8.15}$$

8.4 Invariant Measure for PDMPs

We consider a PDMP $(X_t)_{t\geq 0}$ on E generated by ϕ and N. Let $(Z_t, A_t)_{t\geq 0}$ be an associated CSMP and $(Y,T) = (Y_n, T_n)_{n\geq 0}$ its associated Markov renewal process.

The following lemma is a straightforward consequence of $\mathbb{E}_{x,u}(f(\phi(Z_t, A_t))) = \mathbb{E}_{\phi(x,u)}(f(X_t))$ (Theorem 3.1, Sect. 3.1).

Lemma 8.4 *Let m_0 be an invariant measure for $(Z_t, A_t)_{t\geq 0}$. Define m_ϕ as the image of m_0 by ϕ, that is,*

$$\int_E f(x)\, m_\phi(dx) = \int_{E\times\mathbb{R}_+} f(\phi(x,u))\, m_0(dx,du).$$

Then m_ϕ is invariant for $(X_t)_{t\geq 0}$.

In view of Theorem 8.1, Sect. 8.1 and Lemma 8.4, we immediately get the following theorem.

Theorem 8.5 *Assume that ν is an invariant measure for the driving chain Y and condition (AC) of Proposition 8.2, Sect. 8.1 holds. Define the measure m_ϕ as*

$$\int_E f(x)\, m_\phi(dx) = \int_{E\times\mathbb{R}_+} f(\phi(x,u))\, \bar{F}_x(u)\, \nu(dx)\, du = \int_E \mathbb{E}_x\left(\int_0^{T_1} f(X_u)\, du\right) \nu(dx). \tag{8.16}$$

Then m_ϕ is invariant for the PDMP.

Theorem 8.6 *The Polish space E is assumed to be locally compact. Assume that the driving chain Y is Harris-recurrent with invariant measure ν and define m_ϕ as in (8.16). Then the PDMP is m_ϕ-recurrent and m_ϕ is its unique invariant measure, up to a multiplicative constant.*

Proof We deduce from Theorem 8.4, Sect. 8.2 that $m_0(dx,du) = \bar{F}_x(u)\, \nu(dx)\, du$ is invariant for the CSMP and that the CSMP is m_0-recurrent. Hence m_ϕ is invariant for the PDMP, and condition $m_\phi(A) = m_0(\phi^{-1}(A)) > 0$ leads to $+\infty = \int_0^{+\infty} 1_{\phi^{-1}(A)}(Z_t, A_t)\, dt = \int_0^{+\infty} 1_A(X_t)\, dt$. Therefore, $(X_t)_{t\geq 0}$ is $m_\phi-$recurrent. The uniqueness of the invariant measure up to a multiplicative constant is then a result of Azéma, Duflo, and Revuz [6]. ■

Note that even if the driving chain is positive recurrent, i.e., $\nu(E) < +\infty$, the PDMP is null-recurrent, i.e., $m_\phi(E) = \int_E \mathbb{E}_x(T_1)\, \nu(dx) = +\infty$.

We are now interested in the converse of Theorem 8.5.

Theorem 8.7 *Let π_0 be an invariant measure for the PDMP.*

Assume that $\lim_{t\to+\infty} \int_E \bar{F}_x(t)\, \pi_0(dx) = 0$. *Then there exists a measure ν such that*

$$\int_E f(x)\, \pi_0(dx) = \int_{E\times\mathbb{R}_+} f(\phi(x,u))\, \bar{F}_x(u)\, \nu(dx)\, du \tag{8.17}$$

for every Borel function $f : E \to \mathbb{R}_+$. Moreover, $\nu P = \nu$ where P is the kernel of the driving chain Y.

Note that $\lim_{t\to+\infty} \int_E \bar{F}_x(t)\, \pi_0(dx) = 0$ holds as soon as π_0 is a finite measure.

Proof Define

$$\pi_1(dy, dv) = \int_{\{x \in E\}} N(x, dy, dv)\, \pi_0(dx).$$

The beginning of the proof follows the proof of Proposition 8.4, Sect. 8.1 with π_1 instead of m_1. Indeed,

$$\begin{aligned}\pi_1 * R\,(A\times]t, t+s]) &= \int_E \mathbb{E}_x\left(\sum_{n\geq 1} 1_A(Y_k)\, 1_{]t,t+s]}(T_k)\right) \pi_0(dx)\\ &= \int_E \mathbb{E}_x\left(\sum_{n\geq 1} 1_A(Y_k^t)\, 1_{]0,s]}(T_k^t)\right) \pi_0(dx).\end{aligned}$$

From Corollary 2.2, end of Sect. 2.4, the condition $N_0^{Z_t,A_t}(dy, dv) = N(\phi(Z_t, A_t), dy, dv) = N(X_t, dy, dv)$, and the invariance of π_0 for $(X_t)_{t\geq 0}$, we get $\pi_1 * R\,(A\times]t, t+s]) = \pi_1 * R\,(A\times]0, s])$, and consequently $\pi_1 * R(dx, du) = \nu(dx)\, 1_{\mathbb{R}_+}(u)\, du$ for some measure ν, i.e.,

$$\int_{E\times\mathbb{R}_+} \varphi(x, u)\, \nu(x)\, du = \sum_{n\geq 1} \int_{E\times\mathbb{R}_+} \mathbb{E}_{y,v}(\varphi(Y_n, T_n))\, \pi_1(dy, dv)$$

for every Borel function $\varphi : E \times \mathbb{R}_+ \to \mathbb{R}_+$. Taking $\varphi(x, u) = 1_{[0,t]}(u)\, \bar{F}_x(t-u)\, f \circ \phi(x, t-u)$, we get

$$\begin{aligned}&\int_{E\times\mathbb{R}_+} 1_{[0,t]}(u)\, f(\phi(x, u))\, \bar{F}_x(u)\, \nu(dx)\, du\\ &= \sum_{n\geq 1} \int_{E^2\times\mathbb{R}_+} \mathbb{E}_{y,v}(\varphi(Y_n, T_n))\, N(x, dy, dv)\, \pi_0(dx)\\ &= \sum_{n\geq 1} \int_E \mathbb{E}_x\left(1_{[0,t]}(T_{n+1}) \bar{F}_{Y_{n+1}}(t - T_{n+1})\, f \circ \phi(Y_{n+1}, t - T_{n+1})\right) \pi_0(dx)\\ &= \int_E \mathbb{E}_x\left(1_{\{T_2\leq t\}}\, f(\phi(Z_t, A_t))\right) \pi_0(dx)\\ &= \int_E \mathbb{E}_x\,(f(X_t))\, \pi_0(dx) - \int_E \mathbb{E}_x\left(1_{\{T_2>t\}}\, f(\phi(Z_t, A_t))\right) \pi_0(dx).\end{aligned}$$

Letting t tend to infinity, from the invariance of π_0 for the PDMP and the assumption $\lim_{t\to+\infty} \int_E \mathbb{P}_x(T_1 > t)\, \pi_0(dx) = 0$, we get (8.17) if f is bounded. The result follows for every nonnegative Borel function f by usual techniques.

The proof of $\nu P = \nu$ is as in the proof of Theorem 8.2, Sect. 8.1. ■

Example 8.8 We return to Example 3.4, Sect. 3.1 in which a factory needs a certain commodity. In the model with no jump added we have $Y_n = (S_{T_n}, 0)$, $S_{T_{n+1}} = (S_{T_n} - c(T_{n+1} - T_n))_+ + C_{n+1}$. Thus $(S_{T_n})_{n\geq 0}$ is a Markov chain with probability kernel

P, where $P(x, dy)$ is the probability distribution of $(x - c\xi)_+ + C$ with ξ and C independent random variables, the probability distribution of ξ is the interarrival distribution dF, and the probability distribution of C is μ.

Define $W_n = S_{T_n} - C_n$ $(n \geq 1)$. We have $W_{n+1} = (W_n + C_n - c(T_{n+1} - T_n))_+$. Hence $W = (W_n)_{n\geq 1}$ is a Markov chain with probability kernel P_0, where $P_0(x, dy)$ is the probability distribution of $(x + C - c\xi)_+$. In a GI/G/1 queue, $c = 1$, S_t is the virtual waiting time, and W is called the actual waiting-time process: W_n is the waiting time of customer n, i.e., the time from when he arrives until his service starts in a FIFO discipline (customers are served in order of arrival).

The asymptotic behavior of a Markov chain with probability kernel such as P_0 is known. It is a Lindley chain, i.e., a Markov chain defined as $V_{n+1} = (V_n + U_n)_+$ $(n \geq 0)$, where U_n and V_n are independent for all n and $(U_n)_{n\geq 0}$ are i.i.d. In the example that we consider, we have $U_n = C_n - c(T_{n+1} - T_n)$. Assume $\mathbb{E}(|U_1|) < +\infty$. It can be seen that if $\mathbb{E}(U_1) < 0$, then the probability distribution of V_n converges to some probability distribution ν_0 (Asmussen [5] Chap. III Sect. 7).

That is why from now on we assume $\mathbb{E}(U_1) < 0$, i.e., going back to our example, we assume

$$\mathbb{E}(\mu) < c\,\mathbb{E}(dF) < +\infty, \tag{8.18}$$

where $\mathbb{E}(\mu)$ (respectively $\mathbb{E}(dF)$) is the mean value of an r.v. with probability distribution μ (respectively dF).

When $c = 1$, in queuing terminology condition (8.18) is the well-known condition $\tau < 1$, where $\tau = \mathbb{E}(\mu)/\mathbb{E}(dF)$ is called the traffic intensity.

We go from the case $c > 0$ to the case $c = 1$ by changing the time scale, which is equivalent to replacing the interarrival probability distribution dF by dF_1 defined as $\int f dF_1 = \int f(cu)\, dF(u)$. In any case, let us define τ as $\tau = \mathbb{E}(\mu)/\mathbb{E}(dF_1) = \mathbb{E}(\mu)/(c\mathbb{E}(dF))$. Condition (8.18) is $\tau < 1$.

Since the probability distribution ν_0 is invariant for $(W_n)_{n\geq 1}$, one can check that $\nu_0 * \mu = \nu_0 P_0 * \mu = (\nu_0 * \mu)P$, that is, $\nu = \nu_0 * \mu$ is invariant for the Markov chain $(S_{T_n})_{n\geq 1}$. The measure $\nu(dy)\,\delta_0(du)$ is invariant for the Markov chain $Y = (S_{T_n}, A_{T_n})_{n\geq 1}$, and

$$m_0(dx, du) = m_0(dx_1 dx_2, du) = \bar{F}(u)\,\nu(dx_1)\,\delta_0(dx_2)\,du. \tag{8.19}$$

Theorem 8.5 implies that the measure π_0 defined as

$$\int_{\mathbb{R}_+^2} \varphi(y, v)\,\pi_0(dy, dv) = \int_{\mathbb{R}_+^2} \varphi((y - cu)_+, u)\,\bar{F}(u)\,\nu(dy)\,du \tag{8.20}$$

$$= \int_{\mathbb{R}_+^2} \varphi(0, u)\,1_{\{y\leq cu\}}\,\bar{F}(u)\,\nu(dy)\,du + \int_{\mathbb{R}_+^2} \varphi(y - cu, u)\,1_{\{cu<y\}}\bar{F}(u)\,\nu(dy)\,du$$

is invariant for $(S_t, A_t)_{t\geq 0}$.

Taking $\varphi(y, u) = f(u)$, we obtain the result of Example 8.1, Sect. 8.1. Taking now $\varphi(y, u) = f(y)$, we see that the measure π defined as

$$\int_{\mathbb{R}_+} f(y)\,\pi(dy) = \frac{1}{\mathbb{E}(dF)}\left(\int_{\mathbb{R}_+} \bar{F}(u)\,\nu([0,cu])\,du\right) f(0) + \frac{1}{\mathbb{E}(dF)}\int_{\mathbb{R}_+^2} f(y-cu)\,1_{\{cu<y\}}\bar{F}(u)\,\nu(dy)\,du \tag{8.21}$$

is the probability distribution of S_t for all $t \geq 0$ as soon as the probability distribution of (S_0, A_0) is π_0.

The substitution $w = cu$ in (8.21) gives

$$\int_{\mathbb{R}_+} f(y)\,\pi(dy) = \frac{1}{\mathbb{E}(dF_1)}\left(\int_{\mathbb{R}_+} \bar{F}_1(w)\,\nu([0,w])\,dw\right) f(0) + \frac{1}{\mathbb{E}(dF_1)}\int_{\mathbb{R}_+^2} f(y-w)\,1_{\{w<y\}}\bar{F}_1(w)\,\nu(dy)\,dw, \tag{8.22}$$

which corresponds to the change of scale mentioned above.

Assume now that the probability distribution of S_0 is μ. This is the same as $W_0 = 0$. The Markov chain W is then recurrent with regeneration point $\{0\}$ (Asmussen [5] Chap. VIII Proposition 1.3). Using the generalized Borel–Cantelli lemma, we deduce that the chain $(S_{T_n})_{n\geq 0}$ is μ-recurrent with the same regeneration times as W. Therefore, on the one hand, it has a unique invariant measure, which is therefore ν defined above, and on the other hand, Theorem 8.6 above can be applied: $(S_t, A_t)_{t\geq 0}$ is π_0-recurrent and π_0 is its unique invariant measure (up to a multiplicative constant). Furthermore, ν_0 is the probability distribution of $\sup_{n\geq 0}\sum_{k=0}^{n} U_k$ (Asmussen [5] Chap. III Corollary 7.4).

To have explicit calculations, we must specify μ or dF. Let us denote by $\mathcal{E}_\gamma$ the exponential distribution with parameter γ.

First assume that $\mu = \mathcal{E}_\beta$. At the beginning of Theorem 1.3, Chap. IX of [5], S. Asmussen gives the expression of ν_0 in the case $c = 1$. Replacing dF by dF_1 defined above, we deduce, from Asmussen's result, $\nu_0 = (1-\theta)\,\delta_0 + \theta\,\mathcal{E}_\eta$, where η is the unique positive solution of equation $\int_0^{+\infty} e^{-\eta cu}\,dF(u) = 1 - \eta/\beta$, and $\theta = 1 - \eta/\beta$. Hence

$$\nu = \nu_0 * \mu = (1-\theta)\,\mathcal{E}_\beta + \theta\,\mathcal{E}_\eta * \mathcal{E}_\beta = (1-\theta)\,\mathcal{E}_\beta + \theta\left(\frac{\beta}{\beta-\eta}\mathcal{E}_\eta - \frac{\eta}{\beta-\eta}\mathcal{E}_\beta\right) = \mathcal{E}_\eta.$$

Define $\widehat{dF}$ as the Laplace transform of dF: $\widehat{dF}(s) = \int_{\mathbb{R}_+} e^{-su}\,dF(u)$ $(s \geq 0)$. Equation (8.21) becomes

$$\int_{\mathbb{R}_+} f(y)\,\pi(dy) = \left(1 - \frac{1}{\mathbb{E}(dF)}\frac{1-\widehat{dF}(c\eta)}{c\eta}\right) f(0)$$

$$+\frac{1}{\mathbb{E}(dF)}\frac{1-\widehat{dF}(c\eta)}{c\eta}\int_{\mathbb{R}_+} f(x)\,\eta\,e^{-\eta x}dx$$
$$=(1-\tau)\,f(0)+\tau\int_{\mathbb{R}_+} f(x)\,\eta\,e^{-\eta x}dx,$$

that is, $\pi = (1-\tau)\,\delta_0 + \mathcal{E}_\eta$.

Now assume that μ is a general probability distribution with finite expectation and $dF = \mathcal{E}_\lambda$. Hence $(T_n)_{n\geq 0}$ is a Poisson process with parameter λ. We have seen in Example 3.4, Sect. 3.1 that in this case, there is no need to introduce A_t: $(S_t)_{t\geq 0}$ is a PDMP, and π defined in (8.21) is an invariant probability distribution for this PDMP. Define $\lambda_1 = \lambda/c$. The probability distribution of $(x-c\xi)_+$ is $e^{-\lambda_1 x}\,\delta_0(dy) + h_x(y)\,dy$, where $h_x(y) = 1_{\{0<y<x\}}\lambda_1\,e^{-\lambda_1(x-y)}$. Hence

$$P(x,dy) = e^{-\lambda_1 x}\,\mu(dy) + (h_x * \mu)(dy).$$

Since the Laplace transform on an interval $]s_0,+\infty[$ $(s_0>0)$ of a probability distribution characterizes it, ν is invariant for P iff $\int e^{-sy}\,P(x,dy)\,\nu(dx) = \int e^{-sy}\,\nu(dy)$ for all $s > s_0$. Denoting by $\hat{\nu}$ and $\hat{\mu}$ the respective Laplace transforms of ν and μ, we get from $\nu P = \nu$ that

$$\hat{\nu}(\lambda_1)\,\hat{\mu}(s) + \frac{\lambda_1}{\lambda_1 - s}\hat{\mu}(s)(\hat{\nu}(s) - \hat{\nu}(\lambda_1)) = \hat{\nu}(s),$$

whence

$$\hat{\nu}(s) = \hat{\nu}(\lambda_1)\,\frac{s\,\hat{\mu}(s)}{\lambda_1\,\hat{\mu}(s) + s - \lambda_1}, \qquad s > \lambda_1, \tag{8.23}$$

where the condition $s > \lambda_1$ is to ensure that the denominator is not 0.

From formula (8.22), the mass assigned to δ_0 in π is

$$\frac{1}{\mathbb{E}(dF_1)}\int_{\mathbb{R}_+}\bar{F}_1(w)\,\nu([0,w])\,dw = \lambda_1\int_{\mathbb{R}_+} e^{-\lambda_1 w}\nu([0,w])\,dw = \hat{\nu}(\lambda_1).$$

Let us calculate $\hat{\nu}(\lambda_1)$ by writing $\hat{\nu}(0) = 1$. We have $\hat{\mu}(s) = 1 - \mathbb{E}(\mu)s + s\varepsilon(s)$, $\lim_{s\to 0}\varepsilon(s) = 0$. Hence

$$\hat{\nu}(s) = \hat{\nu}(\lambda_1)\,\frac{1-\mathbb{E}(\mu)s + s\varepsilon(s)}{1-\lambda_1\mathbb{E}(\mu) + \varepsilon_1(s)}.$$

Letting s tend to 0, the condition $\hat{\nu}(0) = 1$ becomes $\hat{\nu}(\lambda_1) = 1 - \lambda_1\,\mathbb{E}(\mu) = 1 - \tau$.

Therefore, when either dF or μ is an exponential distribution, the mass assigned to δ_0 in π is $1-\tau$.

Let us go back to $dF = \mathcal{E}_\lambda$ and $\lambda_1 = \lambda/c$. When $\mu = \mathcal{E}_\beta$, (8.23) gives $\hat{\nu}(s) = a/(\beta - \lambda_1 + s)$ for some a; hence $\nu = \mathcal{E}_{\beta-\lambda_1}$. This is in line with the first particular case studied above, because the equation $\int_0^{+\infty} e^{-\eta c u}\,dF(u) = 1 - \eta/\beta$ gives $\eta = 0$ or $\eta = \beta - \lambda_1$.

As an exercice, keeping $dF = \mathcal{E}_\lambda$, let us take a look at the case in which μ is the probability distribution of $\xi_1 + \xi_2$, where ξ_i $(i = 1, 2)$ has an exponential distribution with parameter β_i. In a queue, it means two successive tasks of exponential distribution for the service. We have $\hat{\mu}(s) = \beta_1\beta_2/(\beta_1 + s)(\beta_2 + s)$, and hence

$$\hat{\nu}(s) = \alpha \frac{\beta_1\beta_2}{s^2 + (\beta_1 + \beta_2 - \lambda_1)s + \beta_1\beta_2 - \lambda_1(\beta_1 + \beta_2)}, \quad \alpha > 0.$$

The condition $\tau < 1$ is $\lambda_1(\beta_1 + \beta_2) < \beta_1 \beta_2$. We deduce

$$s^2 + (\beta_1 + \beta_2 - \lambda_1)s + \beta_1\beta_2 - \lambda_1(\beta_1 + \beta_2) = (s + \alpha_1)(s + \alpha_2), \quad \alpha_1 > \alpha_2 > 0,$$

and as a result, $\nu(du) = \frac{\alpha_1\alpha_2}{\alpha_1 - \alpha_2}(e^{-\alpha_2 u} - e^{-\alpha_1 u})1_{\mathbb{R}_+}(u)\, du$,

$$\pi(dx) = (1 - \tau)\,\delta_0(dx) + \tau \frac{\alpha_1\alpha_2}{\alpha_1 - \alpha_2}\frac{\beta_1\beta_2}{\beta_1 + \beta_2}\left(\frac{1}{\lambda + \alpha_2} e^{-\alpha_2 x} - \frac{1}{\lambda + \alpha_1} e^{-\alpha_1 x}\right) dx.$$

In [51], M.H.A. Davis gives results like Theorem 8.5 and Theorem 8.7 but under stronger assumptions. They concern PDMPs with parameters $\phi, \lambda, Q, \Gamma, q$, where ϕ is governed by an ODE, Γ is a subset of the topological boundary of the state space $\mathring{E}$, and on-site jumps are not allowed. The required correspondence is between the invariant probability distributions of the PDMP and its driving chain. Theorem (34.31) of [51], the analogue of Theorem 8.5 with $\nu(E) = 1$, is proved by checking that $\int_{\mathring{E}} \mathcal{L}f(y)\, m_\phi(dy) = 0$ for f in the domain of the PDMP extended generator, which is fully and explicitly known. The analogue of Theorem 8.7 with $\pi_0(E) = 1$ requires more work. More precisely, let us go back to the notation in Chap. 5. Assuming that π_0 is an invariant probability distribution of $(X_t)_{t\geq 0}$, it is proved in Davis [51], Sect. 34 that the measure σ_Γ^{st} defined as $\sigma_\Gamma^{st}(A) = \frac{1}{t}\int_{\mathring{E}} \underline{\sigma}(x, A \times [0, t])\, \pi_0(dx)$ (A Borel subset of Γ) does not depend on t and $\int_{\mathring{E}} \mathcal{L}f(x)\, \pi_0(dx) + \int_\Gamma \int_{\mathring{E}} (f(y) - f(x))\, q(x; dy)\, \sigma_\Gamma^{st}(dx) = 0$ for all $x \in \mathring{E}$ and suitable functions f (basically, the condition on the boundary for f to be in the domain of the generator is not required). Then it is proved that the probability distribution ν defined as

$$\nu(A) = \frac{\int_{\mathring{E}} \lambda(x)\, Q(x; A)\, \pi_0(dx) + \int_\Gamma q(x; A)\, \sigma_\Gamma^{st}(dx)}{\int_{\mathring{E}} \lambda(x)\, \pi_0(dx) + \sigma_\Gamma^{st}(\Gamma)}$$

is invariant for the driving chain (Theorem (34.21)).

To introduce the time reversal of a Davis PDMP, A. Löpker and Z. Palmowski show more generally in [80] that if π_0 is an invariant probability distribution for the PDMP, then the probability measure μ defined as

$$\mu(A \times B) = \frac{\int_{A\cap\mathring{E}} \lambda(x)\, Q(x; B)\, \pi_0(dx) + \int_{A\cap\Gamma} q(x; B)\, \sigma_\Gamma^{st}(dx)}{\int_{\mathring{E}} \lambda(x)\, \pi_0(dx) + \sigma_\Gamma^{st}(\Gamma)}$$

is invariant for the Markov chain $(X_{T_k-}, X_{T_k})_k$ (Sect. 1.3, Theorem 1).

8.5 Intensities of Stable PDMPs

Let us turn our attention to PDMP stationary intensities ${}^s\rho_{PDMP}$ and ${}^s\varrho$ defined as

$$\int_{E\times\mathbb{R}_+} \varphi(y,v)\,{}^s\rho_{PDMP}(dy,dv) = \int_{E^2\times\mathbb{R}_+} \varphi(y,v)\,\rho(x,dy,dv)\,m_\phi(dx)$$
$$= \sum_{k\geq 1}\int_E \mathbb{E}_x(\varphi(Y_k,T_k))\,m_\phi(dx),$$

$$\int_{E\times\mathbb{R}_+} \varphi(y,v)\,{}^s\varrho(dy,dv) = \int_{E^2\times\mathbb{R}_+} \varphi(y,v)\,\varrho(x,dy,dv)\,m_\phi(dx)$$
$$= \sum_{k\geq 0}\int_E \mathbb{E}_x(\varphi(X_{T_{k+1}-},T_{k+1}))\,m_\phi(dx),$$

where m_ϕ is as in Theorem 8.5, Sect. 8.4, that is,

$$\int_E f(x)\,m_\phi(dx) = \int_{E\times\mathbb{R}_+} f(\phi(x,u))\,m_0(dx,du), \quad m_0(dx,du) = \bar{F}_x(u)\,\nu(dx)\,du,$$

the measure ν being invariant for the driving chain Y.

Recall that to get the steady-state intensities, i.e., assuming $m_\phi(E) < +\infty$, the intensities for which the initial probability distribution of the PDMP is $m_\phi/m_\phi(E)$, the stationary ones must be divided, as for the CSMP, by $\int_{E\times\mathbb{R}_+} \bar{F}_x(u)\,\nu(dx)\,du = \int_E \mathbb{E}_x(T_1)\,\nu(dx)$.

First of all, from (5.4), i.e., $\rho^{x,u} = N_0^{x,u} * R$, we deduce ${}^s\rho_{PDMP}(dy,dv) = \int_{\{(x,u)\in E\times\mathbb{R}_+\}} \rho^{x,u}(dy,dv)\,m_0(dx,du) = m_1 * R\,(dy,dv)$, and therefore

$${}^s\rho_{PDMP}(dy,dv) = {}^s\rho(dy,dv) = \nu(dy)\,1_{\mathbb{R}_+}(v)\,dv. \tag{8.24}$$

It should be noted that

$$\rho^{x,u} = N_0^{x,u} * R = N(\phi(x,u),\,\cdot\,,\,\cdot\,) * R = \rho(\phi(x,u),\,\cdot\,,\,\cdot\,).$$

Do we have $\bar{\rho}^{x,u} = \bar{\rho}(\phi(x,u),\,\cdot\,,\,\cdot\,,\,\cdot\,)$, where $\bar{\rho}^{x,u}$ is as in (5.1), Sect. 5.1 and $\bar{\rho}(x,dy,dv,dw) = \bar{\rho}^{x,0}(dy,dv,dw)$? The answer is no, unless …(see Proposition 8.5 below).

By the way,

$$\int_{E\times\mathbb{R}_+^2} 1_{\{w<v\}}\bar{\rho}(\phi(x,u),dy,dv,dw) = 0, \int_{E\times\mathbb{R}_+^2} 1_{\{w<v\}}\bar{\rho}^{x,u}(dy,dv,dw) = 1_{\{u>0\}}.$$

Actually, on the one hand,

$$\int_{E\times\mathbb{R}_+^2} \psi(y,v,w)\,\bar{\rho}^{x,u}(dy,dv,w) \tag{8.25}$$
$$= \mathbb{E}_{x,u}(\psi(x,u+T_1,T_1)1_{\{T_1<+\infty\}}) + \sum_{k\geq 1}\mathbb{E}_{x,u}(\psi(Y_k,T_{k+1}-T_k,T_{k+1})\,1_{\{T_{k+1}<+\infty\}}),$$

while on the other hand, from $N_0^{x,u} = N(\phi(x,u),\ \cdot\ ,\ \cdot\ )$ we get

$$\int_{E\times\mathbb{R}_+^2} \psi(y,v,w)\,\bar{\rho}(\phi(x,u),dy,dv,w)$$
$$= \mathbb{E}_{\phi(x,u)}(\psi(\phi(x,u),T_1,T_1)1_{\{T_1<+\infty\}}) + \sum_{k\geq 1}\mathbb{E}_{\phi(x,u)}(\psi(Y_k,T_{k+1}-T_k,T_{k+1})\,1_{\{T_{k+1}<+\infty\}}),$$
$$= \mathbb{E}_{x,u}(\psi(\phi(x,u),T_1,T_1)1_{\{T_1<+\infty\}}) + \sum_{k\geq 1}\mathbb{E}_{x,u}(\psi(Y_k,T_{k+1}-T_k,T_{k+1})\,1_{\{T_{k+1}<+\infty\}}).$$

Therefore,

$$\int_{E\times\mathbb{R}_+^2} \psi(y,v,w)\,\bar{\rho}^{x,u}(dy,dv,w) - \int_{E\times\mathbb{R}_+^2} \psi(y,v,w)\,\bar{\rho}(\phi(x,u),dy,dv,w)$$
$$= \mathbb{E}_{x,u}(\psi(x,u+T_1,T_1)1_{\{T_1<+\infty\}}) - \mathbb{E}_{\phi(x,u)}(\psi(\phi(x,u),T_1,T_1)1_{\{T_1<+\infty\}}). \tag{8.26}$$

We are now interested in the intensity of $(Y_k, T_{k+1}-T_k, T_{k+1})_k$ under $\mathbb{P}_{m_\phi}(\,.\,) = \int_{\{x\in E\}}\mathbb{P}_x(\,.\,)\,m_\phi(dx)$.

Proposition 8.5 *(A) Given a Borel function $\psi : E\times\mathbb{R}_+^2 \to \mathbb{R}_+$ such that*

$$\forall (x,u,v)\in E\times\mathbb{R}_+,\ \psi(\phi(x,u),v,v) = \psi(x,u+v,v), \tag{8.27}$$

then

$$\int_E \sum_{k\geq 0}\mathbb{E}_x(\psi(Y_k,T_{k+1}-T_k,T_{k+1})\,1_{\{T_{k+1}<+\infty\}})\,m_\phi(dx)$$
$$= \int_{E\times\mathbb{R}_+^2}\psi(y,v,w)\,{}^s\bar{\rho}(dy,dv,dw) = \int_{E\times\mathbb{R}_+}\psi(y,v,w)\,dF_y(v)\,\nu(dy)\,dw.$$

Since ϕ is a deterministic Markov flow, when $\psi(y,v,w) = \varphi(\phi(y,v),w)$, then (8.27) holds.
If, moreover, $\alpha : E \to \mathbb{R}_+^*$ is a Borel function such that $\alpha(\phi(x,u)) = \alpha(x) - u$ for

all $x \in E$, $u < \alpha(x)$, then (8.27) holds for $\psi(y, v, w) = 1_{\{v<\alpha(y)\}}\, \varphi(\phi(y, v), w)$ and for $\psi(y, v, w) = 1_{\{v=\alpha(y)\}}\, \varphi(\phi(y, \alpha(y)), w)$.

(B) For every Borel function $\psi : E \times \mathbb{R}_+^2 \to \mathbb{R}_+$, we have

$$\int_E \sum_{k\geq 1} \mathbb{E}_x(\psi(Y_k, T_{k+1} - T_k, T_{k+1})\, 1_{\{T_{k+1}<+\infty\}})\, m_\phi(dx) =$$
$$\int_{E\times\mathbb{R}_+^2} \psi(y, v, w)\, 1_{\{v<w\}}\, {}^s\bar{\rho}(dy, dv, dw) = \int_{E\times\mathbb{R}_+} \psi(y, v, w)\, 1_{\{v<w\}}\, dF_y(v)\, \nu(dy)\, dw.$$

Proof The definition of m_ϕ gives

$$\int_E \sum_{k\geq 0} \mathbb{E}_x(\psi(Y_k, T_{k+1} - T_k, T_{k+1})\, 1_{\{T_{k+1}<+\infty\}})\, m_\phi(dx) =$$
$$\int_{E\times\mathbb{R}_+^2} \psi(y, v, w)\, \bar{\rho}(\phi(x, u), dy, dv, dw)\, m_0(dx, du)\,.$$

Hence the first part of Proposition 8.5 is a consequence of (8.26) and (8.13), Sect. 8.3.

For the second part, we deduce from (8.25) that

$$\int_{E\times\mathbb{R}_+^2} \psi(y, v, w)\, 1_{\{v<w\}}\, \bar{\rho}^{x,u}(dy, dv, dw) = \sum_{k\geq 1} E_{x,u}(\psi(Y_k, T_{k+1} - T_k, T_{k+1})\, 1_{\{T_{k+1}<+\infty\}}).$$

Since $N_0^{x,u}(dy, dv) = N(\phi(x, u), dy, dv)$, we have

$$\mathbb{E}_{x,u}(\psi(Y_k, T_{k+1} - T_k, T_{k+1})\, 1_{\{T_{k+1}<+\infty\}}) = \mathbb{E}_{\phi(x,u)}(\psi(Y_k, T_{k+1} - T_k, T_{k+1})\, 1_{\{T_{k+1}<+\infty\}})$$

for all $k \geq 1$. Therefore,

$$\int_E \sum_{k\geq 1} \mathbb{E}_x(\psi(Y_k, T_{k+1} - T_k, T_{k+1})\, 1_{\{T_{k+1}<+\infty\}})\, m_\phi(dx)$$
$$= \int_{E\times\mathbb{R}_+^2} \psi(y, v, w)\, 1_{\{v<w\}}\, {}^s\bar{\rho}(dy, dv, dw),$$

and we apply again (8.13), Sect. 8.3. ■

Corollary 8.1 *The stationary intensity ${}^s\varrho$ is*

$$\int_{E\times\mathbb{R}_+^2} \varphi(y,v)\,{}^s\varrho(dy,dv) := \int_{E\times\mathbb{R}_+^2} \varphi(y,v)\,\varrho(x,dy,dv)\,m_\phi(dx)$$
$$= \int_{E\times\mathbb{R}_+^2} \varphi(\phi(y,v),w)\,dF_y(v)\,\nu(dy)\,dw.$$

Example 8.9 Let us go back to Example 3.4, Sect. 3.1, assuming $\mathbb{E}(dF) < +\infty$. The notation is as in Example 5.9, Sect. 5.3 and Example 8.8, Sect. 8.4. Recall that the PDMP is $(S_t, A_t)_{t\geq 0}$. Here we assume that ν is an invariant probability distribution of the Markov chain $(S_{T_k})_{k\geq 1}$, hence $\nu(dx_1)\,\delta_0(dx_2)$ is an invariant probability distribution for the driving chain $(S_{T_k}, A_{T_k})_{k\geq 1} = (S_{T_k}, 0)_{k\geq 1}, m_0(dx_1dx_2, dv) = \bar{F}(v)\,\nu(dx_1)\,\delta_0(dx_2)\,dv$ and $m_0(E\times\mathbb{R}_+) = \mathbb{E}(dF)$.

We deduce from (5.24), Sect. 5.3 and Corollary 8.1 that in steady state, the mean number of times the stock level is at most a at the time of a delivery that occurs during $]0,t]$ is

$$\mathbb{E}(N_a(t)) = \frac{1}{\mathbb{E}(dF)}\,{}^s\varrho([0,a]\times\mathbb{R}_+\times]0,t])$$
$$= \frac{1}{\mathbb{E}(dF)} \int_{\mathbb{R}_+^3} 1_{\{(y_1-cv)_+\leq a,\,0<w\leq t\}}\,dF(v)\,\nu(dy_1)\,dw$$
$$= \frac{t}{\mathbb{E}(dF)} \int_{\mathbb{R}_+} \nu([0,cv+a])\,dF(v).$$

Now we are interested in the loss of production during a time period such as $]s, s+t]$. Thanks to (5.25), Sect. 5.3, in steady state, we have

$$\mathbb{E}(N^{lp}(t)) = \int_E \sum_{k\geq 0} \mathbb{E}_x(\psi(Y_k, T_{k+1}-T_k, T_{k+1})\,1_{\{T_{k+1}<+\infty\}})\,m_\phi(dx)/\mathbb{E}(dF)$$

with $\psi(y_1, y_2, v, w) = 1_{\{y_1<cv,\,w\leq t\}}$, since the probability distributions of S_t and S_0 are the same. The function ψ does not satisfy (8.27), so we are going to apply the second part of Proposition 8.5. We get

$$\mathbb{E}(dF)\,\mathbb{E}(N^{lp}(t)) = \int_{E\times\mathbb{R}_+^2} 1_{\{(x_1-c(v-w))_+<cw,w\leq t\}}\,1_{\{w<v\}}\,dF(v)\,\nu(dx_1)\,dw$$
$$+ \int_{E\times\mathbb{R}_+^2} 1_{\{y_1<cv,\,w\leq t\}}\,1_{\{v<w\}}\,dF(v)\,\nu(dy_1)\,dw$$
$$= \int_{E\times\mathbb{R}_+^2} 1_{\{y_1<cv,\,w\leq t\}}\,dF(v)\,\nu(dy_1)\,dw.$$

Therefore,

$$\mathbb{E}(N^{lp}(t)) = \frac{t}{\mathbb{E}(dF)} \int_{\mathbb{R}_+} \nu([0, cv[)\, dF(v) = \frac{t}{\mathbb{E}(dF)} \int_{\mathbb{R}_+} \bar{F}(y/c)\, \nu(dy). \quad (8.28)$$

Let us look at how $\mathbb{E}(N^{lp}(t))$ is calculated with the second model, i.e., when we add jumps when the stock becomes empty. The notation is the same as that at the end of Example 3.4, Sect. 3.1 and Example 5.9, Sect. 5.3. Denote by ν_+ a stationary measure of the Markov chain $(S_{T_k^+}, A_{T_k^+})_{k\geq 1}$. It can be written $\nu_+(dx_1, dx_2) = \nu_1(dx_1)\, \delta_0(dx_2) + \delta_0(dx_1)\, \nu_2(dx_2)$, $\nu_1(\{0\}) = 0$. One suspects that ν_1 is invariant for the Markov chain $(S_{T_k})_{k\geq 1}$, which is what we are going to prove. The kernel of the Markov chain $(S_{T_k})_{k\geq 1}$ is P, where $P(x, dy)$ is the probability distribution of $(x - c\xi)_+ + C$, ξ and C are independent, the probability distributions of ξ and C are dF and μ, respectively (Example 3.4, Sect. 3.1). Hence

$$\int_{\mathbb{R}_+} f(y)\, P(x, dy) = \int_{\mathbb{R}_+^2} f(x - cv + y)\, 1_{\{v\leq x/c\}}\, dF(v)\, \mu(dy) + \bar{F}(x/c) \int_{\mathbb{R}_+} f(y)\, \mu(dy).$$

Thanks to (3.6) and (3.7), Sect. 3.1 the transition kernel P_+ of the Markov chain $(S_{T_k^+}, A_{T_k^+})_{k\geq 1}$ is

$$\int_{\mathbb{R}_+^2} \varphi(y_1, y_2)\, P_+((x_1, x_2), dy_1 dy_2) = \int_{\mathbb{R}_+^2} \varphi(x_1 - cv + y_1, 0) 1_{\{v\leq x_1/c\}}\, \mu(dy_1)\, dF_{x_2}(v)$$
$$+ \bar{F}_{x_2}(x_1/c)\, \varphi(0, x_2 + x_1/c), \quad x_1 > 0$$

$$\int_{\mathbb{R}_+^2} \varphi(y_1, y_2)\, P_+((0, x_2), dy_1 dy_2) = \int_{\mathbb{R}_+} \varphi(y_1, 0)\, \mu(dy_1).$$

We have $\nu_+ = \nu_+ P_+$ iff

$$\int_{\mathbb{R}_+} f(x)\, \nu_2(dx) = \int_{\mathbb{R}_+} f(x/c)\, \bar{F}(x/c)\, \nu_1(dx), \quad (8.29)$$

$$\int_{\mathbb{R}_+} f(x)\, \nu_1(dx) = \int_{\mathbb{R}_+} f(x - cv + y)\, 1_{\{v\leq x/c\}}\, dF(v)\, \mu(dy)\, \nu_1(dx) + \nu_2(\mathbb{R}_+) \int_{\mathbb{R}_+} f(y)\, \mu(dy). \quad (8.30)$$

Thanks to (8.29), we get $\nu_2(\mathbb{R}_+) = \int_{\mathbb{R}_+} \bar{F}(x/c)\, \nu_1(dx)$. By entering this value into (8.30), this equation becomes $\nu_1 = \nu_1 P$.

We deduce from (5.27), Sect. 5.3 and (8.24), Sect. 8.5 that

$$m_\phi^+(\mathbb{R}_+^2)\, \mathbb{E}(N^{lp}(t)) = {}^s\rho_+(\{0\} \times \mathbb{R}_+ \times [0,t]) = t\, \nu_2(\mathbb{R}_+) = t \int_{\mathbb{R}_+} \bar{F}(x/c)\, \nu_1(dx). \tag{8.31}$$

We have $m_\phi^+(\mathbb{R}_+^2) = \int_{\mathbb{R}_+^3} \bar{F}_{x_1,x_2}^+(u)\, \nu^+(dx_1, dx_2)\, du$, where

$$dF_{x_1,x_2}^+(dv) = \int_{\{(y_1,y_2)\in\mathbb{R}_+^2\}} N_+((x_1,x_2), dy_1, dy_2, dv)$$

is given by (3.6) and (3.7), Sect. 3.1. We get

$$m_\phi^+(\mathbb{R}_+^2) = \int_{\mathbb{R}_+^2} \bar{F}_{x_1,0}^+(u)\, \nu_1(dx_1)\, du + \int_{\mathbb{R}_+^2} \bar{F}_{0,x_2}^+(u)\, \nu_2(dx_2)\, du,$$

$$dF_{x_1,0}^+(dv) = 1_{\{v \le x_1/c\}}\, dF(v) + \bar{F}(x_1/c)\, \delta_{x_1/c}(dv), \quad dF_{0,x_2}^+(dv) = dF_{x_2}(v).$$

Using (8.29), we obtain $m_\phi^+(\mathbb{R}_+^2) = \mathbb{E}(dF)\, \nu_1(\mathbb{R}_+)$. Hence (8.31) gives (8.28).

The corollary below is an immediate consequence of Corollary 8.1 and of Proposition 5.4, Sect. 5.3.

Corollary 8.2 *Let $(Y_k, T_k)_{k\ge 0}$ be a Markov renewal process with kernel N, ν an invariant measure for the Markov chain $(Y_k)_{k\ge 0}$, and $(X_t)_{t\ge 0}$ a PDMP generated by ϕ and $(Y_k, T_k)_{k\ge 0}$.*

Assume $N(x, dy, dv) = 1_{R_+}(v)\, Q(\phi(x,v); dy)\, dF_x(v)$.

Given B_1 and B_2 two Borel subsets of E, define

$$N_{B_1,B_2}(t) = \sum_{k\ge 0} 1_{B_1}(X_{T_{k+1}-})\, 1_{B_2}(X_{T_{k+1}})\, 1_{[0,t]}(T_{k+1}).$$

Assume $\int_E \mathbb{E}_y(T_1)\, \nu(dy) < +\infty$. In steady state, that is, when the probability distribution of X_0 is m_ϕ defined in (8.16), Sect. 8.4, we have

$$\mathbb{E}(N_{B_1,B_2}(t)) = \int_{E\times\mathbb{R}_+} 1_{B_1}(y)\, Q(y; B_2)\, 1_{[0,t]}(v)\, {}^e\varrho(dy, dv)$$
$$= \frac{t}{\int_E \mathbb{E}_y(T_1)\, \nu(dy)} \int_{E\times\mathbb{R}_+} 1_{B_1}(\phi(y,v))\, 1_{[0,t]}(v)\, Q(\phi(y,v); B_2)\, dF_y(v)\, \nu(dy).$$

In Corollary 8.2 there is no assumption about dF. For a parametrized PDMP we have a more accurate result.

Proposition 8.6 *Let $(X_t)_{t\ge 0}$ be a PDMP with parameters $\phi, \lambda, Q, \alpha, (E_i, q_i)_{i\in I}$, and $(Y_k, T_k)_{k\ge 0}$ the associated Markov renewal process. Given B_1 and B_2 two Borel* *subsets of E, define $N^c_{B_1,B_2}(t)$ and $N^D_{B_1,B_2}(t)$ as the numbers of jumps up to time t from B_1 to B_2 that respectively are not generated by the Dirac measures (purely random jumps) and are generated by the Dirac measures, i.e.,*

$$N^c_{B_1,B_2}(t) = \sum_{k\geq 0} 1_{\{T_{k+1}\leq t, T_{k+1}-T_k<\alpha(Y_k)\}}\, 1_{B_1}(X_{T_{k+1}-})\, 1_{B_2}(X_{T_{k+1}}),$$

$$N^D_{B_1,B_2}(t) = \sum_{k\geq 0} 1_{\{T_{k+1}\leq t, T_{k+1}-T_k=\alpha(Y_k)\}}\, 1_{B_1}(X_{T_{k+1}-})\, 1_{B_2}(X_{T_{k+1}}).$$

Assume $\int_E \mathbb{E}_y(T_1)\,\nu(dy) < +\infty$. Then in steady state, that is, when the probability distribution of X_0 is m_ϕ defined in (8.16), Sect. 8.4, we have

$$\mathbb{E}(N^c_{B_1,B_2}(t)) = \frac{t}{\int_E \mathbb{E}_y(T_1)\,\nu(dy)} \times$$
$$\int_{E\times\mathbb{R}_+} 1_{B_1}(\phi(x,v))\, Q(\phi(x,v);B_2)\, 1_{\{v<\alpha(x)\}}\, \lambda(\phi(x,v))\, e^{-\int_0^v \lambda(\phi(x,s))\,ds}\, \nu(dx)\,dv$$

$$\mathbb{E}(N^D_{B_1,B_2}(t)) = \frac{t}{\int_E \mathbb{E}_y(T_1)\,\nu(dy)} \times$$
$$\int_E 1_{B_1}(\phi(x,\alpha(x))) \sum_{i\in I} 1_{E_i}(x)\, q_i(\phi(x,\alpha(x));B_2)\, e^{-\int_0^{\alpha(x)} \lambda(\phi(x,s))\,ds}\, \nu(dx).$$

Proof Corollary 5.2, end of Sect. 5.3, leads to

$$\mathbb{E}(N^c_{B_1,B_2}(t)) = \int_{(E\times\mathbb{R}_+)^2} \psi_1(y,v,w)\, \bar{\rho}(x,dy,dv,dw)\, m_\phi(dx)$$

with $\psi_1(y,v,w) = 1_{\{w\leq t\}}\, 1_{\{v<\alpha(y)\}} 1_{B_1}(\phi(y,v))\, Q(\phi(y,v);B_2)$ and

$$\mathbb{E}(N^D_{B_1,B_2}(t)) = \int_{(E\times\mathbb{R}_+)^2} \psi_2(y,v,w)\, \bar{\rho}(x,dy,dv,dw)\, m_\phi(dx)$$

with $\psi_2(y,v,w) = 1_{\{w\leq t\}}\, 1_{\{v=\alpha(y)\}} 1_{B_1}(\phi(y,v)) \sum_{i\in I} 1_{E_i}(y) q_i(\phi(y,v);B_2)$.

Part A of Proposition 8.5, Sect. 8.5 gives immediately the result for $\mathbb{E}(N^c_{B_1,B_2}(t))$. On the other hand,

$$\mathbb{E}(N^D_{B_1,B_2}(t)) = \int_E \mathbb{E}_x(\psi_2(x,T_1,T_1)\, 1_{\{T_1<+\infty\}})\, m_\phi(dx)$$
$$+ \int_E \sum_{k\geq 1} \mathbb{E}_x(\psi_2(Y_k, T_{k+1}-T_k, T_{k+1})\, 1_{\{T_{k+1}<+\infty\}})\, m_\phi(dx).$$

Since ϕ is a deterministic Markov flow and $\alpha(\phi(x,u)) = \alpha(x) - u$, we get

$$\int_E \mathbb{E}_x(\psi_2(x, T_1, T_1)\, 1_{\{T_1<+\infty\}})\, m_\phi(dx) = \int_{E\times\mathbb{R}_+} \psi_2(x, v, v)\, dF_x(v)\, m_\phi(dx)$$

$$= \int_E \psi_2(x, \alpha(x), \alpha(x))\, e^{-\int_0^{\alpha(x)} \lambda(x,s)\,ds}\, m_\phi(dx)$$

$$= \int_E 1_{\{\alpha(x)-u\le t\}}\, 1_{B_1}(\phi(x, \alpha(x))\, e^{-\int_0^{\alpha(x)-u} \lambda(\phi(x,u+s))\,ds}$$

$$\times \sum_{i\in I} 1_{E_i}(\phi(x,u))\, q_i(\phi(x,\alpha(x)); B_2)\, 1_{\{u<\alpha(x)\}}\, e^{-\int_0^u \lambda(x,s)\,ds}\, \nu(dx)\, du$$

$$= \int 1_{\{w\le t\}}\, 1_{\{w<\alpha(x)\}}\, 1_{B_1}(\phi(x,\alpha(x))) \sum_{i\in I} 1_{E_i}(x)\, q_i(\phi(x,\alpha(x)); B_2)$$

$$\times\, e^{-\int_0^{\alpha(x)} \lambda(\phi(x,s))\,ds}\, \nu(dx)\, dw.$$

Finally, part B of Proposition 8.5 gives

$$\int_E \sum_{k\ge1} \mathbb{E}_x(\psi_2(Y_k, T_{k+1} - T_k, T_{k+1})\, 1_{\{T_{k+1}<+\infty\}})\, m_\phi(dx)$$

$$= \int 1_{\{w\le t\}}\, 1_{\{\alpha(x)<w\}}\, 1_{B_1}(\phi(x,\alpha(x))) \sum_{i\in I} 1_{E_i}(x)\, q_i(\phi(x,\alpha(x)); B_2)$$

$$\times\, e^{-\int_0^{\alpha(x)} \lambda(\phi(x,s))\,ds}\, \nu(dx)\, dw.$$

This ends the proof. ■

When $(X_t)_{t\ge0}$ is a parametrized PDMP, we have $\varrho = \varrho_c + \underline{\sigma}$ (Proposition 5.5, Sect. 5.3). It is the same for ${}^s\varrho$.

Proposition 8.7 *Let $(X_t)_{t\ge0}$ be a PDMP with parameters $\phi, \lambda, Q, \alpha, (E_i, q_i)_{i\in I}$. Define ${}^s\varrho_c$ and stationary boundary measure ${}^s\sigma$ as*

$$\int_{E\times R_+} \varphi(y,v)\, {}^s\varrho_c(dy, dv) = \int_{E\times R_+} \varphi(y,v)\, \varrho_c(x, dy, dv)\, m_\phi(dx),$$

$$\int_{E\times R_+} \varphi(y,v)\, {}^s\sigma(dy, dv) = \int_{E\times R_+} \varphi(y,v)\, \underline{\sigma}(x, dy, dv)\, m_\phi(dx),$$

where ϱ_c and $\underline{\sigma}$ are as in Proposition 5.5, Sect. 5.3 and m_ϕ is as in (8.16) Sect. 8.4. Then ${}^s\varrho = {}^s\varrho_c + {}^s\sigma$,

$$\int_{E\times\mathbb{R}_+} \varphi(y,v)\,{}^s\varrho_c(dy,dv)$$
$$= \int_{E\times\mathbb{R}_+^2} \varphi(\phi(y,v),w)\,1_{\{v<\alpha(y)\}}\,\lambda(\phi(y,v))\,e^{-\int_0^v \lambda(\phi(y,s))\,ds}dv\,\nu(dy)\,dw,$$

$$\int_{E\times\mathbb{R}_+} \varphi(y,v)\,{}^s\sigma(dy,dv) = \int_{E\times\mathbb{R}_+} \varphi(\phi(y,\alpha(y)),v)\,e^{-\int_0^{\alpha(y)} \lambda(\phi(y,s))\,ds}\,\nu(dy)\,dv.$$

Proof The first formula is a direct consequence of (5.28), Sect. 5.3, $\ell(y,v) = \lambda(\phi(y,v))$, and (8.16), Sect. 8.4.

The second formula comes from (5.29), Sect. 5.3 and part A of Proposition 8.5, Sect. 8.5. ∎

When $m_\phi(E) = \int_E \mathbb{E}_y(T_1)\,\nu(dy) < +\infty$, recall that to get the steady state intensities that are defined as the intensities when the probability distribution of X_0 is $m_\phi/m_\phi(E)$ we must divide the stationary intensities given above by $\int_E \mathbb{E}_y(T_1)\,\nu(dy)$.

8.6 Asymptotic Results

Corollary 2.3, Sect. 2.7 applied with $\varphi = f \circ \phi$ provides for PDMPs a criterion for convergence toward the invariant probability distribution.

Theorem 8.8 *Let $(X_t)_{t\geq 0}$ be a PDMP generated by ϕ and the Markov renewal process $(Y,T) = (Y_n,T_n)_{n\geq 0}$. Assume that the Markov renewal process (Y,T) is nonarithmetic, the driving chain Y is Harris-recurrent with invariant probability distribution ν, and $\int_E \mathbb{E}_y(T_1)\,\nu(dy) < +\infty$.*

Let $f : E \to \mathbb{R}$ be a continuous bounded function. Then

$$\mathbb{E}_x(f(X_t)) \xrightarrow[t\to+\infty]{} \frac{\displaystyle\int_{E\times\mathbb{R}_+} f(\phi(y,u))\,\mathbb{P}_y(T_1>u)\,du\,\nu(dy)}{\displaystyle\int_E \mathbb{E}_y(T_1)\,\nu(dy)} = \frac{\displaystyle\int_E \mathbb{E}_y\left(\int_0^{T_1} f(\phi(y,u))\,du\right)\nu(dy)}{\displaystyle\int_E \mathbb{E}_y(T_1)\,\nu(dy)}$$

for ν-almost all $x \in E$.

Example 8.10 Let us return to Example 8.8, Sect. 8.4 and its notation. We assume that S_0 has the probability distribution μ and τ <1, i.e., $\mathbb{E}(\mu) < c\mathbb{E}(dF)$. We have seen that the Markov chain Y is ν-recurrent, and we have $\int_E \mathbb{E}_y(T_1)\,\nu(dy) =$

$\mathbb{E}(dF)\,\nu(E) = \mathbb{E}(dF) < +\infty$. If dF is nonarithmetic, we can apply Theorem 8.8, and we get for every continuous bonded function $\varphi : \mathbb{R}_+^2 \times \mathbb{R}$ and ν-almost every $x \in E$ that

$$\mathbb{E}_x(\varphi(S_t, A_t)) \xrightarrow[t\to+\infty]{} \int_{\mathbb{R}_+^2} \varphi(y, v)\,\pi_0(dy, dv), \tag{8.32}$$

where π_0 is as in (8.20), Sect. 8.4. Now we further assume that ν_0 can be written as $\nu_0(dx) = p\,\delta_0(dx) + \nu_0'(dx)$ for some $p \in]0, 1]$ and ν_0'. This condition holds as soon as $\int_{\mathbb{R}_+} \bar{F}((x + y)/c)\,\nu(dy) > 0$ for all $x \in E$. Indeed, $\nu_0 = \nu_0 P_0$ gives $\int_{\mathbb{R}_+} f(y)\,\nu_0(dy) = \left(\int_{\mathbb{R}_+} \bar{F}((x + y)/c)\,\mu(dy)\,\nu_0(dx)\right) f(0) + \int_{\mathbb{R}_+} f(y)\,\nu_0'(dy)$. We deduce $\nu = \nu_0 * \mu \geq p\mu$, and hence $0 \leq \mu(\{x : (8.32)\,\textit{does not hold}\}) \leq \nu(\{x : (8.32)\,\textit{does not hold}\}) = 0$. We deduce

$$\mathbb{E}_\mu(\varphi(S_t, A_t)) := \int_{\mathbb{R}_+} \mathbb{E}_x(\varphi(S_t, A_t))\,\mu(dx) \xrightarrow[t\to+\infty]{} \int_{\mathbb{R}_+^2} \varphi(y, v)\,\pi_0(dy, dv).$$

In particular,

$$\begin{aligned}\mathbb{E}_\mu(f(S_t)) \xrightarrow[t\to+\infty]{} &\frac{1}{\mathbb{E}(dF)}\left(\int_{\mathbb{R}_+} \bar{F}(u)\,\nu([0, cu])\,du\right) f(0)\\ &+ \frac{1}{\mathbb{E}(dF)} \int_{\mathbb{R}_+^2} f(y - cu)\,1_{\{cu<y\}}\bar{F}(u)\,\nu(dy)\,du\end{aligned}$$

for every continuous function $f : \mathbb{R}_+ \to \mathbb{R}$.

Theorem 8.8 shows that if T_1 is not large enough, the long-time behavior of a PDMP has nothing to do with the long-time behavior of the flow. Interesting examples are given in the works of Benaïm et al. [9], Lawley et al. [79], and Malrieu [83].

Here only a general result that is an immediate consequence of the semiregenerative property is given. For stronger results obtained in particular cases, see, for instance, Azaïs et al. [7], Benaïm et al. [10], Borovkov and Vere-Jones [16], Browne and Sigman [22], Chafaï et al. [25], Dufour et al. [55], Fontbona et al. [60], Harrison and Resnick [63, 64], Zheng [98].

Chapter 9
Numerical Methods

Monte Carlo simulations are commonly used to get the expectation of quantities related to PDMPs. A rigorous and comprehensive presentation with very interesting examples can be found in De Saporta, Dufour, and Zhang [53].

The aim of this chapter is to provide an alternative method with numerical schemes for parametrized PDMPs. They deliver approximations of the PDMP's marginal distributions and also of the boundary measure σ. We have seen in previous chapters that these are the tools for assessing PDMPs.

Three numerical schemes are given. They are obtained by approximating the generalized Chapman–Kolmogorov equations with finite volume schemes.

In Sect. 9.1 a method for proving their convergence is given. In Sects. 9.2 to 9.4 numerical schemes are introduced. The first scheme is explicit; the two others are implicit. Only the last one is suitable for PDMPs with boundary.

Section 9.5 provides examples. The first example presents the pros and cons of implicit and explicit schemes. The second proves that the cohabitation of spaces with various dimensions does not create a problem. The last one is an example in which the measure σ is interesting in practice.

This chapter was written with Ludovic Goudenège,[1] who in particular managed all the numerical experiments.

Other numerical experiments can be found in Cocozza-Thivent et al. [31] to [40], Eymard and Mercier [58], Eymard et al. [59], Lair et al. [76].

[1]FR3487 - Fédération de Mathématiques de l'École Centrale Paris, France.

C. Cocozza-Thivent, *Markov Renewal and Piecewise Deterministic Processes*, Probability Theory and Stochastic Modelling 100,
https://doi.org/10.1007/978-3-030-70447-6_9

9.1 Method

Throughout this chapter, E is a subset of $\mathbb{R}^d$ or of a hybrid space $\cup_{i\in I}\{i\} \times \mathbb{R}^{d_i}$, I finite. The marginal probability distributions of the PDMP are denoted by $(\pi_t)_{t\geq 0}$.

For a PDMP without boundary, with parameters ϕ, λ, Q, associated with a system of ordinary differential equations with velocity $\mathbf{v}$, we get from (6.9), Sect. 6.2 that:

$$\int_E f(x)\,\pi_t(dx) = \int_E f(x)\,\pi_0(dx) + \int_0^t \int_E \mathbf{v}(x)\cdot\nabla f(x)\,\pi_s(dx)\,ds \quad (9.1)$$
$$+ \int_0^t \int_E \lambda(x) \int_E (f(y) - f(x))\,Q(x;dy)\,\pi_s(dx)\,ds \quad \forall\, f \in C_c^1,$$

where C_c^1 is the set of continuous differentiable functions with compact support.

To explain where the schemes are coming from, let us first assume that $E = \mathbb{R}^d$, $\pi_0(dx) = \tilde{\pi}_0(x)\,dx$, and $\lambda(x)\,Q(x;dy) = a(x,y)\,dy$. Then $\pi_t(x) = u(x,t)\,dx$, and we deduce from equation (9.1) that u satisfies in a weak sense the following hyperbolic equation:

$$\forall x \in \mathbb{R}^d, \quad u(x,0) = \tilde{\pi}_0(x),$$

$$\frac{du}{dt}(x,t) + \operatorname{div}(u(x,t).\mathbf{v}(x)) = \int_{\mathbb{R}^d} a(y,x)\,u(y,t)\,dy - \lambda(x)\,u(x,t)$$

for all $(x,t) \in \mathbb{R}^d \times \mathbb{R}_+$. The schemes set out in Sects. 9.2 and 9.3 are finite-volume schemes that approximate these equations.

Let us now consider a PDMP $(X_t)_{t\geq 0}$ with parameters ϕ, λ, Q, $(E_i, \Gamma_i, q_i)_{i\in I}$, I finite (in the case of a PDMP without boundary, $I = \emptyset$). The Chapman-Kolmogorov equations are

$$\int_E g(x,t)\,\pi_t(dx) = \int_E g(x,0)\,\pi_0(dx) + \int_0^t \int_E \partial_{t,\phi} g(x,s)\,\pi_s(dx)\,ds$$
$$+ \int_0^t \int_E \lambda(x) \int_E (g(y,s) - g(x,s))\,Q(x;dy)\,\pi_s(dx)\,ds$$
$$+ \sum_{i\in I} \int_{\Gamma_i^* \times]0,t]} \int_E (g(y,s) - g_i(x,s))\,q_i(x;dy)\,\sigma_i(dx,ds) \quad (9.2)$$

for "suitable functions g and g_i" (Theorem 6.1, Sect. 6.2). Let us recall that for all $x \in E_i$, $\alpha(x) = \inf\{t > 0 : \phi(x,t) \in \Gamma_i\}$, $\Gamma_i^* = \{\phi(x,\alpha(x)) : x \in E_i, \alpha(x) < +\infty, \int_0^{\alpha(x)} \lambda(\phi(x,v))\,dv < +\infty\}$, and σ_i is as defined in Proposition 5.7, Sect. 5.3. In Sect. 9.4, a finite-volume scheme is given to approximate π_t $(t \geq 0)$ and $(\sigma_i)_{i\in I}$ (I finite).

For a Borel set $A \subset \mathbb{R}^d$ (resp. $A = \{i\} \times A'$, $A' \subset \mathbb{R}^d$), define $|A| = \int_A dx$ (resp. $|A| = \int_{A'} dx$) where dx is the Lebesgue measure. Let $\mathcal{M}$ be an admissible mesh of $\underline{E}$, a space in which the PDMP takes values, consistent with the partition $(E_i)_{i \in I}$ when there is one, i.e.:

- $\mathcal{M}$ is a countable partition of E,
- $\forall K \in \mathcal{M}$, there exists $i \in I$ such that $K \subset E_i$ and $|K| > 0$,
- $|\mathcal{M}| := \sup_{K \in \mathcal{M}} \mathrm{diam(K)} < +\infty$, where $\mathrm{diam(K)} = \sup_{\mathrm{x,y} \in \mathrm{K}} ||\mathrm{x} - \mathrm{y}||$ and $||.||$ is a norm on $\mathbb{R}^{d_i}$.

Let $\delta t > 0$, and if there is a boundary, $\tau > 0$. We set $\mathcal{D} = (\mathcal{M}, \delta t)$ if there is no boundary and $\mathcal{D} = (\mathcal{M}, \delta t, \tau)$ if there is one.
We introduce a family $(p_n^{(K)})_{n \in \mathbb{N}, K \in \mathcal{M}}$ of real values according to a finite-volume scheme, and for all $t > 0$, we define the measure $P_t^{\mathcal{D}}$ as the measure with density $p_n^{(K)}$ on K if $n\,\delta t \leq t < (n+1)\,\delta t$. Under some assumptions, we prove that the measures $P_t^{\mathcal{D}}$ are approximations of the marginal distribution π_t of the PDMP at time t, that is, for all bounded continuous functions $f : E \to \mathbb{R}$,

$$\int_E f(x)\, P_t^{\mathcal{D}}(dx) := \sum_K p_n^{(K)} \int_K f(x)\, dx, \quad (n\delta t \leq t < (n+1)\delta t) \tag{9.3}$$

converges to $\int_E f(x)\, \pi_t(dx) := \mathbb{E}(f(X_t))$ when δt, $|\mathcal{M}|$ (and τ in the case of a boundary) tend to 0. This convergence of probability measures is called weak convergence or narrow convergence or weak* convergence.

In Cocozza-Thivent et al. [40] and Eymard et al. [59], a PDMP's flow is assumed to be governed by a system of ordinary differential equations with velocity $\mathbf{v}$ and there is no boundary. The scheme's convergence is obtained by proving the following results:

1. For all $T > 0$, the family of measures $(P_t^{\mathcal{D}})_{(t \leq T, \mathcal{D})}$ is tight.
2. Let $(t_p)_{p \geq 0}$ be a sequence of real numbers dense in $\mathbb{R}_+$ and $(\mathcal{D}_m)_{m \geq 0}$ a sequence of items such as $\mathcal{D}$. Thanks to Prokhorov's theorem and a diagonal method, there exists a subsequence of $(\mathcal{D}_m)_{m \geq 0}$, denoted also by $(\mathcal{D}_m)_{m \geq 0}$ for the sake of simplicity, such that for all p, $P_{t_p}^{\mathcal{D}_m}$ converges weakly to some probability measure P_{t_p}. The weak convergence of $P_t^{\mathcal{D}_m}$ to some probability measure P_t for all $t > 0$ is then obtained by a uniform (with respect to $(\mathcal{D}_m)_{m \geq 0}$) continuity argument in t.
3. The probability measures $(P_s)_{s \geq 0}$ are solutions of the following equations:

$$\int_E f(x)\, P_t(dx) = \int_E f(x)\, \pi_0(dx) + \int_0^t \int_E \mathbf{v}(x) \cdot \nabla f(x)\, P_s(dx)\, ds \tag{9.4}$$
$$+ \int_0^t \int_E \lambda(x) \int_E (f(y) - f(x))\, Q(x; dy)\, P_s(dx)\, ds \quad \forall f \in C_c^1.$$

We get the result, since it is proved in Cocozza-Thivent et al. [39] that the solutions (P_s) of equations (9.4) are unique and therefore equal to (π_s) (see (9.1)).

In Cocozza-Thivent et al. [41], the PDMP has parameters $\phi, \lambda, Q, \Gamma, q$. In Sect. 9.4 it is generalized for PDMPs with parameters ϕ, λ, Q, $(E_i, \Gamma_i, q_i)_{i\in I}$. Convergence is proved in Appendix C under weaker conditions than in [41]. We give below the main points. The mesh $\mathcal{M}$ is adapted to the partition $(E_i)_{i\in I}$, that is, if $K \in \mathcal{M}$ is such that $K \cap E_i \neq \emptyset$, then $K \subset E_i$.

Let us define the measures $\mu^{\mathcal{D}}$ and $\sigma_i^{\mathcal{D}}$ $(i \in I)$ on $E \times \mathbb{R}_+$ by

$$\int_{E\times\mathbb{R}_+} \varphi(x,s)\,\mu^{\mathcal{D}}(dx,ds) = \sum_{n\in\mathbb{N}} \delta t \sum_{K\in\mathcal{M}} p_{n+1}^{(K)} \int_K \varphi(x, n\delta t)\,dx,$$

$$\int_{E\times\mathbb{R}_+} \varphi(x,s)\,\sigma_i^{\mathcal{D}}(dx,ds) = \sum_{n\in\mathbb{N}} \delta t \sum_{\substack{K\in\mathcal{M},\\ K\subset E_i}} p_{n+1}^{(K)} \frac{1}{\tau} \int_{\{x\in K:\alpha(x)\le\tau\}} \varphi(\phi(x,\alpha(x)), n\delta t)\,dx.$$

The following results give the scheme's convergence:

1. For all $T > 0$, the family of measures $(1_{\{t\le T\}}\,\mu^{\mathcal{D}}(dx,dt))_{\mathcal{D}}$ is tight, and also $\sum_{i\in I}(1_{\{t\le T\}}\,\sigma_i^{\mathcal{D}}(dx,dt))_{\mathcal{D}}$.
2. The limits μ and $\bar\sigma_i$ of all convergent subsequences of measures $1_{\{t\le T\}}\,\mu^{\mathcal{D}_m}$ (dx,dt) and $1_{\{t\le T\}}\,\sigma_i^{\mathcal{D}_m}(dx,dt)$ $(i \in I)$, respectively, satisfy the following equation:

$$\begin{aligned} 0 = &\int_E g(x,0)\,\pi_0(dx) + \int_{E\times\mathbb{R}_+} \partial_{t,\phi} g(x,s)\,\mu(dx,ds) \\ &+ \int_{E\times\mathbb{R}_+} \lambda(x) \int_E (g(y,s) - g(x,s))\,Q(x;dy)\,\mu(dx,ds) \\ &+ \sum_{i\in I} \int_{\Gamma_i^*\times\mathbb{R}_+} \int_E (g(y,s) - g_i(x,s))\,q_i(x;dy)\,\bar\sigma_i(dx,ds) \end{aligned} \tag{9.5}$$

   for "suitable functions g and g_i."
3. The set of "suitable functions g and g_i" is large enough to ensure the uniqueness of measures μ and $\bar\sigma$ satisfying equation (9.5). Therefore, $\mu(dx,ds) = \pi_s(dx)\,ds$ and $\bar\sigma = \sigma$ (see (6.10), Sect. 6.2 or (9.2) with $t \to \infty$).
4. Applying Prokhorov's theorem, we get the convergence of the families of measures $(1_{\{t\le T\}}\,\mu^{\mathcal{D}}(dx,dt))_{\mathcal{D}}$ and $(1_{\{t\le T\}}\,\sigma_i^{\mathcal{D}}(dx,dt))_{\mathcal{D}}$ $(i \in I)$ to $1_{\{t\le T\}}\,\pi_t(dx)\,dt$ and $1_{\{t\le T\}}\,\sigma_i(dx,dt)$, respectively, for all $T > 0$.
5. The convergence of $P_t^{\mathcal{D}}$, defined in (9.3), to π_t for all t is then obtained thanks to the relative compactness of bounded sets in $L^\infty \cap BV$, where BV is the set of functions of bounded variation.

Other numerical schemes have been given in particular cases. An explicit scheme for semi-Markov processes is presented in Cocozza-Thivent and Eymard [32] with the proof of its convergence and some applications. Finite-volume schemes with applications to predictive reliability can be found in Cocozza-Thivent and Eymard [31], Cocozza-Thivent et al. [37], Eymard and Mercier [58], Lair et al. [76].

The following notation is used throughout the following sections. In order to unify the notation of this chapter, it may differ from that of the corresponding articles. For all $(K, L) \in \mathcal{M}^2$,

$$\lambda_{KL} = \int_K \lambda(x)\, Q(x; L)\, dx, \quad \lambda_K = \int_K \lambda(x)\, dx = \sum_{L \in \mathcal{M}} \lambda_{KL}. \tag{9.6}$$

Recall that π_0 is the initial distribution of the PDMP.

9.2 An Explicit Scheme for PDMPs Without Boundary Associated with Ordinary Differential Equations

In this section, we consider a PDMP without boundary, with parameters ϕ, λ, Q. In Cocozza-Thivent et al. [40] it is assumed that $E = \mathbb{R}^d$ or $E = \cup_{i \in I} \{i\} \times \mathbb{R}^{d_i}$ (I finite). Usually in applications, even if the model is defined on a subset of these spaces, the PDMP's parameters can be extended to the whole space without introducing new assumptions. Therefore, the results of this section can be applied.

The flow ϕ is assumed to be governed by an ODE, namely

$$\frac{d}{dt}\phi(x,t) = \mathbf{v}(\phi(x,t)), \quad \phi(x,0) = x.$$

Hypothesis (H1) on the data given in [40] can be written as follows:

1. the function λ is continuous and bounded,
2. the velocity $\mathbf{v}$ is locally Lipschitz continuous, bounded, and sublinear, that is, there exist $V_1 \geq 0$ and $V_2 \geq 0$ such that $||\mathbf{v}(x)|| \leq V_1 ||x|| + V_2$,
3. for every continuous and bounded function $f : E \to \mathbb{R}$, the function $x \to \int f(y)\, Q(x; dy)$ is continuous.

For all $(K, L) \in \mathcal{M}^2$, define

$$m_{KL} = |\{x \in L : \phi(x, \delta t) \in K\}|, \tag{9.7}$$

and the family $(p_n^K)_{n \in \mathbb{N}, K \in \mathcal{M}}$ as follows:

$$|K|\, p_0^{(K)} = \pi_0(K), \tag{9.8}$$

$$|K|\, p_{n+1}^{(K)} = \frac{|K|}{\delta t\, \lambda_K + |K|} \sum_{L\in\mathcal{M}} m_{KL}\, p_n^{(L)} + \delta t \sum_{L\in\mathcal{M}} \frac{\lambda_{LK}}{\delta t\, \lambda_L + |L|} \sum_{M\in\mathcal{M}} m_{LM}\, p_n^{(M)}. \tag{9.9}$$

The term m_{KL} may be replaced with $\tilde{m}_{KL} = |\{x \in L : x + \mathbf{v}(x)\,\delta t \in K\}|$.

When E is a hybrid space, which is the case in [40], a point $x \in E$ is written as $x = (i, y)$, $i \in I$, $y \in \mathbb{R}^{d_i}$. We have $\phi((i, y), t) = (i, \phi_i(y, t))$, where ϕ_i is the solution of

$$\frac{d}{dt}\phi_i(y, t) = \mathbf{v}(i, \phi_i(y, t)), \quad \phi_i(y, 0) = y,$$

and

$$\int_E f\, Q((i, y);\ \cdot\ ) = \sum_{j\in I} \int_{\mathbb{R}^{d_j}} f(j, z)\, \tilde{Q}((i, y); j, dz)$$

for all $(i, j, y) \in I^2 \times \mathbb{R}^{d_i}$, $\tilde{Q}((i, y); j,\ \cdot\ )$ is a nonnegative measure on $\mathbb{R}^{d_j}$ and $\sum_{j\in I} \tilde{Q}((i, y); j, \mathbb{R}^{d_j}) = 1$. Conditions 2 and 3 mean that for all $i \in I$:

- the function $\mathbf{v}(i,\ \cdot\ )$ is locally Lipschitz continuous, bounded, and sublinear,
- for all $j \in I$ and all continuous bounded functions $f : \mathbb{R}^{d_j} \to \mathbb{R}$, the function $y \to \int_{\mathbb{R}^{d_j}} f(z)\, \tilde{Q}((i, y); j, dz)$ is continuous.

We define a mesh $\mathcal{M}_i$ on each $\mathbb{R}^{d_i}$ for all $i \in I$. Formulas (9.6), (9.7), and (9.9) must be understood as follows:

$$\forall\,(i, j) \in I^2,\ \forall\, K \in \mathcal{M}_i,\ \forall\, L \in \mathcal{M}_j,\ \lambda_{KL} = \int_K \lambda(i, y)\tilde{Q}((i, y); j, L)\, dy,$$

$$\forall i \in I,\ \forall\,(K, L) \in \mathcal{M}_i^2,\ m_{KL} = |\{y \in L : \phi_i(y, \delta t) \in K\}|,$$

for all $i \in I$ and for all $K \in \mathcal{M}_i$,

$$|K|\, p_{n+1}^{(K)} = \frac{|K|}{\delta t\, \lambda_K + |K|} \sum_{L\in\mathcal{M}_i} m_{KL}\, p_n^{(L)} + \delta t \sum_{j\in I} \sum_{L\in\mathcal{M}_j} \frac{\lambda_{LK}}{\delta t\, \lambda_L + |L|} \sum_{M\in\mathcal{M}_j} m_{LM}\, p_n^{(M)}.$$

The notation of this book differs from that of [40]. In [40], ϕ is called g; a, b, and μ are used instead of λ and Q. The correspondence between them is the following: $b = \lambda$ and

$$a(i, j, x) = \lambda(i, x)\, \tilde{Q}((i, x); j, \mathbb{R}^{d_j}), \quad \mu(i, j, x)(dy) = \frac{1}{\tilde{Q}((i, x); j, \mathbb{R}^{d_j})} \tilde{Q}((i, x); j, dy),$$

or equivalently, $\lambda(i, x) = \sum_{j\in I} a(i, j, x)$, $\tilde{Q}((i, x); j, dy) = \frac{a(i,j,x)}{\lambda(i,x)}\, \mu(i, j, x)(dy)$.

In [40] the scheme's convergence is obtained under hypothesis (H1) and conditions $\delta t \to 0$, $|\mathcal{M}|/\delta t \to 0$. In some special cases, only hypothesis (H1) and conditions

$\delta t \to 0, |\mathcal{M}| \to 0$ are required (see Remark 5 of [40]). Indeed, condition $|\mathcal{M}|/\delta t \to 0$ is required only to ensure that

$$\frac{1}{\delta t}\int_{t_1}^{t_2}\int_E (\varphi_{\mathcal{M}}(\phi(x,\delta t)) - \varphi(\phi(x,\delta t)))\; P_s^{\mathcal{D}}(dx)\,ds \to 0 \tag{9.10}$$

for every $t_1 \leq t_2$ and continuous function φ with compact support, where

$$\forall K \in \mathcal{M},\ \forall x \in K,\quad \varphi_{\mathcal{M}}(x) = \frac{1}{|K|}\int_K \varphi(y)\,dy.$$

If the mesh and the flow are "well-associated," then (9.10) may vanish as in the following example: $E = \mathbb{R}$, $\mathcal{M} = \{[k\,\delta t, (k+1)\,\delta t[, k \in \mathbb{Z}\}$ and $\phi(x,t) = x + ct$ with $c \in \mathbb{Z}$.

Observe that the velocity $\mathbf{v}$ is not explicitly mentioned in the scheme. Indeed, the results on the scheme's convergence are still true as soon as there exist $A, B \in \mathbb{R}$ such that for all $(x,t) \in E \times \mathbb{R}_+$, $||\phi(x,t) - x|| \leq (||x|| + A)\,(e^{Bt} - 1)$.

9.3 An Implicit Scheme for PDMPs Without Boundary Associated with Ordinary Differential Equations

As in the previous section, a PDMP with parameters ϕ, λ, Q and no boundary is considered. The state space is $E = \mathbb{R}^d$, and the flow ϕ is assumed to be the solution of the ODE

$$\frac{d}{dt}\phi(x,t) = \mathbf{v}(\phi(x,t)),\quad \phi(x,0) = x.$$

The assumptions (H2a) on the data are the following:

1. the function λ is continuous and bounded,
2. the velocity $\mathbf{v}$ is Lipschitz continuous and bounded,
3. for all continuous and bounded functions $f : E \to \mathbb{R}$, the function $x \to \int f(y)\, Q(x; dy)$ is continuous.

For $K, L \in \mathcal{M}$, let ∂K be the topological boundary of K, $K|L = \partial K \cap \partial L$, $\mathbf{n}_{KL}$ the unit normal vector to $K|L$ oriented from K to L. The following additional assumptions on the mesh, denoted by (H2b), are also needed:

1. for all $K \in \mathcal{M}$, the interior of K is an open convex subset of $\mathbb{R}^d$,
2. for all $K, L \in \mathcal{M}$, $K|L$ is included in a hyperplane of $\mathbb{R}^d$, with a strictly positive $(d-1)$-dimensional measure denoted by $|K|L|$,
3. for all $K \in \mathcal{M}$, there exists $\mathcal{N}_K \subset \mathcal{M}$ such that $K \notin \mathcal{N}_K$ and $\partial K = \cup_{L \in \mathcal{N}_K} K|L$,
4. there exists $C_1 \geq 1$ such that for all $K \in \mathcal{M}$,

$$\frac{|\mathcal{M}|}{C_1}\sum_{L\in N_K}|K|L|\leq |K|\leq C_1\,|\mathcal{M}|\sum_{L\in N_K}|K|L|,$$

$$\frac{|\mathcal{M}|}{C_1}\leq \operatorname{diam}(K)\leq |\mathcal{M}|,\quad \frac{|\mathcal{M}|^d}{C_1}\leq |K|\leq C_1\,|\mathcal{M}|^d.$$

Given $\varepsilon \in]0, ||\mathbf{v}||_\infty]$, for all $K \in \mathcal{M}$, $L \in \mathcal{N}_K$, define

$$w_{KL}=\frac{1}{|K|L|}\int_{K|L}\mathbf{v}(x)\cdot \mathbf{n}_{KL}\,ds(x),\quad \tilde{w}_{KL}=\max(|w_{KL}|,\varepsilon)$$

where $ds(x)$ stands for the $(d-1)$-dimensional measure on $K|L$. The scheme is as follows:

$$|K|\,p_0^{(K)}=\pi_0(K), \tag{9.11}$$

$$|K|\,(p_{n+1}^{(K)}-p_n^{(K)})+\delta t\sum_{L\in N_K}|K|L|\left(w_{KL}\frac{p_{n+1}^{(K)}+p_{n+1}^{(L)}}{2}+\frac{\tilde{w}_{KL}}{2}\,(p_{n+1}^{(K)}-p_{n+1}^{(L)})\right)$$
$$=-\delta t\,p_{n+1}^{(K)}+\delta t\sum_{L\in\mathcal{M}}\lambda_{LK}\,p_{n+1}^{(L)}. \tag{9.12}$$

In Eymard et al. [59], the scheme's convergence is proved under hypotheses (H2a), (H2b) and conditions $\delta t \to 0$, $|\mathcal{M}| \to 0$. Observe that the convergence is true for every $\varepsilon \leq ||\mathbf{v}||_\infty$ and that the assumption $|\mathcal{M}|/\delta t \to 0$ is not needed, in contrast to the general case of Sect. 9.2.

9.4 An Implicit Scheme for PDMPs with Boundary

In this section, we give approximations of the PDMP marginal distributions $(\pi_t)_{t\geq 0}$ and boundary measures σ_i (defined in Proposition 5.7, Sect. 5.3) of a PDMP with parameters $\phi, \lambda, Q, (E_i, \Gamma_i, q_i)_{i\in I}$, I finite. For all $x \in E_i$, $\alpha(x) = \inf\{v > 0 : \phi(x, v) \in \Gamma_i\}$.

To simplify the notation, we assume that $E \subset \mathbb{R}^d$. The results can be very easily extended to the case $E_i \subset \mathbb{R}^{d_i}$ and to hybrid spaces.

The partition $(E_i)_{i\in I}$ may be introduced to satisfy continuity assumptions, as explained in Example 3.8, Sect. 3.2.

The results are mainly those of Cocozza-Thivent et al. [41], but some assumptions are relaxed, allowing us to apply the results to more practical cases.

Let $\mathcal{P}(E)$ be the set of probability measures on $\mathbb{R}^d$ with support in E. We assume that the following assumptions, denoted by (H3), are satisfied:

1. each E_i is stable under the flow ϕ until it reaches the boundary, the restriction of the flow ϕ to each E_i is Lipschitz continuous, uniformly in i, until it reaches the boundary, and it can be extended to a continuous function on $\mathbb{R}^d \times \mathbb{R}_+$, i.e.:

   a. $\forall i \in I,\ \forall x \in E_i,\ \forall t \in [0, \alpha(x)[,\ \phi(x,t) \in E_i$,
   b. there exists $L_\phi > 0$ such that for all $i \in I, x_1, x_2 \in E_i, t_1 \leq \alpha(x_1), t_2 \leq \alpha(x_2)$

$$||\phi(x_1,t_1) - \phi(x_2,t_2)|| \leq L_\phi(||x_1 - x_2|| + |t_1 - t_2|),$$

   c. for all $i \in I$, there exists a continuous function $\phi_i : \mathbb{R}^d \times \mathbb{R}_+ \to \mathbb{R}^d$ such that $\forall x \in E_i, \forall t \leq \alpha(x),\ \phi_i(x,t) = \phi(x,t)$;

2. the restriction of α on each E_i is Lipschitz continuous, uniformly in i, and can be extended to a continuous function on $\mathbb{R}^d$, i.e.:

   a. there exists $L_\alpha > 0$ such that

$$\forall i \in I,\ \forall x_1, x_2 \in E_i,\ |\alpha(x_1) - \alpha(x_2)| \leq L_\alpha ||x_1 - x_2||,$$

   b. for all $i \in I$, there exists a continuous function $\alpha_i : \mathbb{R}^d \to \mathbb{R}_+$ such that for all $x \in E_i, \alpha_i(x) = \alpha(x)$;

3. the restriction of λ on each E_i can be extended to a continuous function on $\mathbb{R}^d$, uniformly bounded in i, i.e., there exist $\Lambda > 0$ and for all $i \in I$ a continuous function $\lambda_i : \mathbb{R}^d \to \mathbb{R}_+$ such that for all $i \in I$, for all $x \in E_i$, $\lambda_i(x) = \lambda(x)$, and for all $x \in \mathbb{R}^d$, $||\lambda_i(x)|| \leq \Lambda$,
4. for all $i \in I$, there exists a transition probability $Q_i : \mathbb{R}^d \to \mathcal{P}(E)$ such that:

   a. $\forall x \in E_i, \quad Q_i(x; \cdot) = Q(x; \cdot)$,
   b. for all $j \in I$ and all continuous bounded function $f : E_j \to \mathbb{R}$, the function $x \in \mathbb{R}^d \to \int_{E_j} f(y)\, Q_i(x, dy)$ is continuous,
   c. the function f_Q defined by $f_Q(r) = \sup_{x \in E} \int_{\{y \in E : ||y|| \geq ||x|| + r\}} Q(x; dy)$ satisfies $\lim_{r \to \infty} f_Q(r) = 0$,

5. for all $i \in I$, there exists a transition probability $\hat{q}_i : \mathbb{R}^d \to \mathcal{P}(E)$ such that:

   a. $\forall x \in \Gamma_i, \quad \hat{q}_i(x; \cdot) = q_i(x; \cdot)$,
   b. for all $j \in I$ and all continuous bounded function $f : E_j \to \mathbb{R}$, the function $x \in \mathbb{R}^d \to \int_{E_j} f(y)\, \hat{q}_i(x, dy)$ is continuous,
   c. for all bounded Lipschitz continuous functions $f : E \to \mathbb{R}$, there exists a real number $C_f \geq 0$ such that for all $i \in I$,

$$\forall x, x' \in \Gamma_i, \quad |\int_E f(y)\, q_i(x; dy) - \int_E f(y)\, q_i(x'; dy)| \leq C_f ||x - x'||, \tag{9.13}$$

   d. the function f_q defined by $f_q(r) = \sup_{i \in I} \sup_{x \in \Gamma_i} \int_{\{y \in E : ||y|| \geq ||x|| + r\}} q_i(x; dy)$ satisfies $\lim_{r \to \infty} f_q(r) = 0$,

e. there exist $a > 0$, $B > 0$, $m \geq 1$ such that

$$\sup_{i\in I}\sup_{x\in E_i}\sum_{i_1\in I}\cdots\sum_{i_m\in I}\int_{E_{i_1}\times\cdots\times E_{i_m}} e^{-B(\alpha(x_1)+\cdots+\alpha(x_m))}\, q_i(\phi(x,\alpha(x));dx_1)$$
$$q_{i_1}(\phi(x_1,\alpha(x_1));dx_2)\cdots q_{i_{m-1}}(\phi(x_{m-1},\alpha(x_{m-1}));dx_m) \leq a < 1. \quad (9.14)$$

Implicitly, the x_j involved in (9.14) are such that $\alpha(x_j) < +\infty$, since $\alpha(x_j) = +\infty$ implies $e^{-B\,\alpha(x_j)} = 0$.

The mentions "uniformly in i" are redundant, because I is finite. But as said at the end of this section, it is needed in only one lemma, and we can think that it is unnecessary. If this is the case, the correct assumptions are those given above.

If $m = 1$, hypothesis 5(c) is not needed (see [41]). Condition (H3) 5(e) is the same as (5.22), Sect. 5.2 which is a sufficient condition for the jumps from the boundary to ensure that $\mathbb{E}(N_t) < +\infty$.

Let $\delta t > 0$, $\tau > 0$. Define $\mathcal{M}_i = \{K \in \mathcal{M} : K \subset E_i\}$, and for $(K, L) \in \mathcal{M}^2$,

$$v_{KL} = \frac{1}{\tau}|\{x \in K : \alpha(x) > \tau, \phi(x,\tau) \in L\}|,$$

$$\forall i \in I,\ \forall K \in \mathcal{M}_i,\quad q_{KL} = \frac{1}{\tau}\int_{\{x\in K:\alpha(x)\leq\tau\}} q_i(\phi(x,\alpha(x)); L)\,dx,$$

$$q_K = \frac{1}{\tau}|\{x \in K : \alpha(x) \leq \tau\}| = \sum_{L\in\mathcal{M}} q_{KL}.$$

Note that $v_{KL} \neq 0$ only if there exists $i \in I$ with $K, L \subset E_i$ (see (H3) 1(a)).

We define the family $\left(p_n^{(K)}\right)_{n\in\mathbb{N},K\in\mathcal{M}}$ according to the following finite-volume scheme, which is time-implicit:

$$|K|\,p_0^{(K)} = \pi_0(K), \quad (9.15)$$

$$0 = |K|\frac{p_{n+1}^{(K)} - p_n^{(K)}}{\delta t} + \sum_{L\in\mathcal{M}}\left(v_{KL}\,p_{n+1}^{(K)} - v_{LK}\,p_{n+1}^{(L)}\right)$$
$$+(\lambda_K + q_K)\,p_{n+1}^{(K)} - \sum_{L\in\mathcal{M}} p_{n+1}^{(L)}(\lambda_{LK} + q_{LK}). \quad (9.16)$$

Under assumptions (H3) and conditions $\delta t \to 0$, $\tau \to 0$, $|\mathcal{M}|/\tau \to 0$, the following results give the convergence to the marginal probability distributions π_t and the boundary measures σ_i:

1. for every continuous bounded function $\varphi : E \times \mathbb{R}_+ \to \mathbb{R}$ with compact support in time (i.e., there exists $T > 0$ such that $\varphi(x,t) = 0$ for $(x,t) \in E\times]T,+\infty[$)

and all $i \in I$,

$$\int_{E\times\mathbb{R}_+} \varphi(x,t)\,\mu_i^{\mathcal{D}}(dx,dt) := \sum_{n\in\mathbb{N}} \delta t \sum_{K\in\mathcal{M}_i} p_{n+1}^{(K)} \int_K \varphi(x,n\delta t)\,dx$$

$$\to \int_{E_i\times\mathbb{R}_+} \varphi(x,t)\,\pi_t(dx)\,dt, \tag{9.17}$$

$$\int_{E\times\mathbb{R}_+} \varphi(x,t)\,\sigma_i^{\mathcal{D}}(dx,dt) := \sum_{n\in\mathbb{N}} \delta t \sum_{K\in\mathcal{M}_i} p_{n+1}^{(K)} \frac{1}{\tau} \int_{\{x\in K:\alpha(x)\le\tau\}} \varphi(\phi(x,\alpha(x)),n\delta t)\,dx$$

$$\to \int_{\Gamma_i^*\times\mathbb{R}_+} \varphi(x,t)\,\sigma_i(dx,dt), \tag{9.18}$$

2. given $t > 0$, define n_t such that $n_t\delta t \le t < (n_t+1)\delta t$; then for every continuous and bounded function $f : E \to \mathbb{R}$,

$$\int_{\mathbb{R}^d} f(x)\,P_t^{\mathcal{D}}(dx) = \sum_{K\in\mathcal{M}} p_{n_t+1}^{(K)} \int_K f(x)\,dx \to \int_E f(x)\,\pi_t(dx).$$

Assertion 1 is proved in Appendix C. The proof of assertion 2 is as in [41] Theorem 4.9. The assumption that I is finite seems to be a technical one. It is needed only in Lemma C.7 as explained on Sect. C.3.

As already mentioned, in [41] stronger assumptions are required. First it is assumed that $m = 1$, which is not always convenient. For instance, in Example 3.15, end of Sect. 3.4 condition 6(d) is not satisfied with $m = 1$, but it is with $m = 2$, as proved in Example 5.8, Sect. 5.2. Moreover, I is a singleton in [41].

9.5 Numerical Experiments

In this section we present numerical results for explicit and implicit schemes described in the above sections, and we recall each time whether convergence is ensured by theoretical results. All the numerical experiments were implemented in Python 3.6. When the spaces are unbounded, the theoretical schemes use a mesh made of an infinite number of cells, but of course, in the computer, the meshes are always a finite partition of a truncated space, and the solutions are represented by a finite-dimensional array.

In some applications, the flow is given by an ODE that could be really stiff, i.e., solving it numerically without large errors is difficult and requires a large amount of computation to obtain accurate trajectories. Therefore, Monte Carlo approaches can be very inefficient, since we need to solve the ODE for a given starting point a very large number of times. Moreover, when the initial state is a probability distribution, we need to sample from it. Hence the computational cost of the Monte Carlo approach can be prohibitive for such applications. In contrast, fixing a mesh for the domain, for instance a mesh that could be geometrically adapted to the ODE, and computing the basic quantities such as λ_{KL}, m_{KL}, w_{KL}, v_{KL}, q_{KL} only once is the only time-consuming step. Indeed, this computation could be made offline and in parallel on a very precise mesh, fine enough to solve the problem accurately. Once these computations are made, we can quickly and repeatedly solve the linear systems that appear in the implicit schemes (solving a large linear system is a well-known problem with many available methods). Therefore, this approach is well suited to estimating the sensitivity with respect to the initial state, to compute cost functions such as in Example 5.13, (end of Sect. 5.3) that appear in optimization problems, and also to evaluate very long time behavior (namely the asymptotic distribution), as we will see later.

The aim of this section is to illustrate the pros and cons of schemes previously described. Before applying the methods to examples, we present general features of these schemes. The scheme of Sect. 9.2 is an explicit scheme. An explicit scheme provides an easy way to simulate very complex dynamics with very low CPU time consumption. It has the advantage of not adding a diffusive component to the simulation, since a Dirac distribution moves without deformation. The scheme is easy to understand, since we write explicitly which unit of mass is moving at each iteration from one cell to another. But it has many drawbacks, which explains why it is not used much in practice. The main drawback of an explicit scheme is its poor stability, which is guaranteed for only a small time step δt. It requires having a very small space step $|\mathcal{M}|$ when $\mathcal{M}/\delta t$ has to tend to 0, leading to a very large system of equations. For the same reason, it is really difficult to use an explicit scheme for long-time simulations, which is actually the aim of a large number of numerical simulations: find the asymptotic equilibria of the dynamics. More generally, the other factor that limits the use of large time steps δt is that an explicit scheme may not preserve some physical properties such as the positivity of the $p_n^{(K)}$. This is not the case for our explicit scheme in Sect. 9.2, but it would be the case in an explicit form of the scheme of Sect. 9.4. Generally, explicit schemes should be used only when no other schemes are available.

The schemes of Sects. 9.3 and 9.4 are implicit schemes. They are well adapted to long time simulations, since the step time δt does not have to be small with respect to the space step $|\mathcal{M}|$. However, they do not preserve the Dirac distributions that are approximated by diffusive representations, as will be seen in some of the following examples.

The schemes of Sects. 9.2 and 9.3 are for models without boundary. Nevertheless, we can extend these schemes to models with boundaries by considering a function λ with compact support that is an approximation of the Dirac distribution around the boundaries. However, with such an approximation, the convergence of the schemes is not proved.

A Toy Example

This example is a toy example to compare the three schemes for nonexpert readers. It is Example 1.10, Chap. 1 which is a special case of Example 3.8, Sect. 3.2. Even if this first example is not completely appropriate for the three schemes because of the boundary, we can make many observations about their behavior. Some properties, for instance the diffusive aspect for the implicit schemes, is the same in more complex examples.

A particle is moving on $\mathbb{R}$ with speed 1. When it reaches 0, it jumps uniformly on $[-1, 0[$ with probability p or continues with the same speed with probability $1 - p$. The movement of the particle is described by a PDMP on $\mathbb{R}$ with boundary. Its parameters are ϕ, λ, Q, Γ, q, where

$$\phi(x, t)=x+t, \quad \lambda=0, \quad \Gamma=\{0\}, \quad q(0; dx)=p\, 1_{[-1,0[}(x)\, dx + (1-p)\, \delta_0(dx).$$

Here we take $\mathcal{M} = \mathcal{M}_h := \{[kh, (k+1)h[, k \in \mathbb{Z}\}$, and to have a mesh that is well adapted to our model, we choose $\delta t = d\, h$, $d \in \mathbb{N}^*$.

In Figures 9.1 to 9.4 below, at time $t \in [n\delta t, (n+1)\delta t[$, the line is the graph of the map $x \to \frac{1}{h} \sum_K p_n^{(K)}\, 1_K(x)$, and a vertical arrow in cell K is for a Dirac measure in this cell; its height is $p_n^{(K)}$.

Introducing $E_1 =]-\infty, 0[$, $E_2 = [0, +\infty[$, this example satisfies the assumptions (H3) of Sect. 9.4. Since there is no jump except when the particle reaches the boundary $\Gamma_1 = \{0\}$, we should have $\lambda_{KL} = 0$ and thus $\lambda_K = 0$, for all $(K, L) \in \mathcal{M}^2$. Since there is a boundary, theoretically neither the explicit scheme of Sect. 9.2 nor the implicit scheme of Sect. 9.3 can be used. But as mentioned above, we approximate the obligation for the process to jump at the boundary by taking λ_K constant, instead of tending to 0 with $|\mathcal{M}|$ as in (9.6), for the cells K of the mesh near the boundary $\Gamma_1 = \Gamma$. More precisely, for the schemes of Sects. 9.3 and 9.4 we take

$$\begin{array}{ll} \lambda_{KL} = ph, & \text{if } K = [0, h[, \text{ and } L \cap]-1, 0] \neq \emptyset, \\ \lambda_{KK} = (1-p), & \text{if } K = [0, h[, \\ \lambda_{KL} = 0, & \text{in all other cases.} \end{array}$$

Therefore, $\lambda_K = 1$ if $K = [0, h[$ and 0 elsewhere.

In the three schemes it can be seen that each cell receives a part from the left linked to the transport at speed $v = 1$, except for some cells in the scheme of Sect. 9.4. Moreover, some cells are modified by the behavior at the boundary. This is modeled differently in the three schemes, as we explain below. Instead of writing the formulas, which are easy to derive, we choose to explain the differences with words; this makes it more accessible even without prior familiarity with numerical methods.

With the scheme of Sect. 9.2, the cell $[0, h[$ loses a fraction $p\frac{\delta t}{\delta t + h} \sim_{\frac{h}{\delta t} \to 0} p$ for the cell $[-dh, (-d+1)h[$, which is immediately uniformly distributed between the cells in $[-1, 0[$ (recall $\delta t = dh$).

With the scheme of Sect. 9.3, each cell in $[-1, 0[$ receives a fraction ph of the cell $[0, h[$, and therefore the cell $[0, h[$ loses a fraction p.

With the scheme of Sect. 9.4, each cell in $[-1, 0[$ receives a fraction ph of the cells close to the boundary, and this fraction comes from the cells located in $[-\tau, 0[$, τ being an additional parameter. The cells located in $[0, \tau[$ do not receive mass due to the transport at speed 1 (see the definition of v_{KL}). Finally, the cell $[0, h[$ receives a fraction $1 - p$ from the cells located in $[-\tau, 0[$, which corresponds to the jump at point 0.

Now we present the results of simulations. The initial distribution is δ_{-2} and $p = 0.3$ for the three schemes. In order to be in the "well-associated" case, (see Sect. 9.2), we need $h = 1/m$, $m \in \mathbb{N}^*$. For our example, we have chosen $\delta t = 0.01$ and $h = 0.01$.

Results for the Explicit Scheme of Sect. 9.2 (**Fig.** 9.1):
The Dirac measure is moving at speed $v = 1$ without diffusion/dispersion/deformation until it reaches 0. At this time, a fraction $(1 - p) \frac{\delta t}{\delta t + h}$ crosses the boundary and a fraction $p \frac{\delta t}{\delta t + h}$ jumps uniformly on $[-1, 0[$. We can see that our choice of δt and h is not optimal, since $\frac{\delta t}{\delta t + h} = \frac{1}{2}$, which is not close to 1, as is required above when we compare the behaviors at the boundary. Nevertheless, the results are reasonable, as can be seen by comparing them to the Monte Carlo simulations of Fig. 9.3 below.

Little by little, the part of the distribution that is in $\mathbb{R}_-$ crosses the boundary, as illustrated by all the figures of this section.

Results for the Implicit Scheme of Sect. 9.3 (**Fig.** 9.2):
We see that the Dirac measure is not conserved, since there is diffusion due to the implicit scheme, as announced before. Instead of a Dirac measure, we see a function with compact support with its mass conserved.

At time $t = 3.5$ we see a regular distribution, which does not seem really close to the distribution drawn in Fig. 9.1 at the same time. Indeed, the function that is the approximation of the Dirac measure and the "true" density are added. Since the convergence of this scheme is not proved in the case of dynamics with a boundary,

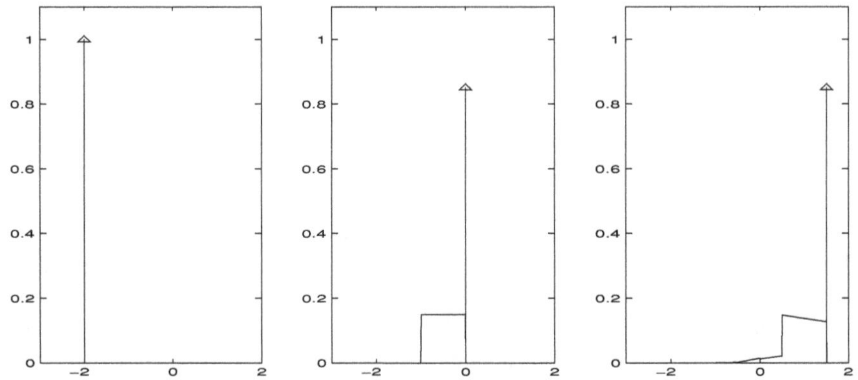

Fig. 9.1 Toy example with the scheme of Sect. 9.2 at time $t = 0$, $t = 2$ and $t = 3.5$

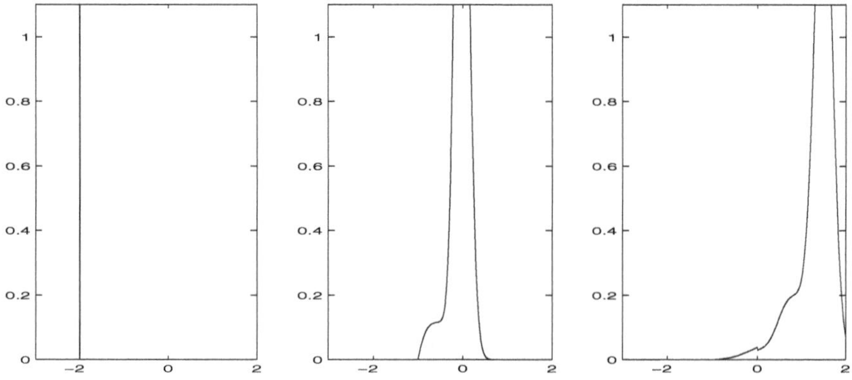

Fig. 9.2 Toy example with scheme of Sect. 9.3 at time $t = 0$, $t = 2$, and $t = 3.5$

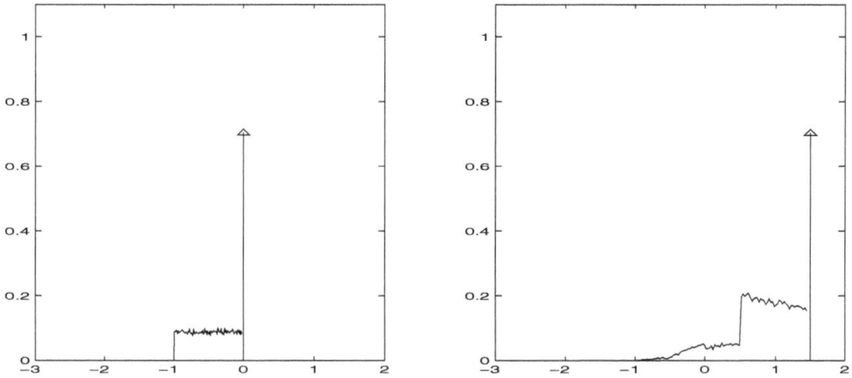

Fig. 9.3 Toy example with Monte Carlo procedure at time $t = 2$ and $t = 3.5$ with 10^6 particles, a time step $\delta t = 0.01$, and a reconstruction of the density by a histogram with 100 bins on $[-1, t-2]$

this difference could be a side effect of the nonconvergence of the schemes. Another explanation could be that h is not small enough. In order to be completely clear, we have provided a Monte Carlo simulation with 10^6 particles. The result is given in Fig. 9.3, and we see that it is closer to that of the explicit scheme than to that of the implicit scheme. The principal reason is the Dirac distribution, which is well represented by the Monte Carlo procedure.

Results for the Implicit Scheme of Sect. 9.4 **(Fig.** 9.4**):**
For this scheme there is a new parameter τ, which should be chosen according to the mesh (as δt was chosen for the two previous schemes). Precisely when $\tau/h \in \mathbb{N}^*$, we get a mesh that is adapted to the time variable τ used to localize the boundary. In contrast, the rate $\delta t/h$ is no longer important. We take $\delta t = 0.01$, $h = 0.01$, as mentioned above, and $\tau = 0.01$.

In Fig. 9.4 we see that the Dirac measure is not conserved, since there is diffusion due to the implicit scheme, and the figure is not really close to the distribution drawn

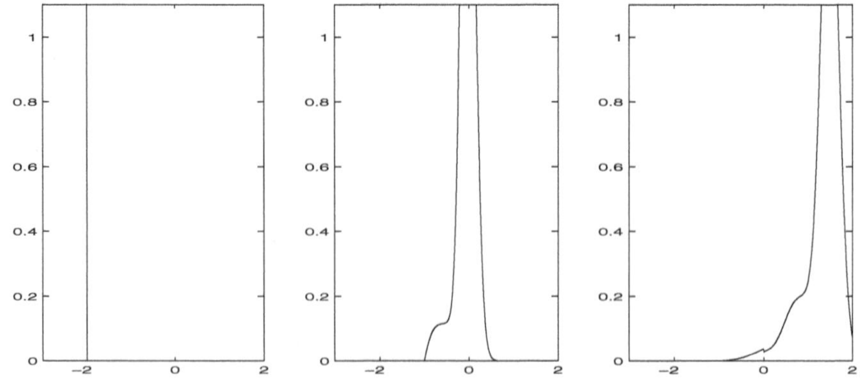

Fig. 9.4 Toy example with scheme of Sect. 9.4 at iterations $t = 0$, $t = 2$, and $t = 3.5$

in Fig. 9.1 but very close to the distribution drawn in Fig. 9.2. Since the convergence of the scheme that gives Fig. 9.4 has been proved, it seems to confirm that the same artifact seen also in Fig. 9.2 is due to the diffusion of the Dirac measure and not to a convergence problem of the scheme of Sect. 9.3.

An Example with Spaces of Various Dimensions

Let us go back to Example 3.10, Sect. 3.4. This is an example in a two-dimensional space, but the space $A = \mathbb{R} \times \{0\}$ is a one-dimensional space. This example has been chosen to show that we can mix spaces of various dimensions. Moreover, since there are only orthogonal speeds, we consider Cartesian meshes in order to simplify the computation of flux. Since $\lambda = 0$, there is no jump, except at the boundary. As in the case of the toy example, the explicit scheme of Sect. 9.2 and the implicit scheme of Sect. 9.3 cannot be used, but they can be adapted. We do not consider these extensions here, but only the scheme of Sect. 9.4, which works well.

We take $\mathcal{M} = \cup_{i=1}^{3} \mathcal{M}_i$, where

$$\mathcal{M}_1 := \{[kh, (k+1)h[\times]mh, (m+1)h], k \in \mathbb{Z}, m \in \mathbb{N}\},$$
$$\mathcal{M}_2 := \{[kh, (k+1)h[\times[-mh, (-m+1)h[, k \in \mathbb{Z}, m \in \mathbb{N}^*\},$$
$$\mathcal{M}_3 := \{[kh, (k+1)h[\times\{0\}, k \in \mathbb{Z}\},$$

$\delta t = 0.1$, $\tau = 0.05$ and $h = 0.05$, so τ is a multiple of h. Let $d = \tau/h \in \mathbb{N}$; we get $v_{KL} = h^2/\tau$ if one of the following conditions is satisfied:

- $K, L \in \mathcal{M}_1$ and for some $k \in \mathbb{Z}$ and $m \in \mathbb{N}$ we have $K = [kh, (k+1)h[\times]mh, (m+1)h]$ and $L = [kh, (k+1)h[\times](m-d)h, (m+1-d)h]$,
- $K, L \in \mathcal{M}_2$ and for some $k \in \mathbb{Z}$ and $m \in \mathbb{N}^*$ we have $K = [kh, (k+1)h[\times[-mh, (-m+1)h[$ and $L = [kh, (k+1)h[\times[(-m+d)h, (-m+d+1)h[$,
- $K, L \in \mathcal{M}_3$ and for some $k \in \mathbb{Z}$ we have $K = [kh, (k+1)h[$ and $L = [(k+d)h, (k+d+1)h[$,

and $v_{KL} = 0$ elsewhere. The computation could be more complex if τ is not a multiple of h.

Moreover, $\lambda_{KL} = 0$ for all $K, L \in \cup_{i=1}^{3}\mathcal{M}_i$ and

$$q_{KL} = \begin{cases} (1-p_1)h^2\mu_1(L)/\tau, & \text{if } K \subset \mathbb{R}\times]0,\tau] \text{ and } L \notin \mathcal{M}_3, \\ (1-p_2)h^2\mu_2(L)/\tau, & \text{if } K \subset \mathbb{R}\times[-\tau,0[\text{ and } L \notin \mathcal{M}_3, \\ p_1 h^2/\tau, & \text{if } K \subset \mathbb{R}\times]0,\tau] \text{ and } L \in \mathcal{M}_3, \text{ is below } K, \\ p_2 h^2/\tau, & \text{if } K \subset \mathbb{R}\times[-\tau,0[\text{ and } L \in \mathcal{M}_3, \text{ is above } K, \\ 0, & \text{in all other cases;} \end{cases}$$

hence $q_K = h^2/\tau$ if $K \in \mathcal{M}_1$ and $K \subset \mathbb{R}\times]0,\tau]$, or if $K \in \mathcal{M}_2$ and $K \subset \mathbb{R}\times[-\tau,0[$; $q_K = 0$ if not.

In Fig. 9.5, we see the evolution of the measure $0.3\,\delta_{(0,3)} + 0.7\,\delta_{(2,-1)}$ at different times when $p_1 = p_2 = 1$. The right axis from 0 to 5 is for the first variable, and the left axis from -5 to 5 is for the second variable. As mentioned before, we glue meshes $\mathcal{M}_1$, $\mathcal{M}_2$, and $\mathcal{M}_3$ on the space $\mathbb{R}^2$.

With these parameters, the trajectory is purely deterministic and very easy to get explicitly. The goal of the simulation is to see the effect of the scheme. As expected, the Dirac measures are first moving toward E_3, and then they stay on E_3 and are moving to the right. But the Dirac measures are not conserved; there is a diffusion due to the implicit scheme. At time $t = 1.5$, neither the Dirac measure starting from (3, 0) nor the Dirac measure starting from (2, −1) has arrived in E_3, but we see in Fig. 9.5 that some part of the Dirac measure starting from (2, −1) has started to turn from E_2 to E_3. However, this shows the capacity of the scheme to treat smooth transfers of mass despite some discontinuities for the global flow ϕ on $\mathbb{R}^2$. At time $t = 4$, almost all the mass has been transferred to E_3. Due to the diffusion, the two initial data could be mixed. Choosing smaller parameters can reduce this phenomenon.

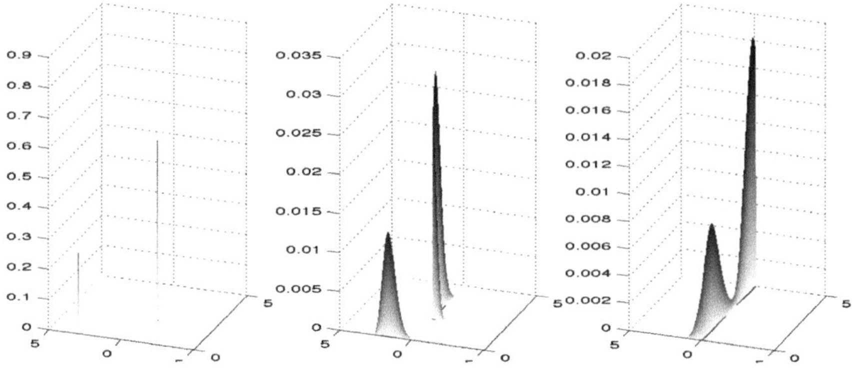

Fig. 9.5 Example 3.10 with the scheme of Sect. 9.4 at times $t = 0$, $t = 1.5$, and $t = 4$

An Example with an Interesting Boundary Measure

We take again Example 3.12, Sect. 3.4 which models a stock (or the virtual waiting time of a GI/G1/1 queue). We choose the last model, which is the model without hybrid space but with "boundary" (i.e., where jumps when the stock becomes empty are added), to get the mean number of times the factory must be stopped because the stock is empty. From a theoretical point of view, the only well-adapted scheme is the scheme of Sect. 9.4. Thus we consider only that one, even though we suspect that the others may be implemented, since we don't get an ODE merely because $s \to \phi(x, s)$ is not differentiable at one point.

We consider the case in which $c = 1$, μ is the uniform distribution on $[0, 1]$, and the hazard rate of the interarrival distribution is $\bar{\lambda}(x_2) = 2\,x_2$. Any other rate can be chosen, and the algorithm does not depend essentially on this function. In particular, the assumption of boundedness of rates is not essential in numerical simulations.

We take $E = E_0 \cup E_1$ with $E_0 = \{0\} \times \mathbb{R}_+$, $E_1 = \mathbb{R}_+^* \times \mathbb{R}_+$, $\Gamma_1 = \{0\} \times \mathbb{R}_+$, $\Gamma_0 = \emptyset$. The subset E_0 is considered a one-dimensional set. We take $\mathcal{M}_0 := \{[mh, (m+1)h[\colon m \in \mathbb{N}\}$, $\mathcal{M}_1 := \{]kch, (k+1)ch] \times [mh, (m+1)h[\ : (k, m) \in \mathbb{N}^2\}$. For all $(m, n) \in \mathbb{N}^2$, we define $p_n^{(m)} := p_n^{(K)}$ for the cell $K = [mh, (m+1)h[\in \mathcal{M}_0$. In the same way, for all $(k, m, n) \in \mathbb{N}^3$ we define $p_n^{(k,m)} := p_n^{(K)}$ for the cell $K =]kch, (k+1)ch] \times [mh, (m+1)h[\in \mathcal{M}_1$. We do not give details about the computation of v_{KL}, λ_{KL}, and q_{KL}, since it is relatively obvious.

We want to get the distribution $\bar{\pi}_t$ of the variable S_t ($t \in \mathbb{R}_+$). It is the first marginal of the probability distribution π_t of $X_t = (S_t, D_t)$. We have $\bar{\pi}_t(dx_1) = f_t(x_1)\,dx_1 + p_t\,\delta_0(dx_1)$. For $t \in [n\delta t, (n+1)\delta t[$, the function f_t at a point $x_1 \in]kch, (k+1)ch]$ is approximated by $\sum_{m\in\mathbb{N}} |K|\ p_{n+1}^{(k,m)} = \sum_{m\in\mathbb{N}} ch^2\ p_{n+1}^{(k,m)}$. The weight p_t is approximated by $\sum_{m\in\mathbb{N}} |K|\ p_{n+1}^{(m)} = \sum_{m\in\mathbb{N}} h\ p_{n+1}^{(m)}$. The values $|K|$ for the sums approximating respectively f_t and p_t are not of the same order, since one mesh is on $\mathbb{R}_+^2$ and the other on $\mathbb{R}_+$.

In Fig. 9.6, we see the approximation of $\bar{\pi}_t$ for $t = 3$ hours and $t = 10$ hours. The initial probability distribution is $\delta_{3,0}$. The height of the vertical arrow at point $x_1 = 0$ is the approximation of p_t. The line approximates the graph of f_t. Although the scheme is implicit, there is no diffusion for the approximation of the Dirac measure, since its known localization has been taken into account.

The approximation of $\bar{\pi}_t$ allows us to estimate, for instance, the probability that the number of commodities of the factory is above a threshold C at time t, namely $\int_C^{+\infty} f_t(x_1)\,dx_1$, or the probability that the factory is empty at time t, i.e., p_t.

However, the loss of production $LP(t)$, that is, the mean number of times there is no available commodity for the factory during the time interval $[0, t]$, and the frequency of annual loss of production $FLP(t) := 8766\,LP(t)/t$ (the time unit will be an hour and 8766 is the number of hours in a year) cannot be obtained from π_t but from the boundary measure σ. We have $LP(t) = \sigma(\Gamma_1 \times [0, t])$ (Example 5.1, Sect. 5.3).

In another example, in Eymard and Mercier [58], where no boundary is considered, it is suggested to replace $FLP(t)$ for large t with $FRP(t) = 8766\,RP(t)/t$, where $RP(t)$ is the mean number of times the production is recovered in the time

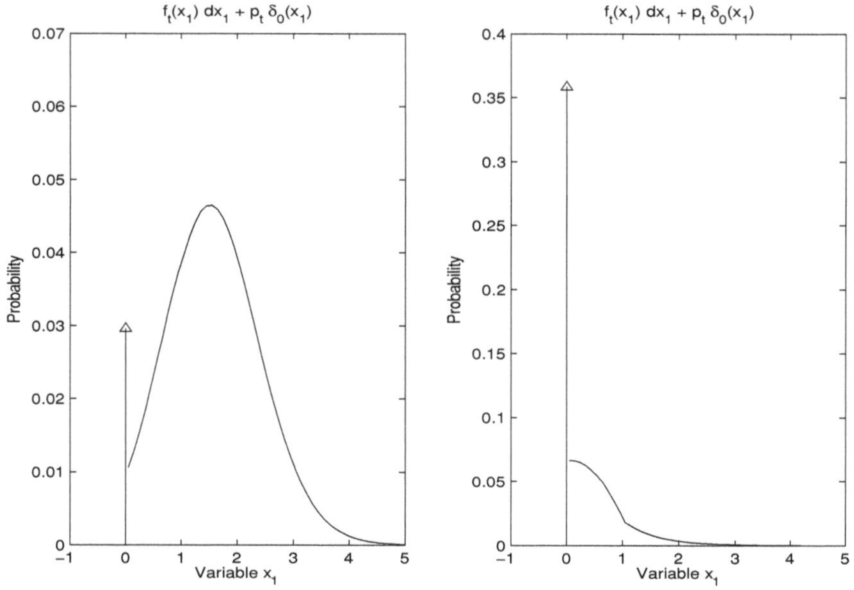

Fig. 9.6 Example 3.12 with scheme 9.4 at $t = 3$ and $t = 10$

interval $[0, t]$. Hence $RP(t)$ is the mean number of times the process jumps from E_0 to E_1, and hence $RP(t) = \int_0^t \int_{\mathbb{R}_+} \bar{\lambda}(x_2)\, \pi_s(\{0\}, dx_2)\, ds$ (Corollary 5.2, Sect. 5.3). Clearly $|LP(t) - RP(t)| \leq 1$. Hence $LP(t)$ and $RP(t)$, which tend to infinity, are equivalent as $t \to \infty$ but are perceptibly different when t is not large enough.

In Fig. 9.7, the evolution of these quantities is shown for small t, and also the evolution of $IRP(t) := (RP(t + \delta t) - RP(t))/\delta t$ and $ILP(t) := (LP(t + \delta t) - LP(t))/\delta t$, which are the instantaneous rates of growing of the quantities RP and LP. We see that the quantities IRP and ILP reach their asymptotic values very quickly. This shows that the quantities RP and LP are asymptotically linearly growing. But the quantities FLP and FRP have not reached their asymptotic values after 20 hours.

In Fig. 9.8, the evolution of these quantities is represented during 20 days, i.e., 480 hours. In Fig. 9.9 the same quantities are drawn with a logarithmic time scale $\log_{10}(1 + t)$.

The scheme is well adapted for computing these values as t tends to infinity, since the parameter δt can be taken large. Precisely, at the beginning of the scheme we take $\delta t = 0.1$. It is taken small to approximate well the transitory dynamics. When some asymptotic behavior seems to have been reached after many iterations of the scheme, we take δt bigger, namely $\delta t = 2.4$, to approximate the asymptotic dynamics and go faster to infinity in time, without disrupting the numerical approximation.

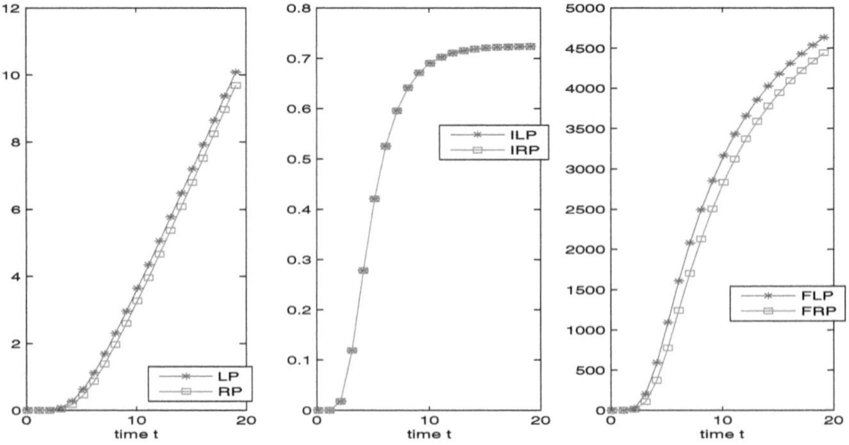

Fig. 9.7 Left: $LP(t)$ and $RP(t)$, Center: $ILP(t)$ and $IRP(t)$, Right: $FLP(t)$ and $FRP(t)$ for small t

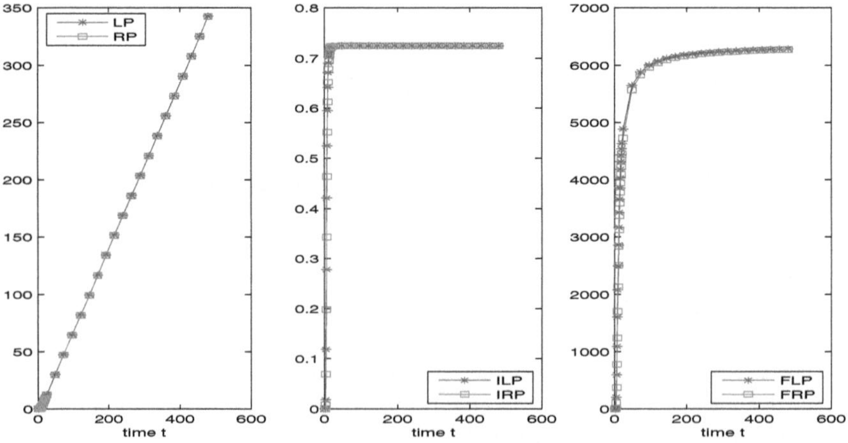

Fig. 9.8 Left: $LP(t)$ and $RP(t)$, Center: $ILP(t)$ and $IRP(t)$, Right: $FLP(t)$ and $FRP(t)$ for large t

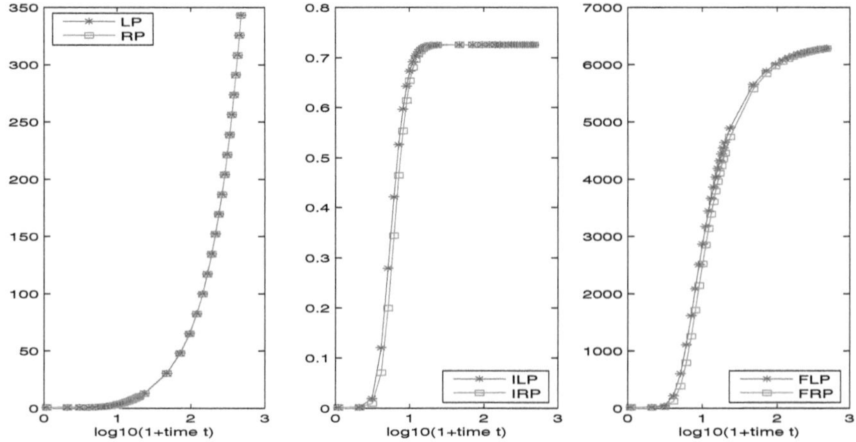

Fig. 9.9 Left: LP and RP, Center: ILP and IRP, Right: FLP and FRP with respect to logarithmic time scale $log(1 + t)$

Chapter 10
Switching Processes

This chapter introduces switching processes (SP). Their construction follows the construction of a PDMP, except that the deterministic flow ϕ is replaced by a stochastic process ζ, wrongly called "the" intrinsic process. More specifically, for each n, $\phi(Y_n, .)$ is replaced by a process whose probability distribution is that of ζ given $\zeta(0) = Y_n$. A PDMP is a switching process.

To better understand the model by reading the examples and consequently the definition of a switching process below, it can be assumed that the intrinsic process is a Markov process even if it is not necessary to get results of Sects. 10.2 to 10.4.

Example 10.1 In a nominal situation, the evolution of an item's state (price, wear and tear, ...) is described by a Markov process ζ. Moreover, shocks occur at times T_n $(n \geq 1)$. The magnitude of a shock is random and may depend on the state of the item right before the shock. The process $(T_n)_{n\geq 1}$ can be a Poisson process or more generally a renewal process. The probability distribution of $T_{n+1} - T_n$ given the evolution until time T_n may also have a hazard rate that is a function of the item's state as for the earthquake model of Example 1.3, Chap. 1 when tectonic forces between earthquakes are described by a Markov process. These different possibilities will be considered in Examples 10.7, Sect. 10.1 and 10.9, Sect. 10.5.

Example 10.2 We return to Example 3.11, Sect. 3.4 of an item whose failure rate is temperature-dependent, and assume that at the end of each repair, the item is returned to the nominal temperature τ_0. Then the item's behavior is modeled by an alternating renewal process. The probability distribution $dF_1(t)$ (respectively $dF_0(t)$) of the duration of trouble-free operation (respectively of the duration of repair) is

$$dF_1(v) = \bar{\lambda}(z(v))\, e^{-\int_0^v \bar{\lambda}(z(s))\, ds}\, 1_{\mathbb{R}_+}(v)\, dv,$$
$$\text{(respectively } dF_0(v) = \bar{\lambda}(v)\, e^{-\int_0^v \bar{\mu}(s)\, ds}\, 1_{\mathbb{R}_+}(v)\, dv),$$

C. Cocozza-Thivent, *Markov Renewal and Piecewise Deterministic Processes*, Probability Theory and Stochastic Modelling 100,
https://doi.org/10.1007/978-3-030-70447-6_10

where z is the solution of $z'(t) = \mathbf{v_1}(z(t))$, $z(0) = \tau_0$.

Suppose now that this item is a mechanical component whose failure rate does not depend on the temperature but depends on the wear of the item, and that this wear is modeled by a gamma process ξ (see Example 10.10, Sect. 10.6 for the definition). Given the degradation process ξ, the probability distribution of the duration of trouble-free operation is

$$\bar{\lambda}(\xi(v))\, e^{-\int_0^v \bar{\lambda}(\xi(s))\,ds}\, 1_{\mathbb{R}_+}(v)\, dv,$$

i.e., its hazard rate is $\bar{\lambda}(\xi(v))$.

Recall that a nonnegative random variable with hazard rate h can be simulated as follows. Let $H(t) = \int_0^t h(s)\,ds$, κ the left-continuous generalized inverse of H, that is, $\kappa(t) = \inf\{s : H(s) \geq t\}$, and U an r.v. with uniform probability distribution on $[0, 1]$. Then the hazard rate of $\kappa(-\log U)$ is h (see Sect. 3.3).

Let $(\xi^{(n)})_{n\geq 1}$ be i.i.d. processes with the same probability distribution as ξ, let $(U_n)_{n\geq 1}$ be i.i.d. random variables with uniform probability distribution on $[0, 1]$, and assume that $(\xi^{(n)})_{n\geq 1}$, $(U_n)_{n\geq 1}$ are independent. Let κ_i be the left-continuous generalized inverse of $t \to \int_0^t \bar{\lambda}(\xi^{(i)}(s))\,ds$ and κ that of $t \to \int_0^t \mu(s)\,ds$.

Assume that the component is new at time 0 and that $\int_0^{+\infty} \bar{\lambda}(\xi(s))\,ds = +\infty$ a.s. and $\int_0^{+\infty} \mu(s)\,ds = +\infty$.

(A) The behavior of the component can be described by an alternating renewal process $(T_n)_{n\geq 1}$. Its simulation can be achieved by taking $T_0 = 0$ and

$$T_{2n+1} - T_{2n} = \kappa_{n+1}(-\log U_{2n+1})\ n \geq 0, \quad T_{2n} - T_{2n-1} = \kappa(-\log U_{2n})\ n \geq 1.$$

To get more information, especially a Markov process, we can define $X_t \in \{e_1, e_0\} \times \mathbb{R}_+$ as

$$X_t = \begin{cases} (e_1, \xi^{(n+1)}(t - T_n)), & T_{2n} \leq t < T_{2n+1}\ (n \geq 0) \\ (e_0, t - T_n), & T_{2n+1} \leq t < T_{2n+2}\ (n \geq 0). \end{cases}$$

(B) Now suppose that preventive maintenance is carried out when wear reaches the threshold a. Define $\xi^{(i)}$, κ_i, κ as in (A), e_2 as the state of preventive maintenance, $\bar{\mu}$ as the repair rate for the preventive maintenance, and $\bar{\kappa}$ as the left-continuous generalized inverse of $t \to \int_0^t \bar{\mu}(s)\,ds$. The behavior of the component and of its wear can be simulated by $X_t = (I_t, \eta_t) \in \{e_1\} \times \mathbb{R} \cup \{e_0, e_2\} \times \mathbb{R}_+$ defined as follows:

- $T_0 = 0$, $I_0 = e_1$,
- $\alpha^{(1)} = \inf\{t : \xi^{(1)}(t) \geq a\}$, $V_1 = \kappa_1(-\log U_1)$, $T_1 = \min(\alpha^{(1)}, V_1)$,
- for $t < T_1$, $X_t = (e_1, \xi^{(1)}(t))$,
- if $T_1 = \alpha^{(1)}$ then $I_{T_1} = e_2$, $T_2 = T_1 + \bar{\kappa}(-\log U_2)$.
  Otherwise, $I_{T_1} = e_0$, $T_2 = T_1 + \kappa(-\log U_2)$,
- for $T_1 \leq t < T_2$, $X_t = (I_{T_1}, t - T_1)$.

For $n \geq 1$:

- $\alpha^{(n+1)} = \inf\{t : \xi^{(n+1)}(t) \geq a\}$, $V_{n+1} = \kappa_{n+1}(-\log U_{2n+1})$, $W_{n+1} = \min(\alpha^{(n+1)}, V_{n+1})$, $T_{2n+1} = T_{2n} + W_{n+1}$,
- for $T_{2n} \leq t < T_{2n+1}$, $X_t = (e_1, \xi^{(n+1)}(t - T_n))$,
- if $W_{n+1} = \alpha^{(n+1)}$, then $I_{T_{2n+1}} = e_2$, $T_{2n+2} = T_{2n+1} + \bar{\kappa}(-\log U_{2n+2})$. Otherwise, $I_{T_{2n+1}} = e_0$, $T_{2n+2} = T_{2n+1} + \kappa(-\log U_{2n+2})$,
- for $T_{2n+1} \leq t < T_{2n+2}$, $X_t = (I_{T_{2n+1}}, t - T_{2n+1})$.

Example 10.3 The price of an agricultural product is modeled by Markov processes whose parameters depend on climatic conditions of three types: normal (type 1), drought (type 2), erratic (type 3). Under type i conditions, the price is modeled by a stochastic process ξ^{θ_i} with parameter θ_i ($i \in \{1, 2, 3\}$). The change in climatic condition is described by a zero-delayed Markov renewal process $(\bar{Y}_n, T_n)_{n\geq 0}$ with values in $\{1, 2, 3\}$. At time 0, the climatic condition is normal and the price is ξ_0. Define C_t (respectively ξ_t) as the climatic condition (respectively the price) at time t and let $X_t = (C_t, \xi_t)$. The process $(X_t)_{t\geq 0}$ can be simulated as follows:

- simulate $(\bar{Y}_n, T_n)_{n\geq 1}$,
- simulate a process $\xi^{(1)}$ such that its probability distribution given $(\bar{Y}_n, T_n)_{n\geq 1}$ is the same as that of ξ^{θ_1} given $\xi^{\theta_1}(0) = \xi_0$,
- for $t < T_1$, take $C_t = 1$, $\xi_t = \xi^{(1)}(t)$, $Y_1 = (\bar{Y}_1, \xi^{(1)}(T_1))$,
- simulate a process $\xi^{(2)}$ such that its probability distribution given $(\bar{Y}_n, T_n)_{n\geq 1}$, $\xi^{(1)}$, $Y_1 = (i_1, x_1)$ is the probability distribution of $\xi^{\theta_{i_1}}$ given $\xi^{\theta_{i_1}}(0) = x_1$,
- for $T_1 \leq t < T_2$, take $C_t = i_1$, $\xi_t = \xi^{(2)}(t - T_1)$, $Y_2 = (\bar{Y}_2, \xi^{(2)}(T_2 - T_1))$,
- simulate a process $\xi^{(3)}$ such that its probability distribution given $(\bar{Y}_n, T_n)_{n\geq 1}$, $\xi^{(1)}$, $\xi^{(2)}$, $Y_2 = (i_2, x_2)$ is the probability distribution of $\xi^{\theta_{i_2}}$ given $\xi^{\theta_{i_2}}(0) = x_2$,
- for $T_2 \leq t < T_3$, take $C_t = i_2$, $\xi_t = \xi^{(3)}(t - T_2)$, $Y_3 = (\bar{Y}_3, \xi^{(3)}(T_3 - T_2))$,

and so on.

These processes make it possible to model many interesting situations (see, for instance, the good introduction of Bect [8] including a historical overview). In the last section we make the link between these processes and processes that are now being addressed by experts on diffusion processes. Our aim is to show the contribution of our approach through CSMPs for such processes. We hope that this will motivate further research.

In applications, switching processes are often hybrid processes, i.e., processes taking values in a space that is the product of a discrete space and a continuous one. However, in order not to complicate the notation and also for the sake of generality, we assume that they are taking values in a general state space E.

In Sect. 10.1, we give a heuristic construction of the process, avoiding measurability questions.

In Sect. 10.2 we prove that a vector of independent parametrized switching processes is a switching process; therefore, it holds also for PDMPs.

In Sect. 10.3, we give conditions for a switching process to be decomposed as defined in Sect. 4.2. Such processes generalize the results for PDMPs.

In Sect. 10.4 we introduce formulas connected with the semiregenerative property.

In Sect. 10.5 conditions for a switching process to be a Markov process are given.

The last two sections are devoted to Kolmogorov equations. We assume that the jumps added to the intrinsic process are not caused by a boundary. In Sect. 10.6 the intrinsic process is assumed to be a time-homogeneous semimartingale. In Sect. 10.7 the results are briefly extended to non-time-homogeneous intrinsic processes such as Itô–Lévy processes and therefore to diffusion processes.

10.1 The General Model

Let $\zeta = (\zeta(t))_{t\geq 0}$ be a cad-lag process, taking values in E, called the intrinsic process.

Let $M(\zeta; dy, dv)$ be the restriction to $E \times \mathbb{R}_+$ of a probability distribution on $E \times \mathbb{R}_+ \cup \{(\Delta, +\infty)\}$ that depends on ζ and with the suitable measurability properties. We call M a random semi-Markov kernel on E (associated with ζ). We assume $M(\zeta; E \times \{0\}) = 0$.

Example 10.4 Let $\lambda : E \to \mathbb{R}_+$ be a Borel function such that for every $x \in E$, $\mathbb{P}(\exists\, \varepsilon(x) > 0 : \int_0^{\varepsilon(x)} \lambda(\zeta(v))\, dv < +\infty \mid \zeta(0) = x) = 1$, a random variable $\alpha(\zeta)$ with values in $]0, +\infty]$ depending on the process ζ, and Q, q probability kernels on E. Define

$$\begin{aligned} M(\zeta; dy, dv) &= 1_{\{v<\alpha(\zeta)\}}\, \lambda(\zeta(v))\, e^{-\int_0^v \lambda(\zeta(s))\, ds}\, Q(\zeta(v); dy)\, dv \\ &\quad + 1_{\{\alpha(\zeta)<+\infty\}}\, e^{-\int_0^{\alpha(\zeta)} \lambda(\zeta(s))\, ds}\, q(\zeta(\alpha(\zeta)); dy)\, \delta_{\alpha(\zeta)}(dv). \end{aligned}$$

Definition 10.1 A switching process (SP) on E generated by ζ and M with initial state $x_0 \in E$ is a process $(X_t)_{t\geq 0}$ that can be described as follows:

1. let $\zeta^{(1)}$ be a process whose probability distribution is the conditional probability distribution of ζ given $\zeta(0) = x_0$,
2. the conditional probability distribution of (Y_1, T_1) given $\zeta^{(1)}$ is $M(\zeta^{(1)}; dy, dv) + (1 - M(\zeta^{(1)}; E \times \mathbb{R}_+))\, \delta_{\Delta,+\infty}(dy, dv)$,
3. for $t < T_1$, $X_t = \zeta^{(1)}(t)$ and $X_{T_1} = Y_1$.
   Assuming that $\zeta^{(1)}, \ldots, \zeta^{(n)}, Y_1, T_1, \ldots, Y_n, T_n$ $(n \geq 1)$ have been defined,
4. let $\zeta^{(n+1)}$ be a process such that its conditional probability distribution given $\zeta^{(1)}, \ldots, \zeta^{(n)}, Y_1, T_1, \ldots, Y_n = y_n, T_n$ is the conditional probability distribution of ζ given $\zeta(0) = y_n$,
5. the conditional probability distribution of $(Y_{n+1}, T_{n+1} - T_n)$ given $\zeta^{(1)}, \ldots, \zeta^{(n+1)}, Y_1, T_1, \ldots, Y_n, T_n$ is $M(\zeta^{(n+1)}; dy, dv) + (1 - M(\zeta^{(n+1)}; E \times \mathbb{R}_+))\, \delta_{\Delta,+\infty}(dy, dv)$,
6. for $T_n \leq t < T_{n+1}$, $X_t = \zeta^{(n+1)}(t - T_n)$ and $X_{T_{n+1}} = Y_{n+1}$.

The times T_n $(n \geq 1)$ are called the SP-jump times.

The process ζ may have its own jump times, they are not considered to be SP-jump times.

When M is as in Example 10.4, we say that $(X)_{t\geq 0}$ is a parametrized switching process; it is generated by ζ, with parameters λ, Q, α, q. When $\alpha(\zeta) = \inf\{t > 0 : \zeta(t) \in B\}$, it is with parameters λ, Q, B, q. When $\alpha(\zeta) = +\infty$, it is without boundary, and its parameters are λ, Q.

The above definition is appropriate for a time-homogeneous process ζ. The inhomogeneous case is mentioned in Sect. 10.7.

Throughout this chapter we assume that $\lim_{n\to+\infty} T_n = +\infty$.

Example 10.5 Let $\phi : E \times \mathbb{R}_+ \to E$ be a Borel function such that $\phi(x, 0) = x$ for all $x \in E$, and $(Z_t, A_t)_{t\geq 0}$ a CSMP with kernel N. The process $(\phi(Z_t, A_t))_{t\geq 0}$ is a switching process generated by ζ and M, where $\zeta(t) = \phi(\zeta(0), t)$ and $M(\zeta; dy, dv) = N(\zeta(0), dy, dv)$.

A PDMP with parameters ϕ, λ, Q, B, q is a switching process generated by ζ defined just above, with parameters λ, Q, B, q. Moreover, Example 10.4 can be easily extended to the case of a PDMP with parameters $\phi, \lambda, Q, \alpha, (E_i, q_i)_{i\in I}$.

Example 10.6 The processes $(X_t)_{t\geq 0}$ of Example 10.2 are switching processes. Define λ and Q as $\lambda(e_1, x) = \bar{\lambda}(x)$, $Q((e_1, x); dy) = \delta_{(e_0,0)}(dy)$, $\lambda(e_0, x) = \mu(x)$, $Q((e_0, x); dy) = \delta_{(e_1,\tau_0)}(dy)$.

1. In case A, define the intrinsic process $\zeta = (\zeta^1, \zeta^2)$ as $\zeta^1(t) = \zeta^1(0)$ $(t \geq 0)$ and $\zeta^2 = \xi$ if $\zeta^1 = e_1$, $\zeta^2(t) = \zeta^2(0) + t$ $(t \geq 0)$ if $\zeta^1 = e_0$. Then $(X_t)_{t\geq 0}$ is a parametrized switching process without boundary, generated by ζ, with parameters λ, Q.

2. In case B, define the intrinsic process $\zeta = (\zeta^1, \zeta^2)$ as $\zeta^1(t) = \zeta^1(0)$ $(t \geq 0)$, $\zeta^2 = \xi$ if $\zeta^1 = e_1$, $\zeta^2(t) = \zeta^2(0) + t$ $(t \geq 0)$ if $\zeta^1 \in \{e_0, e_2\}$. Then $(X_t)_{t\geq 0}$ is a parametrized switching process generated by ζ, with parameters λ, Q, B, q, where $B = \{e_1\} \times [a, +\infty[$, $q((e_1, x); dy) = \delta_{(e_2,0)}(dy)$. Note that we cannot take $B = \{e_1\} \times \{a\}$, since ξ is not assumed to be continuous.

If the failure rate depends also on the age of the component, then the switching process and the intrinsic process take values in $\{e_1\} \times \mathbb{R} \times \mathbb{R}_+ \cup \{e_0\} \times \mathbb{R}_+$ when there is no preventive maintenance and in $\{e_1\} \times \mathbb{R} \times \mathbb{R}_+ \cup \{e_0, e_2\} \times \mathbb{R}_+$ when there is preventive maintenance. On $\zeta^1 = \{e_1\}$, ζ^2 is replaced by $\tilde{\zeta}^2 = (\xi, \zeta^2)$, we have $Q((e_0, x_1, x_2); dy) = \delta_{e_1,\tau_0,0}(dy)$, $B = \{e_1\} \times [a, +\infty[\times\mathbb{R}_+$, and moreover, $Q((e_2, x_1, x_2); dy) = \delta_{e_1,\tau_0,0}(dy)$ in case of preventive maintenance.
If the preventive maintenance occurs as soon as the temperature is at least a or the component reaches the age of c, then $B = \{e_1\} \times [a, +\infty[\times[c, +\infty[$. We can also take $B = \{e_1\} \times [a, +\infty[\times\{c\}$.

Example 10.7 Let us go back to the shock processes of Example 10.1. Denote by X_t the item's state at time t. Define $Q(x; dy)$ as the probability distribution of the item's state right after a shock given that the item's state is x right before it.

(A) Assume that $(T_n)_{n\geq 0}$ is a renewal process with interarrival distribution dF. Then $(X_t)_{t\geq 0}$ is a switching process generated by ζ and M defined as $M(\zeta; dy; dv) = Q(\zeta(v); dy)\, dF(v)$.

(B) Assume that the probability distribution of $T_{n+1} - T_n$ $(n \geq 0)$ given $X_s, s \leq T_n$, has a hazard rate λ that is a function of the item's state. Then $(X_t)_{t \geq 0}$ is a switching process generated by ζ, with parameters λ, Q.

This is also the case for the earthquake model of Example 1.3, Chap. 1 when tectonic forces between earthquakes are described by the intrinsic process.

10.2 Vector of Independent Parametrized Switching Processes

Proposition 10.1 *Let $(X_{k,t})_{t \geq 0}$ $(1 \leq k \leq n)$ be independent parametrized switching processes on E_k, generated by ζ_k, with parameters λ_k, Q_k, α_k, q_k, respectively.*

Then $X = (X_1, \ldots, X_n)$ is a parametrized switching process generated by $\zeta = (\zeta_1, \ldots, \zeta_n)$, with parameters λ, Q, α, q, where ζ_k $(1 \leq k \leq n)$ are independent,

$$\lambda(x_1, \ldots, x_n) = \sum_{k=1}^{n} \lambda_i(x_i), \quad \alpha(\zeta) = \min\{\alpha_k(\zeta_k) : 1 \leq k \leq n\},$$

$$Q((x_1, \ldots, x_n); dy_1 \ldots dy_n) = \sum_{k=1}^{n} \frac{\lambda_k(x_k)}{\lambda(x_1, \ldots, x_n)} Q_k(x_k; dy_k) \prod_{\substack{1 \leq j \leq n \\ j \neq k}} \delta_{x_j}(dy_j), \tag{10.1}$$

$$q((x_1, \ldots, x_n); dy_1 \ldots dy_n) = \prod_{i:\alpha_i(\zeta_i)=\alpha(\zeta)} q_i(x_i; dy_i) \prod_{j:\alpha_j(\zeta_j)>\alpha(\zeta)} \delta_{x_j}(dy_j). \tag{10.2}$$

The proof is by induction. It is a direct application of the following lemma.

Lemma 10.1 *For $k = 1, 2$, let ζ_k be a stochastic process taking values in E_k, and Z_k and U_k random variables taking values in E_k and $\mathbb{R}_+$ respectively. Assume that (ζ_1, Z_1, U_1) and (ζ_2, Z_2, U_2) are independent and that the conditional probability distribution of (Z_k, U_k) $(k = 1, 2)$ given ζ_k is M_k defined as M in Example 10.4, Sect. 10.1 with λ_k, Q_k, α_k, q_k instead of λ, Q, α, q.*

Define

$$T = U_1 \wedge U_2, \quad \xi_k = \begin{cases} \zeta_k(T), & T < U_k, \\ Z_k, & T = U_k. \end{cases} \quad \xi = (\xi_1, \xi_2),$$

Then the conditional probability distribution of (ξ, T) given $\zeta = (\zeta_1, \zeta_2)$ is $M(\zeta, dy_1 dy_2, dv)$, where M is as in Example 10.4, Sect. 10.1 with

$$\lambda(x_1, x_2) = \lambda_1(x_1) + \lambda_2(x_2), \quad \alpha(\zeta) = \min(\alpha_1(\zeta_1), \alpha_2(\zeta_2)),$$

$$Q((x_1, x_2); dy_1 dy_2) = \frac{\lambda_1(x_1)}{\lambda_1(x_1) + \lambda_2(x_2)} Q_1(x_1; dy_1)\, \delta_{x_2}(dy_2) + \frac{\lambda_2(x_2)}{\lambda_1(x_1) + \lambda_2(x_2)} \delta_{x_1}(dy_1)\, Q_2(x_2; dy_2),$$

$$q((x_1, x_2); dy_1 dy_2) = \begin{cases} q_1(x_1; dy_1)\, \delta_{x_2}(dy_2), & \alpha_1(\zeta_1) < \alpha_2(\zeta_2), \\ \delta_{x_1}(dy_1)\, q_2(x_2; dy_2), & \alpha_2(\zeta_2) < \alpha_1(\zeta_1) \\ q_1(x_1; dy_1)\, q_2(x_2; dy_2), & \alpha_1(\zeta_1) = \alpha_2(\zeta_2). \end{cases}$$

Proof Given a Borel function $f : E_1 \times \mathbb{E}_2 \times \mathbb{R}_+$, write

$$\begin{aligned} \mathbb{E}(f(\xi_1, \xi_2, T) \mid \zeta_1, \zeta_2) &= \mathbb{E}\left(1_{\{U_1 < U_2\}} f(Z_1, \zeta_2(U_1), U_1) \mid \zeta_1, \zeta_2)\right) \\ &+ \mathbb{E}\left(1_{\{U_2 < U_1\}} f(\zeta_1(U_2), Z_2, U_2) \mid \zeta_1, \zeta_2)\right) + \mathbb{E}\left(1_{\{U_1 = U_2\}} f(Z_1, Z_2, U_1) \mid \zeta_1, \zeta_2)\right) \\ &= \int_{E_1 \times \mathbb{R}_+} f(y, \zeta_2(v), v)\, 1_{\{v < \alpha_2(\zeta_2)\}}\, e^{-\int_0^v \lambda_2(\zeta_2(s))\, ds}\, M_1(\zeta_1, dy, dv) \\ &+ \int_{E_2 \times \mathbb{R}_+} f(\zeta_1(v), y, v)\, 1_{\{v < \alpha_1(\zeta_1)\}}\, e^{-\int_0^v \lambda_1(\zeta_1(s))\, ds}\, M_2(\zeta_2, dy, dv) \\ &+ 1_{\{\alpha_1(\zeta_1) = \alpha_2(\zeta_2) < +\infty\}} e^{-\int_0^{\alpha_1(\zeta_1)} \lambda_1(\zeta_1(s))\, ds} e^{-\int_0^{\alpha_2(\zeta_2)} \lambda_2(\zeta_2(s))\, ds} \\ &\times \int_{E_1 \times E_2} f(y_1, y_2, \alpha_1(\zeta_1))\, q_1(\zeta_1(\alpha_1(\zeta_1)); dy_1)\, q_2(\zeta_2(\alpha_2(\zeta_2)); dy_2). \end{aligned}$$

Then the calculations are straightforward. ■

Corollary 10.1 *Let $(X_{k,t})_{t \geq 0}$ $(1 \leq k \leq n)$ be independent PDMPs with parameters $\phi_k, \lambda_k, Q_k, \Gamma_k, q_k$ respectively. Then $X = (X_1, \ldots, X_n)$ is a PDMP with parameters $\phi, \lambda, Q, \Gamma, q$ defined as*

$$\phi((x_1, \ldots, x_n), t) = (\phi_1(x_1, t), \ldots, \phi_n(x_n, t)), \quad \lambda(x_1, \ldots, x_n) = \sum_{k=1}^{n} \lambda_k(x_k),$$

$$\Gamma = \{(x_1, \ldots, x_n) : \exists k \text{ such that } x_k \in \Gamma_k\},$$

Q and q as in (10.1) and (10.2) respectively.

10.3 Process Decomposition

The decomposition of a switching process can be performed as for PDMPs (Sect. 4.2). In this short section we only state the results. The proof of Proposition 10.2 below follows that of Proposition 4.3, Sect. 4.2. It does not require new ideas.

Proposition 10.2 *Let $A \subset E$ be a Borel set, $\gamma : E \to \mathbb{R}_+$ a Borel function, M and M_0 two random semi-Markov kernels satisfying the following conditions:*

$$\forall t \geq 0 \quad 1 - M(\zeta; E \times [0,t]) = e^{-\int_0^t \gamma(\zeta(s))\,ds}\,(1 - M_0(\zeta; E \times [0,t]))\,, \quad (10.3)$$

$$1_A(y)\,M(\zeta; dy, dv) = e^{-\int_0^v \gamma(\zeta(s))\,ds}\,M_0(\zeta; dy, dv). \quad (10.4)$$

Let X be a switching process generated by ζ and M such that $\mathbb{P}(X_0 \in A) = 1$. Let X^0 be a switching process generated by ζ and M_0 with the same initial probability distribution as X. Define

$$\tau = \inf\{t \geq 0 : X_t \in A^c\}, \quad \tau^0 = \inf\{t \geq 0 : X_t^0 \in A^c\}.$$

Then

$$\mathbb{P}(\tau > t) = \mathbb{E}\left(1_{\{t<\tau^0\}}\,e^{-\int_0^t \gamma(X_s^0)\,ds}\right).$$

Corollary 10.2 *Let $A \subset E$ be a Borel set and $X = (X_t)_{t\geq 0}$ a parametrized switching process generated by ζ, with parameters λ, Q, α, q.*

Let X^0 be a parametrized switching process generated by ζ, with parameters $\lambda_0, Q_0, \alpha, q$, where

$$\lambda_0(x) = \lambda(x)\,Q(x; A), \quad Q_0(x; dy) = 1_A(y)\,Q(x; dy)/Q(x; A).$$

Define

$$\tau = \inf\{t \geq 0 : X_t \in A^c\}, \quad \tau^0 = \inf\{t \geq 0 : X_t^0 \in A^c\}, \quad \gamma(x) = \lambda(x)\,Q(x; A^c).$$

Then

$$\mathbb{P}(\tau > t) = \mathbb{E}\left(1_{\{t<\tau^0\}}e^{-\int_0^t \gamma(X_s^0)ds}\right).$$

10.4 Semiregenerative Property, Link with the CSMP

As a consequence of the switching process construction, we get the following result.

Proposition 10.3 *Let $(X_t)_{t\geq 0}$ be a switching process generated by ζ and M, $(T_n)_{n\geq 1}$ its SP-jump times, and $Y_n = X_{T_n}$. Define the renewal kernel N as*

$$N(x, dy, dv) = \mathbb{E}(M(\zeta; dy, dv) \mid \zeta(0) = x),$$

namely

$$\int_{E\times\mathbb{R}_+} \varphi(y, v)\,N(x, dy, dv) = \mathbb{E}\left(\int_{E\times\mathbb{R}_+} \varphi(y, v)\,M(\zeta; dy, dv) \mid \zeta(0) = x\right)$$

for any Borel function $\varphi : E \times \mathbb{R}_+ \to \mathbb{R}_+$.

The process $(Y_n, T_n)_{n\geq 1}$ is a Markov renewal process with kernel N, and the probability distribution of (Y_1, T_1) given $X_0 = x$ is $N(x, \cdot, \cdot)$.

The process $(X_t)_{t\geq 0}$ is a semiregenerative process, and $(Y_n, T_n)_{n\geq 1}$ is its embedded Markov renewal process.

Let $(Z_t, A_t)_{t\geq 0}$ be the CSMP associated with the zero-delayed Markov renewal process (Y, T). It is called an embedded CSMP of the switching process X.

Proposition 10.4 *Let $(X_t)_{t\geq 0}$ be a switching process generated by ζ and M, let $(Z_t, A_t)_{t\geq 0}$ be an embedded CSMP, and $(Y_n, T_n)_{n\geq 1}$ the associated Markov renewal process. Define $N_t = \sum_{n\geq 1} 1_{\{T_n \leq t\}}$.*

Given a Borel function $g : \mathbb{R}_+ \times E \to \mathbb{R}_+$ and $t > 0$, we have

$$\mathbb{E}(g(t, X_t) \mid Z_t, A_t, T_{N_t}) = \psi(Z_t, A_t, T_{N_t}) \quad a.s.,$$

where

$$\bar{F}_y(v)\,\psi(y, v, s) = \mathbb{E}\left(g(s + v, \zeta(v))\,(1 - M(\zeta; E \times [0, v])) \mid \zeta(0) = y\right),$$

$$\bar{F}_y(v) = \mathbb{E}(1 - M(\zeta; E \times [0, v]) \mid \zeta(0) = y),$$

that is,

$$\psi(y, v, s) = \mathbb{E}(g(s + v, \zeta(v)) \mid T_1 > v, \zeta(0) = y).$$

In particular,

$$\mathbb{E}(g(t, X_t)) = \mathbb{E}(\psi(Z_t, A_t, T_{N_t})).$$

If $(X_t)_{t\geq 0}$ is a parameter switching process with parameters λ, Q, α, q, then we have

$$\bar{F}_y(v)\,\psi(y, v, s) = \mathbb{E}\left(g(s + v, \zeta(v))\,1_{\{v<\alpha(\zeta)\}}\,e^{-\int_0^v \lambda(\zeta(w))\,dw} \mid \zeta(0) = y\right),$$

$$\bar{F}_y(v) = \mathbb{E}\left(1_{\{v<\alpha(\zeta)\}}\,e^{-\int_0^v \lambda(\zeta(w))\,dw} \mid \zeta(0) = y\right).$$

Proof Let $\psi_0 : E \times \mathbb{R}_+^2 \to \mathbb{R}_+$ be a Borel function. Recall that $T_0 = 0$, $Y_0 = X_0$. We have

$$\mathbb{E}(\psi_0(Z_t, A_t, T_{N_t})\,g(t, X_t)) = \sum_{n\geq 0} \mathbb{E}(\psi_0(Y_n, t - T_n, T_n)\,g(t, \zeta^{(n+1)}(t - T_n))\,1_{\{N_t = n\}}).$$

From

$$\begin{aligned}&\mathbb{E}(g(t, \zeta^{(n+1)}(t-T_n))\, 1_{\{T_{n+1}-T_n>t-T_n\}} \mid Y_n, T_n)\\&= \mathbb{E}(g(t-T_n+T_n, \zeta^{(n+1)}(t-T_n))(1-M(\zeta^{(n+1)}, E\times[0,t-T_n])) \mid Y_n, T_n)\\&= \psi(Y_n, t-T_n, T_n)\, \bar{F}_{Y_n}(t-T_n)\end{aligned}$$

we get

$$\mathbb{E}(\psi_0(Z_t, A_t, T_{N_t})\, g(t, X_t)) = \mathbb{E}(\psi_0(Z_t, A_t, T_{N_t})\, \psi(Z_t, A_t, T_{N_t})),$$

which completes the proof. ∎

Since a switching process is a semiregenerative process, the results given at the end of Sect. 2.7 can be applied. Under the assumptions of Theorem 2.3, Sect. 2.7 or Corollary 2.4, Sect. 2.7, we get

$$\begin{aligned}\mathbb{E}(f(X_t)) \xrightarrow[t\to\infty]{} & \frac{\int_E \mathbb{E}\left(\int_0^{T_1} f(\zeta(v))\, dv \mid \zeta(0)=x)\, dv\right) \nu(dx)}{\int_E \mathbb{E}(T_1 \mid \zeta(0)=x)\, \nu(dx)}\\ = & \frac{\int_E \mathbb{E}\left(\int_{\mathbb{R}_+} f(\zeta(v))\, (1-M(\zeta; E\times[0,v]))\, dv \mid \zeta(0)=x\right) \nu(dx)}{\int_E \mathbb{E}\left(\int_{\mathbb{R}_+} (1-M(\zeta; E\times[0,v]))\, dv \mid \zeta(0)=x\right) \nu(dx)}\end{aligned}$$

for ν-almost all $x \in E$, where ν is "the" invariant measure of the Markov chain Y.

10.5 Conditions to Get a Markov Process

Theorem 10.1 *Let M be a random semi-Markov kernel. Assume*

i. ζ is a Markov process,
ii. for all $s \in \mathbb{R}_+$, $M(\zeta; E\times[0,s])$ is measurable for the σ-field generated by $\zeta(v), v \le s$,
iii. for all $s > 0$ and every Borel function $\varphi : E\times\mathbb{R}_+ \to \mathbb{R}_+$, we have

$$\begin{aligned}&\int_{E\times\mathbb{R}_+} 1_{\{s<v\}} \varphi(y, v-s)\, M(\zeta; dy, dv)\\&= (1-M(\zeta; E\times[0,s])) \int_{E\times\mathbb{R}_+} \varphi(y, v)\, M(\zeta(s+\cdot\,); dy, dv). \quad (10.5)\end{aligned}$$

Then the switching process X generated by ζ and M is a Markov process.

This theorem generalizes Theorem 3.1, Sect. 3.1. Indeed, return to Example 10.5, Sect. 10.1 and its notation: the Markov property for the process ζ is $\phi(x, s+t) = \phi(\phi(x,s), t)$ for all x, s, t and condition iii is $N_0^{x,s} = N(\phi(x,s), ., .)$ for all x, s.

The proof of Theorem 10.1 is based on the following lemma.

Lemma 10.2 *Let $f : E \to \mathbb{R}_+$ be a Borel function. Assume that the assumptions of Theorem 10.1 hold and define $\zeta^{(1)}$ and T_1 as in Definition 10.1, Sect. 10.1. Then for all $(s,t) \in \mathbb{R}_+^2$,*

$$\mathbb{P}(T_1 > t+s \,|\, \zeta^{(1)}) = \mathbb{P}(T_1 > s \,|\, \zeta^{(1)})\,(1 - M(\zeta^{(1)}(s + \cdot\,); E \times [0,t])), \quad (10.6)$$

and for all $(v, v_1) \in \mathbb{R}_+^2$,

$$\begin{aligned}\mathbb{E}(1_{\{T_1>v\}} f(X_{v+v_1}) \,|\zeta^{(1)}(s), s \le v) &= \mathbb{P}(T_1 > v \,|\zeta^{(1)}(s), s \le v)\\ &\quad \times \mathbb{E}(f(X_{v_1}) \,|X_0 = \zeta^{(1)}(v)).\end{aligned}$$

Proof Formula (10.6) is an immediate consequence of condition iii, and it provides

$$\begin{aligned}&\mathbb{E}(1_{\{T_1>v+v_1\}} f(\zeta^{(1)}(v+v_1)) \,|\, \zeta^{(1)})\\ &= f(\zeta^{(1)}(v+v_1))\,\mathbb{P}(T_1 > v \,|\, \zeta^{(1)})\,(1 - M(\zeta^{(1)}(v + \cdot\,); E \times [0, v_1])).\end{aligned}$$

Therefore, it follows from conditions i and ii that

$$\mathbb{E}(1_{\{T_1>v+v_1\}} f(X_{v+v_1}) \,|\, \zeta^{(1)}(s), s \le v) = \mathbb{P}(T_1 > v \,|\, \zeta^{(1)}(s), s \le v)\,\varphi(\zeta^{(1)}(v), v_1), \quad (10.7)$$

where

$$\begin{aligned}\varphi(y_1, v_1) &= \mathbb{E}(f(\zeta^{(1)}(v_1))\,(1 - M(\zeta^{(1)}; E \times [0, v_1])) \,|\, \zeta^{(1)}(0) = y_1)\\ &= \mathbb{E}(f(X_{v_1})\,1_{\{v_1<T_1\}} \,|\, X_0 = y_1).\end{aligned}$$

On the other hand, from the construction of X, more precisely the semiregenerative property, we get

$$1_{\{T_1 \le v+v_1\}}\,\mathbb{E}(f(X_{v+v_1}) \,|\, Y_1, T_1, \zeta^{(1)}) = 1_{\{T_1 \le v+v_1\}}\,\varphi_1(Y_1, v + v_1 - T_1),$$

where $\varphi_1(x, u) = \mathbb{E}(f(X_u) \,|\, X_0 = x)$. Hence, by assumption iii, we have

$$\begin{aligned}&\mathbb{E}(1_{\{v<T_1\le v+v_1\}} f(X_{v+v_1}) \,|\, \zeta^{(1)})\\ &= \int_{E\times\mathbb{R}_+} 1_{\{v<u\le v+v_1\}}\,\varphi_1(x, v+v_1-u)\,M(\zeta^{(1)}; dx, du)\\ &= \mathbb{P}(T_1 > v \,|\, \zeta^{(1)}) \int_{E\times\mathbb{R}_+} 1_{\{u\le v_1\}}\mathbb{E}(f(X_{v_1-u}) \,|\, X_0 = x)\,M(\zeta^{(1)}(v + \cdot\,); dx, du).\end{aligned}$$

Consequently, since $\zeta^{(1)}$ is a Markov process and assumption ii holds, we get

$$\mathbb{E}(1_{\{v<T_1\leq v+v_1\}} f(X_{v+v_1}) \mid \zeta^{(1)}(s), s \leq v) = \mathbb{P}(T_1 > v \mid \zeta^{(1)}(s),\ s \leq v)\, \psi(\zeta^{(1)}(v), v_1), \tag{10.8}$$

where

$$\psi(y_1, v_1) = \mathbb{E}\left(\int 1_{\{0<u\leq v_1\}} \mathbb{E}(f(X_{v_1-u}) \mid X_0 = x)\, M(\zeta^{(1)}; dx, du) \mid \zeta^{(1)}(0) = y_1\right)$$
$$= \mathbb{E}(1_{\{T_1\leq v_1\}} f(X_{v_1}) \mid X_0 = y_1),$$

thanks again to the semiregenerative property. Summing (10.7) and (10.8) completes the proof of the lemma. ■

Now we turn to the proof of Theorem 10.1 with the notation of Definition 10.1, Sect. 10.1.

Let $k \geq 1$, $0 < t_1 < \cdots < t_k < t_{k+1}$, and let $f_1, \ldots, f_k$ be Borel functions from E to $\mathbb{R}_+$. Define $t_0 = 0$, $f_0 = 1$. We have

$$\mathbb{E}(f_1(X_{t_1}) \ldots f_k(X_{t_k})\, f_{k+1}(X_{t_{k+1}})) = \sum_{n\geq 0} \sum_{m=0}^{k-1} A_{n,m},$$

where

$$A_{n,m} = \mathbb{E}(f_1(X_{t_1}) \ldots f_k(X_{t_k})\, f_{k+1}(X_{t_{k+1}})\, 1_{\{t_{k-m-1}<T_n\leq t_{k-m}<t_k<T_{n+1}\}}).$$

Conditioning on $Y_1, \ldots, Y_n$, $(\zeta^{(p)})_{p\leq n}$, we get

$$A_{n,m} = \mathbb{E}(f_1(X_{t_1}) \ldots f_{k-m-1}(X_{t_{k-m-1}})\, 1_{\{t_{k-m-1}<T_n\}} \varphi_m(Y_n, T_n)),$$

where

$$\varphi_m(x, u) = 1_{\{u\leq t_{k-m}\}} \mathbb{E}\left(f_{k-m}(\zeta^{(1)}(t_{k-m} - u)) \ \ldots\ f_k(\zeta^{(1)}(t_k - u))\right.$$
$$\left.\times\ 1_{\{T_1>t_k-u\}}\, f_{k+1}(X_{t_{k+1}-u}) \mid \zeta^{(1)}(0) = x\right).$$

Consequently,

$$\mathbb{E}(f_1(X_{t_1}) \ldots f_k(X_{t_k})\, f_{k+1}(X_{t_{k+1}}))$$
$$= \sum_{n\geq 0} \sum_{m=0}^{k-1} \mathbb{E}\left(f_1(X_{t_1}) \ldots f_{k-m-1}(X_{t_{k-m-1}})\, 1_{\{t_{k-m-1}<T_n\}}\, \varphi_m(Y_n, T_n)\right). \tag{10.9}$$

Moreover, by conditioning on $\zeta^{(1)}(s)$, $s \leq t_k - u$, Lemma 10.2 provides

$$\varphi_m(x, u) = 1_{\{u\leq t_{k-m}\}} \mathbb{E}(1_{\{T_1>t_k-u\}}\, f_{k-m}(X_{t_{k-m}-u}) \ \ldots\ f_k(X_{t_k-u})\, \psi_k(X_{t_k-u}) | X_0 = x)$$

where $\psi_k(x) = \mathbb{E}(f_{k+1}(X_{t_{k+1}-t_k})|X_0 = x)$.

Now taking $f_{k+1} = 1$, $f_k = g_k$ in (10.9), we get

$$\mathbb{E}(f_1(X_{t_1}) \dots f_{k-1}(X_{t_{k-1}}) g_k(X_{t_k}))$$
$$= \sum_{n\geq 0} \sum_{m=0}^{k-1} \mathbb{E}\left(f_1(X_{t_1}) \dots f_{k-m-1}(X_{t_{k-m-1}})\, 1_{\{t_{k-m-1}<T_n\}}\, \tilde{\varphi}_m(Y_n, T_n)\right),$$

where

$$\tilde{\varphi}_m(x, u) = 1_{\{u \leq t_{k-m}\}} \mathbb{E}(1_{\{T_1 > t_k - u\}} f_{k-m}(X_{t_{k-m}-u}) \dots f_{k-1}(X_{t_{k-1}-u}) g_k(X_{t_k-u})|X_0 = x).$$

Finally, taking $g_k(x) = f_k(x)\,\psi_k(x)$, we have $\tilde{\varphi}_m = \varphi_m$ for all m. Thus

$$\mathbb{E}(f_1(X_{t_1}) \dots f_k(X_{t_k})\, f_{k+1}(X_{t_{k+1}})) = \mathbb{E}(f_1(X_{t_1}) \dots f_k(X_{t_k})\, \psi_k(X_{T_k})).$$

This completes the proof of Theorem 10.1.

Corollary 10.3 *A parametrized switching process generated by a Markov process ζ, with parameters λ, Q, α, q, is a Markov process as soon as $\alpha(\zeta)$ is a stopping time for the filtration $(\mathcal{F}_t)_{t\geq 0}$ generated by the process ζ, i.e., $\{\alpha(\zeta) \leq t\} \in \mathcal{F}_t$ for all $t \geq 0$, and $\alpha(\zeta) = s + \alpha(\zeta(s + \cdot))$ on $\{\alpha(\zeta) > s\}$.*

In particular, a switching process generated by a Markov process ζ, with parameters λ, Q, B, q, is a Markov process.

As a consequence of Corollary 10.3 and Example 10.6, Sect. 10.1, the processes $(X_t)_{t\geq 0}$ of Example 10.2 are Markov processes. Now let us take a look at Example 10.3, Chap. 10 and Example 10.7, Sect. 10.1.

Example 10.8 Let us go back to Example 10.3 and its notation. Define N as the kernel of the Markov renewal process $(\bar{Y}_n, T_n)_{n\geq 0}$. We assume that N is proper, that is, $N(i, \{1, 2, 3\} \times \mathbb{R}_+) = 1$ for any $i \in \{1, 2, 3\}$. We define the intrinsic process $\zeta = (\zeta^1, \zeta^2)$ taking values in $\{1, 2, 3\} \times \mathbb{R}_+$ as follows: $\zeta^1(t) = \zeta^1(0)$ for all t, $\zeta^2 = \xi^{\theta_i}$ if $\zeta^1 = i$. Then X is a switching process generated by ζ and M defined as $M(\zeta; dy_1, dy_2, dv) = N(i, dy_1, dv)\, \delta_{\zeta^2(v)}(dy_2)$ if $\zeta^1 = i$ $(i \in \{1, 2, 3\})$.

The process ζ is a Markov process, since processes ξ^{θ_i} are assumed to be. Moreover, $M(\zeta; E \times [0, s])$ is deterministic and condition ii holds. But Condition iii does not hold in the general case; indeed, if we take $\varphi(y, v) = 1_{]t,+\infty[}(v)$, the first member of (10.5) is $\mathbb{P}(T_1 > t + s)$ when the second member is $\mathbb{P}(T_1 > s)\,\mathbb{P}(T_1 > t)$; hence a necessary condition for (10.5) to hold is that the interarrival probability distributions of the Markov renewal process $(\bar{Y}_n, T_n)_{n\geq 0}$ be exponential. This is quite natural, since $X_t = (C_t, \xi_t)$ does not contain the memory of the elapsed time in the current climatic state. Therefore, we add A_t defined as $A_t = t - T_n$ if $T_n \leq t < T_{n+1}$. The process $(\widetilde{X}_t)_{t\geq 0} = (C_t, \xi_t, A_t)_{t\geq 0}$ is a switching process generated by $\tilde{\zeta} = (\tilde{\zeta}^1, \tilde{\zeta}^2, \tilde{\zeta}^3)$ and $\widetilde{M}$ defined for $\tilde{\zeta}^1(0) = i$, $\tilde{\zeta}^3(0) = u$ as follows:

$$\forall t \geq 0: \quad \tilde{\zeta}^1(t) = i, \ \tilde{\zeta}^2(t) = \xi^{\theta_i}(t), \ \tilde{\zeta}^3(t) = u + t,$$

$$\widetilde{M}(\zeta; dy_1 dy_2 dw, dv) = N_0^{i,u}(dy_1, dv)\, \delta_{\xi^{\theta_i}(v)}(dy_2)\, \delta_0(dw).$$

The two members of (10.5) are then equal to

$$\frac{1}{\mathbb{P}(T_1 > u \mid Y_0 = i)} \sum_{j \in \{1,2,3\}} \int_{\mathbb{R}_+} 1_{\{s+u<v\}}\, \varphi(j, \xi^{\theta_i}(v-u), 0, v-u-s)\, N(i, \{j\}, dv);$$

hence Condition iii holds and $(\widetilde{X}_t)_{t \geq 0}$ is a Markov process.

Example 10.9 Let us look at Example 10.7, end of Sect. 10.1. In case B, Corollary 10.3 states that $(X_t)_{t \geq 0}$ is a Markov process.

In case A, if $(T_n)_{n \geq 0}$ is a Poisson process, then $(X_t)_{t \geq 0}$ is a parametrized switching process without boundary, hence a Markov process. If $(T_n)_{n \geq 0}$ is not a Poisson process, the problem is the same as in Example 10.8, and we have to consider $\widetilde{X}_t = (X_t, A_t)$ to get a Markov process. Then the intrinsic process is $\widetilde{\zeta} = (\zeta, \zeta_e)$ with $\zeta_e(t) = \zeta_e(0) + t$, and $\widetilde{X}$ is a switching process generated by $\widetilde{\zeta}$ and $\widetilde{M}$ defined as

$$\widetilde{M}(\widetilde{\zeta}; dy, dw, dv) = Q(\zeta(v); dy)\, dF_0^u(v)\, \delta_0(dw) \text{ on } \zeta_e(0) = u,$$

where $\int_{\mathbb{R}_+} f(v)\, dF_0^u(v) = \frac{1}{\bar{F}(u)} \int_{\{v:v>u\}} f(v-u)\, dF(v)$. The two members of (10.5) are then equal to

$$\frac{1}{\bar{F}(u)} \int 1_{\{s+u<v\}} \varphi(y, 0, v-u-s)\, Q(\zeta(v-u); dy)\, dF(v),$$

which finally proves that $\widetilde{X}$ is indeed a Markov process.

10.6 Case of a Semimartingale Intrinsic Process

The goal of this section is to get Kolmogorov equations for the switching process X from the generator of the intrinsic process ζ when the added jumps are not caused by a boundary. We focus on the methodology, not on the minimal assumptions. The equations we get are the ones we expect.

We assume $E \subset \mathbb{R}^d$. The proofs depend on Itô's formula, which is recalled below.

Define $C^{1,2}$ as the class of functions $g = g(t, x) : \mathbb{R}_+ \times \mathbb{R}^d \to \mathbb{R}$ that are continuously differentiable with respect to the time variable t and twice continuously differentiable with respect to the spatial variable x.

Itô's formula.

Given a filtration on a probability space, consider a semimartingale $X = (X^{(1)}, \cdots, X^{(d)})$ taking values in $\mathbb{R}^d$, that is, $X^{(i)} = X_0^{(i)} + M^{(i)} + V^{(i)}$, where $M^{(i)}$ is a local martingale and $V^{(i)}$ an adapted finite variation process $(1 \leq i \leq d)$.

Let $g \in C^{1,2}$. Then a.s. for all $t \in \mathbb{R}_+$,

$$g(t, X_t) = g(0, X_0) + \int_0^t \frac{\partial g}{\partial t}(s, X_{s-})\, ds + \sum_{i=1}^d \int_0^t \frac{\partial g}{\partial x_i}(s, X_{s-})\, dV_s^{(i)}$$

$$+ \sum_{i=1}^d \int_0^t \frac{\partial g}{\partial x_i}(s, X_{s-})\, dM_s^{(i)} + \frac{1}{2} \sum_{i=1}^d \sum_{j=1}^d \int_0^t \frac{\partial^2 g}{\partial x_i \partial x_j}(s, X_{s-})\, d < M^{(i)}, M^{(j)} >_s^c$$

$$+ \sum_{0<s\leq t} \left(g(s, X_s) - g(s, X_{s-}) - \sum_{i=1}^d \frac{\partial g}{\partial x_i}(s, X_{s-})\, \Delta X_s^{(i)} \right).$$

Generally, the intrinsic process ζ is a semimartingale, and then on applying Itô's formula, we can see that $g(t, \zeta(t)) - g(0, \zeta(0)) - \int_0^t \mathcal{A}_0 g(v, \zeta(v))\, dv$ is a martingale, for some operator $\mathcal{A}_0$, under suitable conditions on g. This section is based on such a martingale property.

Given $g : \mathbb{R}_+ \times E \to \mathbb{R}$, define $\tau_s g(u, x) = g(s + u, x)$.

Theorem 10.2 *Assume*

$$M(\zeta; dy, dv) = \lambda(\zeta(v))\, e^{-\int_0^v \lambda(\zeta(w))\, dw}\, Q(\zeta(v); dy)\, dv$$

and that λ is bounded.

Let $(\mathcal{D}(\mathcal{A}_0), \mathcal{A}_0)$ be an operator on functions $g : \mathbb{R}_+ \times E \to \mathbb{R}$ that are bounded on $[0, t] \times E$ for all $t > 0$. Assume that for all $g \in \mathcal{D}(\mathcal{A}_0)$:

1. *the process $t \to g(t, \zeta(t))$ is given by*

$$g(t, \zeta(t)) = g(0, \zeta(0)) + \int_0^t \mathcal{A}_0 g(v, \zeta(v)) dv + M_t^g,$$

   where M_t^g is a martingale,
2. *the function $\mathcal{A}_0 g$ is bounded on $[0, t] \times E$ for all $t > 0$,*
3. *for all $s > 0$, $\tau_s g \in \mathcal{D}(\mathcal{A}_0)$ and $\mathcal{A}_0 \tau_s g = \tau_s \mathcal{A}_0 g$.*

Define X as a switching process generated by ζ and M, and the operator $\tilde{\mathcal{A}}_0$ on $\mathcal{D}(\mathcal{A}_0)$ as

$$\tilde{\mathcal{A}}_0 g(v, x) = \mathcal{A}_0 g(v, x) + \lambda(x) \int_E (g(v, y) - g(v, x))\, Q(x; dy).$$

Then

$$\mathbb{E}(g(t, X_t)) = \mathbb{E}(g(0, X_0)) + \int_0^t \mathbb{E}(\tilde{\mathcal{A}}_0 g(s, X_s))\, ds$$

for all $g \in \mathcal{D}(\mathcal{A}_0)$.

Proof The notation is as in Propositions 10.3 and 10.4, Sect. 10.4. Define also $\mathbb{E}_x(\,\cdot\,) = \mathbb{E}(\,\cdot \mid \zeta(0) = x)$ $(x \in E)$; we use this notation when only the process ζ appears in the formulas.

To apply Proposition 10.4 and Corollary 6.1, Sect. 6.1, we need some formulas to check their assumptions.

First, from Itô's formula we get for $g \in \mathcal{D}(\mathcal{A}_0)$,

$$e^{-\int_0^t \lambda(\zeta(w))\, dw} g(t, \zeta(t)) = g(0, \zeta(0)) + \int_0^t e^{-\int_0^s \lambda(\zeta(w))\, dw} \mathcal{A}_1 g(s, \zeta(s))\, ds + \int_0^t e^{-\int_0^s \lambda(\zeta(w))\, dw}\, dM_s^g,$$

where $\mathcal{A}_1 g(s, x) = \mathcal{A}_0 g(s, x) - \lambda(x)\, g(s, x)$ (recall that $\{v : \zeta(v_-) \neq \zeta(v)\}$ is countable, since ζ is assumed to be cad-lag). Thus for all $z \in E$,

$$\mathbb{E}_z\left(e^{-\int_0^t \lambda(\zeta(w))\, dw} g(t, \zeta(t))\right) - g(0, z) = \mathbb{E}_z\left(\int_0^t e^{-\int_0^v \lambda(\zeta(w))\, dw} \mathcal{A}_1 g(v, \zeta(v))\, dv\right). \tag{10.10}$$

Second, write $N(x, dy, dv) = 1_{\mathbb{R}_+}(v)\, dF_x(v)\, \beta(x, v; dy)$. It follows from Proposition 10.3, Sect. 10.4 that

$$1_{\mathbb{R}_+}(v)\, dF_x(v) = \int_{\mathbb{R}^d} N(x, dy, dv) = \mathbb{E}_x(\lambda(\zeta(v))\, e^{-\int_0^v \lambda(\zeta(w))\, dw})\, dv,$$

and thus $1_{\mathbb{R}_+}(v)\, dF_x(v) = \ell(x, v)\, e^{-\int_0^v \ell(x,w)\, dw}\, dv$ for some ℓ (Proposition A.6, Sect. A.3) and

$$\ell(z, v) = \frac{\mathbb{E}_z\left(e^{-\int_0^v \lambda(\zeta(w))\, dw} \lambda(\zeta(v))\right)}{\mathbb{E}_z\left(e^{-\int_0^v \lambda(\zeta(w))\, dw}\right)} \leq ||\lambda||_\infty.$$

Consequently, $\mathbb{E}(\sum_{n \geq 0} 1_{\{T_n \leq t\}}) < +\infty$ (Corollary 5.1, end of Sect. 5.2).

Now we turn to the heart of the proof.

From Proposition 10.4, Sect. 10.4 we get $\mathbb{E}(g(t, X_t)) = \mathbb{E}(\psi(Z_t, A_t, T_{N_t}))$, where $\psi(z, v, s) = \mathbb{E}_z\left(e^{-\int_0^v \lambda(\zeta(w))\, dw} g(s + v, \zeta(v))\right) / \bar{F}_z(v)$. The function ψ is bounded on $\mathbb{R}^d \times \mathbb{R}_+ \times [0, t]$, since $\bar{F}_z(v) = \mathbb{E}_z(e^{-\int_0^v \lambda(\zeta(w))\, dw}) \geq e^{-||\lambda||_\infty v}$. From (10.10) applied to $\tau_s g$, we deduce that the function $v \to \psi(z, v, s)$ is absolutely continuous. Its derivative $\partial_2 \psi$ is bounded on $\mathbb{R}^d \times [0, A] \times [0, B]$ for all $A > 0$, $B > 0$. Corollary 6.1, Sect. 6.1 yields

$$\mathbb{E}(\psi(Z_t, A_t, T_{N_t})) = \mathbb{E}(\psi(Z_0, 0, 0)) + \mathbb{E}\left(\int_0^t \tilde{L}\psi(Z_s, A_s, T_{N_s})\right) ds, \tag{10.11}$$

where $\tilde{L}\psi(z, v, s) = \partial_2\psi(z, v, s) + \int_{\mathbb{R}^d} (\psi(z_1, 0, s + v) - \psi(z, v, s))\, \ell(z, v)\, \beta(z, v; dz_1)$.

We now turn our attention to the second term of (10.11). For $(z, s) \in \mathbb{R}^d \times \mathbb{R}_+$, the function $v \to \bar{F}_z(v)\, \tilde{L}\psi(z, v, s)$ is Lebesgue integrable on $[0, t]$. An integration by parts, the relation $dF_z(v) = \bar{F}_z(v)\, \ell(z, v)\, dv$, the assumption $\mathcal{A}_0\tau_s g = \tau_s\mathcal{A}_0 g$, and Formula (10.10) applied to $\tau_s g$ give

$$\int_0^t \bar{F}_z(v)\tilde{L}\psi(z, v, s)\, dv = \int_0^t \mathbb{E}_z\left(e^{-\int_0^v \lambda(\zeta(w))\, dw}\, \tilde{\mathcal{A}}_0 g(s + v, \zeta(v))\right) dv. \quad (10.12)$$

Define ψ_1 as $\psi_1(z, v, s) = \mathbb{E}_z\left(e^{-\int_0^v \lambda(\zeta(w))\, dw}\, \tilde{\mathcal{A}}_0 g(s + v, \zeta(v))\right) / \bar{F}_z(v)$. From (10.12) we get $\tilde{L}\psi(z, v, s)\, dv = \psi_1(z, v, s)\, dv$ for all $(z, s) \in \mathbb{R}^d \times \mathbb{R}_+$. Hence if $T_n \le t < T_{n+1}$, we have

$$\int_0^t \tilde{L}\psi(Z_s, A_s, T_{N_s})\, ds = \sum_{k=0}^{n-1} \int_{T_k}^{T_{k+1}} \tilde{L}\psi(Y_k, s - T_k, T_k)\, ds$$
$$+ \int_{T_n}^t \tilde{L}\psi(Y_n, s - T_n, T_n)\, ds = \int_0^t \psi_1(Z_s, A_s, T_{N_s})\, ds.$$

Applying again Proposition 10.4, we get

$$\mathbb{E}\left(\int_0^t \tilde{L}\psi(Z_s, A_s, T_{N_s})\right) ds = \int_0^t \mathbb{E}(\tilde{\mathcal{A}}_0 g(s, X_s))\, ds.$$

The result follows. ∎

Corollary 10.4 *Let $(\mathcal{D}(\mathcal{A}), \mathcal{A})$ be an operator on bounded functions $f : E \to \mathbb{R}$. Assume that for all $f \in \mathcal{D}(\mathcal{A})$:*

1. *the process $t \to f(\zeta(t))$ is given by*

$$f(\zeta(t)) = f(\zeta(0)) + \int_0^t \mathcal{A}f(\zeta(v))\, dv + M_t^f,$$

where M_t^f is a martingale,

2. *the function $\mathcal{A}f$ is bounded on E.*

Let X be a parametrized switching process without boundary, generated by ζ, with parameters λ, Q. Assume that λ is bounded. Define the operator $\tilde{\mathcal{A}}$ on $\mathcal{D}(\mathcal{A})$ as

$$\tilde{\mathcal{A}}f(x) = \mathcal{A}f(x) + \lambda(x) \int_E (f(y) - f(x))\, Q(x; dy).$$

Then

$$\mathbb{E}(f(X_t)) = \mathbb{E}(f(X_0)) + \int_0^t \mathbb{E}(\tilde{\mathcal{A}}f(X_s))\, ds$$

for all $f \in \mathcal{D}(\mathcal{A})$.

Theorem 10.2 can be applied when the intrinsic process ζ is a Lévy process. Before giving the corresponding corollary (Corollary 10.5), let us briefly recall some results on Lévy processes, which come mainly from Pascucci [86].

A process ζ is a Lévy process taking values in $\mathbb{R}^d$ if

1. $\zeta(0) = 0$ a.s.,
2. ζ has independent stationary increments, i.e., for all $0 \leq s < t$, $\zeta(t) - \zeta(s)$ is independent of the σ field generated by $\zeta(u)$, $u \leq s$, and $\zeta(t) - \zeta(s)$ and $\zeta(t-s)$ have the same probability distribution,
3. the process ζ is stochastically continuous, that is, for all $\varepsilon > 0, t > 0$, we have

$$\lim_{h \to 0} \mathbb{P}(||\zeta(t+h) - \zeta(t)|| \geq \varepsilon) = 0.$$

Given a Lévy process ζ taking values in $\mathbb{R}^d$, there exists a σ-finite random measure J, called the jump measure, on the Borel sets of $\mathbb{R}_+ \times \mathbb{R}^d$, such that

i. $J(\{0\} \times \mathbb{R}^d) = 0$,
ii. $J(I \times A) = \text{Card}\{s \in I : \Delta\zeta(s) \in A\}$ for all Borel sets $I \subset \mathbb{R}_+$ and $A \subset \mathbb{R}^d$. If I is bounded and $0 \notin \bar{A}$ (where $\bar{A}$ is the topological closure of A), then $J(I \times A) < +\infty$ a.s.

Hence a Lévy process has only a finite number of "large" jumps in a finite time interval.

The Lévy measure of a Lévy process ν is defined as $\nu(A) = \mathbb{E}(J([0, 1] \times A))$.

Given a Borel set $A \subset \mathbb{R}^d$ such that $0 \notin \bar{A}$ and $\nu(A) < +\infty$, the process $t \to J_t(A) := J([0, t] \times A)$ is a Poisson process with intensity $\nu(A)$. Therefore, the process $t \to \tilde{J}_t(A) := J_t(A) - t\,\nu(A)$ is a martingale. More generally, let $f : \mathbb{R}_+ \times \mathbb{R}^d \to \mathbb{R}$ be a Borel function for which there exists $\varepsilon > 0$ such that $\int_0^t \int_{\{x:||x|| \leq \varepsilon\}} |f(s, x)|\, \nu(dx)\, ds < +\infty$. Then

$$\int_0^t \int_{\mathbb{R}^d} f(s, x)\, J(ds, dx) := \sum_{0<s\leq t, \Delta\zeta(s) \neq 0} f(s, \Delta\zeta(s)) < +\infty \quad a.s.$$

Define $\tilde{J}$ as $\tilde{J}(ds, dx) = J(ds, dx) - ds\, \nu(dx)$. If $\int_0^t \int_{\mathbb{R}^d} |f(s, x)|\, \nu(dx)\, ds < +\infty$ for all $t > 0$, then the process $M_t = \int_0^t \int_{\mathbb{R}^d} f(s, x)\, \tilde{J}(ds, dx)$ is a martingale.

Recall that given a correlation matrix C, there exists a matrix Σ such that $C = \Sigma\Sigma^T$ (Σ^T is the transposed matrix of Σ). We write $\Sigma = \sqrt{C}$.

Theorem 10.3 (Lévy-Itô decomposition) *Let* ζ *be a Lévy process taking values in* $\mathbb{R}^d$ *with jump measure* J *and Lévy measure* ν. *Then*

$$\int_{\{x:||x||\geq 1\}} \nu(dx) < +\infty, \quad \int_{\{x:||x||<1\}} ||x||^2 \nu(dx) < +\infty.$$

Moreover, there exist a d-dimensional standard Wiener process W, independent of J, a correlation matrix C, and for all $R > 0$ a constant $\mu_R \in \mathbb{R}^d$ such that

$$\zeta(t) = \mu_R\, t + \sqrt{C}\, W_t + \int_0^t \int_{\{x:||x||\geq R\}} x\, J(ds, dx) + \int_0^t \int_{\{x:||x||<R\}} x\, \tilde{J}(ds, dx). \tag{10.13}$$

The triplet (μ_R, C, ν) is the R-triplet of the Lévy process.

If $\int_{\{x:||x||\geq 1\}} ||x||\, \nu(dx) < +\infty$, we also have

$$\zeta(t) = \mu_\infty t + \sqrt{C}\, W_t + \int_0^t \int_{\mathbb{R}^d} x \tilde{J}(ds, dx) = \mu_\infty t + \sqrt{C}\, W_t + \sum_{0<s\leq t} \Delta\zeta(s) - t\, \mathbb{E}(||\Delta\zeta(1)||)$$

for some $\mu_\infty \in \mathbb{R}^d$. The triplet (μ_∞, C, ν) is called the ∞-triplet of the Lévy process.

Example 10.10 A gamma process (with parameters $a, b > 0$) is a Lévy process ζ taking values in $\mathbb{R}_+$ such that the probability distribution of $\zeta(t)$ $(t > 0)$ is $b^{at} x^{at-1} e^{-bx} 1_{\mathbb{R}_+}(x)\, dx / \Gamma(at)$. Its Levy measure is $\nu(dx) = \frac{a}{x} e^{-bx} 1_{\mathbb{R}_+}(x)\, dx$. It is a purely jump process, that is, $\mu_R = \mu_\infty = 0, C = 0$.

Corollary 10.5 *Let ζ be a Lévy process taking values in $\mathbb{R}^d$ with jump measure J and 1-triplet (μ, C, ν). Write $\mu = (\mu_1, \ldots, \mu_d)$.*

Define $\mathcal{D}(\mathcal{A}_0)$ as the set of bounded functions $g = g(t, x) : \mathbb{R}_+ \times \mathbb{R}^d \to \mathbb{R}$ of class $C^{1,2}$ with derivatives bounded on $[0, t] \times \mathbb{R}^d$ for all $t > 0$. For $g \in \mathcal{D}(\mathcal{A}_0)$, define

$$\mathcal{A}_0 g(s, x) = \frac{\partial g}{\partial s}(s, x) + \sum_{i=1}^d \mu_i \frac{\partial g}{\partial x_i}(s, x) + \frac{1}{2} \sum_{i=1}^d \sum_{j=1}^d C_{i,j} \frac{\partial^2 g}{\partial x_i \partial x_j}(s, x) + \int_{\mathbb{R}^d} \left(g(s, x + y) - g(s, x) - \sum_{i=1}^d \frac{\partial g}{\partial x_i}(s, x)\, y_i 1_{\{||y||<1\}} \right) \nu(dy),$$

$$\tilde{\mathcal{A}}_0 g(s, x) = \mathcal{A}_0 g(s, x) + \lambda(x) \int_{\mathbb{R}^d} (g(s, y) - g(s, x))\, Q(x; dy).$$

Let X be a parametrized switching process without boundary, generated by ζ, with parameters λ, Q. Assume that λ is bounded.

Then for all $g \in \mathcal{D}(\mathcal{A}_0)$,

$$\mathbb{E}(g(t, X_t)) = \mathbb{E}(g(0, X_0)) + \int_0^t \mathbb{E}(\tilde{\mathcal{A}}_0 g(s, X_s))\, ds.$$

Proof From Itô's formula we get

$$g(t,\zeta(t)) = g(0,0) + \int_0^t \mathcal{A}_0 g(s,\zeta(s))\,ds + \sum_{i=1}^d \int_0^t \frac{\partial g}{\partial x_i}(s,\zeta(s_-)) \sum_{j=1}^d \sigma_{i,j}\,dW_s^j$$
$$+ \int_0^t \int_{\mathbb{R}^d} \Big(g(s,\zeta(s_-)+x) - g(s,\zeta(s_-)) \Big) \tilde{J}(ds,dx). \quad (10.14)$$

On the one hand, the partial derivatives $\partial g/\partial x_i$ are bounded; thus the processes $t \to \int_0^t \partial g/\partial x_i(s,\zeta(s_-))\,dW_s^j$ $(1 \le i,j \le d)$ are martingales. On the other hand, Taylor's formula yields

$$\mathbb{E}\left(\int_0^t \int_{\mathbb{R}^d} |g(s,\zeta(s_-)+x) - g(s,\zeta(s_-)|^2 \nu(dx)\,ds \right) \le 4||g||^2\, t\, \nu(\{x : ||x|| \ge 1\})$$
$$+ C_t\, t \int_{\{x:||x||<1\}} ||x||^2\, \nu(dx) \; < +\infty,$$

where C_t is a constant that depends only on t and $\sup_{1\le i\le d} \sup_{s\le t, x\in\mathbb{R}^d} |\frac{\partial g}{\partial x_i}(s,x)|$. Therefore, $\int_0^t \int_{\mathbb{R}^d} (g(s,\zeta(s_-)+x) - g(s,\zeta(s_-)))\, \tilde{J}(ds,dx)$ is a martingale.

In the same way, it can be proved that $\mathcal{A}_0 g$ is bounded on $[0,t] \times \mathbb{R}_+$. Hence Theorem 10.2 can be applied. ■

Taking $g(s,z) = u(t-s,z)$, we get the Feynman–Kac formula.

Corollary 10.6 (Feynman–Kac formula) *Let ζ be a Lévy process taking values in $\mathbb{R}^d$ with jump measure J and 1-triplet (μ, C, ν). Let X be a parametrized switching process without boundary, generated by ζ, with parameters λ, Q. Assume that λ is bounded.*

Let $u = u(t,x) : \mathbb{R}_+ \times \mathbb{R}^d \to \mathbb{R}$ be a bounded function of class $C^{1,2}$ with derivatives bounded on $[0,t] \times \mathbb{R}^d$ for all $t > 0$. Assume that u satisfies

$$\frac{\partial u}{\partial t}(t,x) = \mathcal{A}u(t,x) + \lambda(x) \int_{\mathbb{R}^d} (u(t,y) - u(t,x))\, Q(x;dy), \quad u(0,x) = h(x),$$

where

$$\mathcal{A}u(t,x) = \sum_{i=1}^d \mu_i \frac{\partial u}{\partial x_i}(t,x) + \frac{1}{2} \sum_{i=1}^d \sum_{j=1}^d C_{i,j} \frac{\partial^2 u}{\partial x_i \partial x_j}(t,x)$$
$$+ \int_{\mathbb{R}^d} \left(u(t,x+y) - u(t,x) - \sum_{i=1}^d \frac{\partial u}{\partial x_i}(t,x)\, y_i 1_{\{||y||<1\}} \right) \nu(dy).$$

Then $u(t,x) = \mathbb{E}_x(h(X_t))$ for all $t > 0$.

10.7 Case of an Inhomogeneous Intrinsic Process

When ζ is the solution of an inhomogeneous stochastic differential equation or is an inhomogeneous Itô–Lévy process, we feel that the switching process is not correctly defined after a jump T_n $(n \geq 1)$, since it goes again as if T_n were the initial time. Besides, Theorem 10.2, Sect. 10.6 cannot be applied, because the condition $\mathcal{A}_0 \tau_s g = \tau_s \mathcal{A}_0 g$ does not hold. This is why we are going to give a definition of inhomogeneous switching processes, adapted to inhomogeneous intrinsic process, and we might as well ask for λ and Q to depend on time.

Let $\lambda : \mathbb{R}_+ \times E \to \mathbb{R}_+$ be a Borel function and Q a probability kernel from $\mathbb{R}_+ \times E$ to E. Define

$$M(\zeta; dy, dv) = \lambda(v, \zeta(v))\, e^{-\int_0^v \lambda(w,\zeta(w))\,dw}\, Q(v, \zeta(v); dy)\, dv. \tag{10.15}$$

The inhomogeneous switching process is defined as follows. Let $\zeta^{s,x} = (\zeta^{s,x}(t))_{t\geq 0}$ be a process whose probability distribution is the probability distribution of $\zeta(s + \cdot\,)$ given $\zeta(s) = x$. Assume that we have defined $\zeta^{(1)}, \ldots, \zeta^{(n)}, Y_1, T_1, \ldots, Y_n, T_n$. Let $\zeta^{(n+1)}$ be a process whose probability distribution given $\zeta^{(1)}, \ldots, \zeta^{(n)}, Y_1, T_1, \ldots, Y_n = x, T_n = s$ is the probability distribution of $\zeta^{s,x}$. The probability distribution of $(Y_{n+1}, T_{n+1} - T_n)$ given $\zeta^{(1)}, \ldots, \zeta^{(n+1)}, Y_1, T_1, \ldots, Y_n = x, T_n = s$ is $M(\zeta^{(n+1)}; dy, dv)$. For $T_n \leq t < T_{n+1}$, define $X_t = \zeta^{(n+1)}(t - T_n)$ and $X_{T_{n+1}} = Y_{n+1}$.

Define $\tilde{M}$ from M as follows: given a process $\xi = (\xi^{(1)}, \xi^{(2)})$ taking values in $\mathbb{R}_+ \times \mathbb{R}^d$, $\tilde{M}(\xi; ds, dy, dv) = M(\xi^{(2)}; dy, dv)\, \delta_{\xi^{(1)}(v)}(ds)$. Let $\tilde{\zeta}$ be a process whose probability distribution given $\tilde{\zeta}(0) = (s, x)$ is the probability distribution of the process $t \to (s + t, \zeta^{s,x}(t))$. Then the process $\tilde{X} = (t, X_t)_{t\geq 0}$ is a switching process generated by $\tilde{\zeta}$ and $\tilde{M}$.

For an inhomogeneous switching process, $(Y_n, T_n)_{n\geq 1}$ is generally not a Markov renewal process. However, the process $(\tilde{Y}_n, T_n)_{n\geq 1}$, where $\tilde{Y}_n = (T_n, Y_n)$, is a Markov renewal process with kernel

$$\tilde{N}((s, x); ds_1 dx_1, dv) = \mathbb{E}(M(\zeta^{s,x}; dx_1, dv))\, \delta_{s+v}(ds_1).$$

Applying Corollary 10.4, Sect. 10.6 to the switching process generated by $\tilde{\zeta}$ and $\tilde{M}$, we get the following proposition.

Proposition 10.5 *Assume that M is defined as in (10.15), λ is bounded, and X is an inhomogeneous switching process generated by ζ and M, as described right above.*

Let $(\mathcal{D}(\mathcal{A}_0), \mathcal{A}_0)$ be an operator on bounded functions $g : \mathbb{R}_+ \times \mathbb{R}^d \to \mathbb{R}$. Assume that for all $g \in \mathcal{D}(\mathcal{A}_0)$:

1. *the process $t \to g(t, \zeta(t))$ is given by*

$$g(t, \zeta(t)) = g(0, \zeta(0)) + \int_0^t \mathcal{A}_0 g(v, \zeta(v))\, dv + M_t^g,$$

where M_t^g is a martingale,

2. *the function $\mathcal{A}_0 g$ is bounded on E.*

Then

$$\mathbb{E}(g(t, X_t)) = \mathbb{E}(g(0, X_0)) + \int_0^t \mathbb{E}(\tilde{\mathcal{A}}_0 g(s, X_s))\, ds,$$

where

$$\tilde{\mathcal{A}}_0 g(s, x) = \mathcal{A}_0 g(s, x) + \lambda(s, x) \int_{\mathbb{R}^d} (g(s, y) - g(s, x))\, Q(s, x; dy).$$

Example 10.11 When $d = 1$ and the intrinsic processes ζ is an Itô–Lévy process solution of

$$d\zeta(t) = b(t, \zeta(t_-))\, dt + \sigma(t, \zeta(t_-))\, dW_t + \int a(t, \zeta(t_-), x)\, \tilde{J}(dt, dx),$$

we get (Pascucci [86]), under suitable conditions on g:

$$\begin{aligned} g(t, \zeta(t)) = g(0, \zeta(0)) &+ \int_0^t \mathcal{A}_0 g(s, \zeta(s))\, ds + \int_0^t \sigma(s, \zeta(s_-)) \frac{\partial g}{\partial x}(s, \zeta(s_-))\, dW_s \\ &+ \int_0^t \int (g(s, \zeta(s_-)) + a(s, \zeta(s_-), y)) - g(s, \zeta(s_-)))\, \tilde{J}(ds, dy), \end{aligned}$$

where

$$\begin{aligned} \mathcal{A}_0 g(s, x) = &\frac{\partial g}{\partial s}(s, x) + b(s, x) \frac{\partial g}{\partial x}(s, x) + \frac{1}{2} \sigma^2(s, x) \frac{\partial^2 g}{\partial x^2} g(s, x) \\ &+ \int_{\mathbb{R}} \left(g(s, x + a(s, x, y)) - g(s, x) - a(s, x, y) \frac{\partial g}{\partial x}(s, x) \right) \nu(dy). \end{aligned}$$

Thus Proposition 10.5 can be applied.

The generalization to $d > 1$ is left to the reader.

Appendix A
Tools

In this appendix, we recall what a conditional expectation and a conditional distribution are, we give some definitions and results for the readers who are not familiar with absolute continuous functions and more generally functions of bounded variations, and with the hazard rate of a positive random variable with a probability density function. These notions are of great use throughout this book. The end of Sect. A.3 is an introduction to Sect. A.4. Section A.4 is only useful in Chap. 7 which is not essential in this book.

A.1 Conditional Expectation, Conditional Probability Distribution

We briefly give the definitions and basic properties of conditional expectation and conditional probability distribution. We accept their existence.

Equalities and inequalities between random variables are true almost surely. We do not mention this below. In addition, "unique" means that two random variables satisfying the condition are equal almost surely.

Given a σ-field $\mathcal{B}$ and a real random variable X such that $\mathbb{E}(|X|) < +\infty$, the conditional expectation $\mathbb{E}(X|\mathcal{B})$ is the unique random variable such that for all $B \in \mathcal{B}$, $\mathbb{E}(1_B X) = \mathbb{E}(1_B \mathbb{E}(X|\mathcal{B}))$. We have:

- $\mathbb{E}(\mathbb{E}(X|\mathcal{B})) = \mathbb{E}(X)$,
- if X is $\mathcal{B}$-measurable, then $\mathbb{E}(X|\mathcal{B}) = X$,
- $\mathbb{E}(XY|\mathcal{B}) = Y\,\mathbb{E}(X|\mathcal{B})$ for every bounded $\mathcal{B}$-measurable random variable Y,
- if $X \geq 0$, then $\mathbb{E}(X|\mathcal{B}) \geq 0$,
- if $\mathbb{E}(|X_1|) < +\infty$, $\mathbb{E}(|X_2|) < +\infty$, $\lambda_1, \lambda_2 \in \mathbb{R}$, then $\mathbb{E}(\lambda_1 X_1 + \lambda_2 X_2|\mathcal{B}) = \lambda_1$ $E(X_1|\mathcal{B}) + \lambda_2 \mathbb{E}(X_2|\mathcal{B})$,
- if $\mathcal{B}_1 \subset \mathcal{B}_2$, then $\mathbb{E}(X|\mathcal{B}_1) = \mathbb{E}(\mathbb{E}(X|\mathcal{B}_2)|\mathcal{B}_1)$.

C. Cocozza-Thivent, *Markov Renewal and Piecewise Deterministic Processes*, Probability Theory and Stochastic Modelling 100,
https://doi.org/10.1007/978-3-030-70447-6

When $X \geq 0$, $\mathbb{E}(X|\mathcal{B})$ exists but takes values in $[0, +\infty]$. All the above properties hold assuming only $X \geq 0$, $Y \geq 0$, $\lambda_1 \geq 0$, $\lambda_2 \geq 0$.

Assume that $\mathcal{B} = \sigma(X)$, i.e., that $\mathcal{B}$ is the σ-field generated by X, and that X takes values in $(E, \mathcal{E})$. Then every real random variable $\mathcal{B}$-measurable is equal to $f(X)$ for some measurable function $f : E \to \mathbb{R}$.

By definition, $\mathbb{E}(Y|X) = \mathbb{E}(Y|\sigma(X))$; hence $\mathbb{E}(Y|X) = f(X)$ for some measurable function $f : E \to \mathbb{R}$ and $f(X)$ is the unique real random variable such that $\mathbb{E}(Yg(X)) = \mathbb{E}(f(X)\, g(X))$ for "every" measurable function $g : E \to \mathbb{R}$. "Every" means $g \geq 0$ if $Y \geq 0$, and g bounded if not but $\mathbb{E}(|Y|) < +\infty$.

Assume that X takes values in $(E, \mathcal{E})$ and Y in $(F, \mathcal{F})$. A (regular) conditional probability distribution of Y given $X = x$ $(x \in E)$ is $n(x, dy)$, where n is a probability kernel from E to F, as defined Sect. 2.1, such that $\mathbb{E}(f(Y)|X) = \int_F f(y)\, n(X, dy)$ for every measurable function $f : F \to \mathbb{R}_+$. We write $\mathbb{E}(f(Y)|X = x) = \int_F f(y) n(x, dy)$. Thanks to the monotone class theorem, we have

$$E(\varphi(X, Y)|X) = \int_F \varphi(X, y)\, n(X, dy)$$

for every measurable function $\varphi : E \times F \to \mathbb{R}_+$.

A.2 Functions of Bounded Variation

In this section we consider only functions and measures on $(\mathbb{R}_+, \mathcal{B}(\mathbb{R}_+))$.

Given a cad-lag (right continuous with left limits) nondecreasing function $f : \mathbb{R}_+ \to \mathbb{R}$, there exists one and only one nonnegative measure μ on $\mathcal{B}(\mathbb{R}_+)$ such that

$$\forall t \geq 0, \quad \mu([0, t]) = f(t). \tag{A.1}$$

Conversely, if μ is a nonnegative measure on $\mathcal{B}(\mathbb{R}_+)$, then the function $f(t) = \mu([0, t])$ is nondecreasing and cad-lag. Thus a nondecreasing cad-lag function $f : \mathbb{R}_+ \to \mathbb{R}$ is equivalent to a nonnegative measure μ on $\mathcal{B}(\mathbb{R}_+)$. The mapping is given by (A.1), or equivalently,

$$\mu(\{0\}) = f(0), \qquad \forall 0 \leq a < b : \ \mu(]a, b]) = f(b) - f(a), .$$

We write $\mu = df$.

Example A.1 1. if $f = 1_{[a,+\infty[}$ $(a \geq 0)$, then $df = \delta_a$,

2. if $f(t) = a + bt$, then $df = a\,\delta_0 + b\,dt$,
3. given a nonnegative measure μ on $\mathbb{R}_+$ and a Borel function $g : \mathbb{R}_+ \to \mathbb{R}_+$ μ-integrable on $[0, t]$ for all $t \geq 0$, define $f(t) = \int_0^t g(s)\, \mu(ds)$; then $df(t) = g(t)\, \mu(dt)$.

Definition A.1 The function $f : \mathbb{R}_+ \to \mathbb{R}$ is said to be of bounded variation on $\mathbb{R}_+$ if for all $t > 0$,

$$V_f(t) = \sup \left\{ \sum_{i=0}^{n-1} |f(t_{i+1}) - f(t_i)| : n \geq 1,\ 0 = t_0 < t_1 < \ldots < t_n = t \right\} < +\infty. \tag{A.2}$$

The function V_f is its total variation.

Given an interval $I \subset \mathbb{R}_+$, the function $f : I \to \mathbb{R}$ is said to be of bounded variation on I if (A.2) holds for all $n \geq 1$ and $t_0 < \ldots < t_n$ belonging to I.

Remark A.1 Every function f of bounded variation on $[a, b] \subset \mathbb{R}_+$ can be extended to a function of bounded variation on $\mathbb{R}_+$ by taking $f(s) = f(a)$ for $0 \leq s \leq a$ and $f(s) = f(b)$ for $s \geq b$.

Every monotonic function is of bounded variation, and the following theorem is a kind of converse result.

Theorem A.1 (Jordan decomposition) (Billingsley [12] Chap. 6 Sect. 31)

If $f : \mathbb{R}_+ \to \mathbb{R}$ is a function of bounded variation, then there is a pair of cad-lag nondecreasing functions Φ_1 and Φ_2 such that $f - f(0) = \Phi_1 - \Phi_2$, and we have $df = f(0)\,\delta_0 + d\Phi_1 - d\Phi_2$.

The pair (Φ_1, Φ_2) is unique up to addition of a constant, and there is one and only one pair (Φ_1, Φ_2), called the canonical decomposition of f, such that moreover, $V_f = \Phi_1 + \Phi_2$.

The measure $|df|$ is defined as $|df| = |f(0)|\,\delta_0 + d\Phi_1 + d\Phi_2$, where (Φ_1, Φ_2) is the canonical decomposition of f.

Proposition A.1 (Rudin [89] Chap. 7 Exercise 13 p. 157)

A function of bounded variation is differentiable at Lebesgue-almost every point of its domain, and its derivative is Lebesgue integrable.

Given a function $f : \mathbb{R}_+ \to \mathbb{R}$ of bounded variation, df is the unique signed measure μ on $\mathcal{B}(\mathbb{R}_+)$ such that

$$\forall t > 0, \quad f(t) = \mu([0, t]),$$

or equivalently,

$$\mu(\{0\}) = f(0), \quad \forall\, 0 \leq a < b, \quad f(b) - f(a) = \mu(]a, b]).$$

Conversely, given a signed measure μ on $\mathcal{B}(\mathbb{R}_+)$ that is finite on every compact set, the function $f(t) = \mu([0, t])$ is of bounded variation.

Proposition A.2 (Fubini's theorem) *Given $f_1, f_2 : \mathbb{R}_+ \to \mathbb{R}$ two functions of bounded variation and $g : \mathbb{R}_+^2 \to \mathbb{R}$ a Borel function, we have*

$$\int_{\mathbb{R}_+}\left(\int_{\mathbb{R}_+}|g(s_1,s_2)|\,|df_1|(s_1)\right)|df_2|(s_2)=\int_{\mathbb{R}_+}\left(\int_{\mathbb{R}_+}|g(s_1,s_2)|\,|df_2|(s_2)\right)|df_1|(s_1).$$

If $\int_{\mathbb{R}_+}(\int_{\mathbb{R}_+}|g(s_1,s_2)|\,|df_1|(s_1))|df_2|(s_2)<+\infty$, *then*

$$\begin{aligned}\int_{\mathbb{R}_+}\left(\int_{\mathbb{R}_+}g(s_1,s_2)\,df_1(s_1)\right)df_2(s_2) &= \int_{\mathbb{R}_+}\left(\int_{\mathbb{R}_+}g(s_1,s_2)\,df_2(s_2)\right)df_1(s_1)\\ &:= \int_{\mathbb{R}_+^2}g(s_1,s_2)\,df_1(s_1)\,df_2(s_2).\end{aligned}$$

Given a cad-lag function f, define $f(s_-)=\lim_{\substack{u\to s,\\ u<s}} f(u)$, $\Delta f(s)=f(s)-f(s_-)$.

Proposition A.3 (Integration by parts) (Brémaud [20] Appendix A4 Theorem T2)
Let $f,g:\mathbb{R}_+\to\mathbb{R}$ *be two functions of bounded variation. Then*

$$\begin{aligned}&f(t)\,g(t)\\ &=f(0)\,g(0)+\textstyle\int_{]0,t]}g(s)\,df(s)+\int_{]0,t]}f(s_-)\,dg(s)\\ &=f(0)\,g(0)+\textstyle\int_{]0,t]}g(s_-)\,df(s)+\int_{]0,t]}f(s_-)\,dg(s)+\sum_{s\le t}\Delta f(s)\Delta g(s).\end{aligned}$$

Corollary 1 *(Cocozza-Thivent [29] Corollary A6)*
Let $f:\mathbb{R}_+\to\mathbb{R}$ *be a continuous function of bounded variation and* $a\in\mathbb{R}$. *Then*

$$df^n(s)=n\,f^{n-1}(s)\,df(s),\quad de^{af}(s)=a\,e^{af(s)}\,df(s).$$

Proposition A.4 (Dacunha-Castelle and Duflo [48] Proposition 6.2.11, Brémaud [20] *A4 Theorem T4, Cocozza-Thivent [29] Appendice A Proposition A7)*
Let $f:\mathbb{R}_+\to\mathbb{R}$ *be a cad-lag function of bounded variation such that* $f(0)=0$ *and* $a\in\mathbb{R}$. *Then the equation*

$$\forall t\ge 0\quad z(t)=z(0)+a\int_{]0,t]}z(s_-)\,df(s)$$

has one and only one solution satisfying $\sup_{s\le t}|z(s)|<+\infty$ *for all* $t\ge 0$. *This solution is given by*

$$\forall t\ge 0\quad z(t)=z(0)\left(\prod_{s\le t}(1+a\Delta f(s))\right)e^{af^c(t)},$$

where $f^c(t)=f(t)-\sum_{s\le t}\Delta f(s)$ *is the continuous part of* f.
In particular, if f *is continuous and* $f(0)=0$, *then* z *defined as*

$$z(t)=z(0)\,e^{af(t)}$$

is the unique solution of

$$\forall t \geq 0 \quad z(t) = z(0) + a \int_0^t z(s)\, df(s)$$

such that $\sup_{s \leq t} |z(s)| < +\infty$ *for all* $t \geq 0$.

Definition A.2 A function $f : [a, b] \to \mathbb{R}$ is absolutely continuous on $[a, b]$ if for all $\varepsilon > 0$, there exists $\delta > 0$ such that whenever a finite sequence of disjoint subintervals $]a_i, b_i[\subset [a, b]$ satisfies $\sum_{i=1}^{n}(b_i - a_i) < \delta$, then $\sum_{i=1}^{n} |f(b_i) - f(a_i)| < \varepsilon$.

A function $f : \mathbb{R} \to \mathbb{R}$ (resp. $f : \mathbb{R}_+ \to \mathbb{R}$) is absolutely continuous if it is absolutely continuous on every interval $[a, b] \subset \mathbb{R}$ (resp. $[a, b] \subset \mathbb{R}_+$).

Clearly, a Lipschitz continuous function is absolutely continuous.

Remark A.2 If f_1 and f_2 are two bounded functions that are absolutely continuous on $[a, b] \subset \mathbb{R}$, then $f_1 f_2$ is absolutely continuous on $[a, b]$. If, moreover, $|f_2(s)| \geq c > 0$ for all $s \in [a, b]$, then f_1/f_2 is absolutely continuous on $[a, b]$.

Theorem A.2 (Rudin [89] pp. 144–149)
A function $f : \mathbb{R} \to \mathbb{R}$ *is absolutely continuous iff*

1. f *has a derivative* f' *Lebesgue-almost everywhere,*
2. f' *is Lebesgue integrable on all* $[a, b] \subset \mathbb{R}$,
3. *for all* $[a, b] \subset \mathbb{R}$, $f(b) - f(a) = \int_a^b f'(s)\, ds$.

Throughout the book, given an absolutely continuous function f, we denote by f' the function defined as the derivative of f at the points where the derivative exists and equal to 0 at the other points. Similarly, if $\varphi : E \times \mathbb{R}_+ \to \mathbb{R}$ is such that for all $x \in E$ the function $s \to \varphi(x, s)$ is absolutely continuous, we denote by $\partial_2\varphi(x, s)$ the function defined as $(\partial\varphi/\partial s)(x, s)$ at the points where this derivative exists and equal to 0 at the other points.

Theorem A.3 (Rudin [89] Chap. 7 Exercise 13 p. 157)
A right continuous function f *is absolutely continuous iff*

1. *it is of bounded variation,*
2. *the associated measure* df *is absolutely continuous with respect to the Lebesgue measure.*

A.3 Hazard Rate, Hazard Measure

In this section T is a random variable taking values in $\bar{\mathbb{R}}_+ = \mathbb{R}_+ \cup \{+\infty\}$, and F is its cumulative distribution function, that is, $F(t) = \mathbb{P}(T \leq t)$ $(t \geq 0)$. Define $\bar{F} = 1 - F$.

We start by assuming that T has a probability density function f, i.e., the probability distribution of T is $f(v)\,dv + \left(1 - \int_0^{+\infty} f(s)\,ds\right) \delta_{+\infty}(dv)$.

Definition A.3 If T has a probability density function f, its hazard rate h is defined as

$$h(v) = \begin{cases} \dfrac{f(v)}{\bar{F}(v)}, & \bar{F}(v) \neq 0, \\ 0, & \bar{F}(v) = 0. \end{cases} \tag{A.3}$$

The probability density function f is defined Lebesgue-almost everywhere. Therefore the same holds for the hazard rate. Nevertheless, in many cases, the probability density function is continuous on its support and the hazard rate is defined by taking, in (A.3), f as this continuous function.

The name "rate" comes from the following proposition.

Proposition A.5 (Cocozza-Thivent [29] Proposition 1.1)
Assume that T has a continuous probability density function. Then the hazard rate h satisfies

$$h(v) = \lim_{\Delta \to 0_+} \frac{1}{\Delta} \mathbb{P}(v < T \leq v + \Delta | T > v)$$

for all $v > 0$ such that $\mathbb{P}(T > v) > 0$.

Example A.2 The hazard rate of the exponential distribution with parameter λ is equal to λ. It does not depend on the time, which is a way to understand why the exponential distribution is memoryless. The density function of the Weibull distribution with parameters α, β is $f(v) = \frac{\beta}{\alpha} (v/\alpha)^{\beta-1} e^{-(v/\alpha)^{\beta}}$, and its hazard rate is $h(v) = \frac{\beta}{\alpha}(v/\alpha)^{\beta-1}$.

Proposition A.6 *Let T be a random variable with values in $\bar{\mathbb{R}}_+$, and $h : \mathbb{R}_+ \to \mathbb{R}$ a Borel function. The following assertions are equivalent:*

1. *the random variable T has a probability density function and its hazard rate is h,*
2. *there exists $\varepsilon > 0$ such that $\int_0^{\varepsilon} h(s)\,ds < +\infty$ and $\mathbb{P}(T > t) = \exp\left(- \int_0^t h(s)\,ds\right)$ holds for all $t > 0$.*

If these conditions are fulfilled, then the probability distribution of T can be written as

$$h(v)\, e^{-\int_0^v h(s)\,ds}\, dv + e^{-\int_0^{+\infty} h(s)\,ds}\, \delta_{+\infty}(dv),$$

and T takes values in $\mathbb{R}_+$ iff $\int_0^{+\infty} h(s)\,ds = +\infty$.

Proof Assuming that assertion 1 holds, the function $z(t) = \mathbb{P}(T > t)$ is such that $z(t) = 1 - \int_0^t z(s)h(s)\,ds$ holds for all $t > 0$. Hence $z(t) = \exp\left(- \int_0^t h(s)\,ds\right)$ (Proposition A.4, Sect. A.2), and assertion 2 follows.

Conversely, assuming that assertion 2 holds, assertion 1 follows from Corollary 1, Sect. A.2. ■

In the general case of a random variable T with values in $]0, +\infty]$ and probability distribution dF, the hazard rate is replaced by the hazard measure defined as

$$dH(v) = \frac{1}{1 - F(v_-)} dF(v),$$

where $F(v_-) = \lim_{u \to v, u < v} f(u) = \mathbb{P}(T < v)$.

If T has a probability density function and hazard rate h, then $dH(v) = h(v)\, dv$.

Proposition A.7 (Cocozza-Thivent [29] Proposition 1.15)

Let dH be the hazard measure of T. Then

$$\bar{F}(t) = \mathbb{P}(T > t) = (1 - \mathbb{P}(T = 0)) \prod_{s \le t} (1 - \mathrm{d}H(s))\, e^{-\int_0^t dH^c(s)},$$

where $H(t) = dH(]0, t])$ and $H^c(t) = H(t) - \sum_{s \le t} \Delta H(s)$ is the continuous part of H.

In particular, if $\mathbb{P}(T = t) = 0$ for all $t \in \mathbb{R}_+$, then

$$\forall t > 0 \quad \bar{F}(t) = \mathbb{P}(T > t) = e^{-\int_0^t dH(s)}.$$

Proposition A.8 (Dacunha-Castelle and Duflo [48] Chap. 6 Sect. 6.2.3)

Let T be a random variable with values in $]0, +\infty]$ and hazard rate dH. Let $\mathcal{F}_t$ be the σ-algebra generated by the random variables $1_{\{T \le s\}}$, $(s \le t)$. Define

$$M_t = 1_{\{T \le t\}} - \int_{]0,t]} 1_{\{s \le T\}}\, dH(s). \tag{A.4}$$

Then $(M_t)_{t \ge 0}$ is a martingale with respect to the filtration $(\mathcal{F}_t)_{t \ge 0}$.

More generally, let X be an r.v. with values in $(E, \mathcal{E})$, $\mu_{X,T}$ the probability distribution of (X, T), and $\mathcal{F}_t$ the σ-algebra generated by the random variables $1_{\{T \le s, X \in A\}}$, $(s \le t$, $A \in \mathcal{E})$. Define

$$\tilde{p}(dx, ds) = \frac{1}{\mathbb{P}(T \ge s)} \mu_{X,T}(dx, ds), \tag{A.5}$$

and for $A \in \mathcal{E}$ and $t \in \mathbb{R}_+$,

$$M_t^A = 1_{\{T \le t\}}\, 1_{\{X \in A\}} - \int_{A \times]0,t]} \tilde{p}(dx, ds).$$

Then $(M_{t \wedge T}^A)_{t \ge 0}$ is a martingale with respect to the filtration $(\mathcal{F}_t)_{t \ge 0}$.

The following section provides a generalization of this result.

A.4 Compensator of a Marked Point Process

Definition A.4 Let $(E, \mathcal{E})$ be a measurable space and $\Delta \notin E$. A marked point process with values in E is a sequence $(Y_n, T_n)_{n\geq 1}$ of random variables such that

- $(T_n)_{n\geq 1}$ is a nondecreasing sequence of random variables with values in $\bar{\mathbb{R}}_+$,
- for all $n \geq 1$, $\mathbb{P}(T_n < T_{n+1}, T_n < +\infty) = \mathbb{P}(T_n < +\infty)$,
- for all $n \geq 1$, $Y_n \in E \cup \{\Delta\}$ and $\mathbb{P}(Y_n = \Delta, T_n = +\infty) = \mathbb{P}(T_n = +\infty) = \mathbb{P}(Y_n = \Delta)$.

In this section we further assume that $\mathbb{P}(\lim_{n\to\infty} T_n = +\infty) = 1$.

Define the random measure dp on $E \times \mathbb{R}_+$ as

$$dp(z, s) := p(dz, ds) := \sum_{n\geq 1: T_n < +\infty} \delta_{Y_n, T_n}(dz, ds)$$

and the random measure $d\tilde{p}$ on $E \times \mathbb{R}_+$ as follows:

- on $]0, T_1]$, $d\tilde{p}$ is as in (A.5) with $\mu_{X,T}$ replaced by the probability distribution of (Y_1, T_1),
- on $]T_k, T_{k+1}]$ $(k \geq 1)$, $d\tilde{p}$ is as in (A.5) with $\mu_{X,T}$ replaced by the conditional probability distribution of (Y_{k+1}, T_{k+1}) given $(Y_1, T_1, \ldots, Y_k, T_k)$.

Denote by $\mathcal{F}_t$ the σ-algebra generated by the random variables $\sum_{n\geq 1} 1_{\{T_n \leq s, Y_n \in A\}}$, $(s \leq t,\ A \in \mathcal{E})$.

Proposition A.9 (Dachuna-Castelle and Duflo [48] Chap. 6 Sect. 6.2.3)
With the previous notation, $d\tilde{p}$ is the compensator of dp with respect to the filtration $(\mathcal{F}_t)_{t\geq 0}$, that is, for all $A \in \mathcal{E}$, $t \to \int_{A\times]0,t]} d\tilde{p}(z, s)$ is predictable and

$$M_t^A = \int_{A\times]0,t]} dp(z, s) - \int_{A\times]0,t]} d\tilde{p}(z, s)$$

is a local martingale with respect to the filtration $(\mathcal{F}_t)_{t\geq 0}$.

Notation Given $h : E \times \mathbb{R}_+ \times \Omega \to \mathbb{R}$, we introduce the following notation:

- $h \in L^1(p)$ if $\mathbb{E}(\int |h(z, s, \omega)|\, p(dz, ds)) < +\infty$.
- $h \in L^1(\tilde{p})$ if $\mathbb{E}(\int |h(z, s, \omega)|\, \tilde{p}(dz, ds)) < +\infty$,
- $h \in L^1_{loc}(p)$ (resp. $L^1_{loc}(\tilde{p})$) if there exists a sequence σ_n of stopping times increasing to infinity such that $h 1_{[0,\sigma_n[} \in L^1(p)$ (resp. $L^1(\tilde{p})$).

Proposition A.10

$$L^1(p) = L^1(\tilde{p}), \quad L^1_{loc}(p) = L^1_{loc}(\tilde{p}).$$

If $h \in L^1_{loc}(p)$ *is predictable, then*

$$M_t = \int_{E\times\mathbb{R}_+} h(z,s,\omega) 1_{\{s\leq t\}}\, p(dz,ds) - \int_{E\times\mathbb{R}_+} h(z,s,\omega) 1_{\{s\leq t\}}\, \tilde{p}(dz,ds)$$

is a local martingale.

If furthermore $(z,s,\omega) \to h(z,s,\omega)\, 1_{\{s\leq t\}} \in L^1(p)$ *for all* $t \in \mathbb{R}_+$, *then* M_t *is a martingale.*

A left continuous process is predictable. Hence given a cad-lag process $(X_t)_{t\geq 0}$, define $X_{t-} = \lim_{s\to t, s<t} X_s$. Then the process $(X_{t-})_{t>0}$ is predictable.

Appendix B
Interarrival Distributions with Several Dirac Measures

The aim of this chapter is to answer the following question: how to cope when the interarrival distributions are mixtures of absolutely continuous probability distributions and countable numbers of Dirac measures, that is,

$$1_{\mathbb{R}_+}(v)\, dF_x(v) = q_0(x)\, f_0(x, v)\, dv + \sum_{i \geq 1} 1_{\{\alpha_i(x) < +\infty\}}\, q_i(x)\, \delta_{\alpha_i(x)}(dv), \qquad \text{(B.1)}$$

where for all $x \in E$, we have $q_i(x) > 0$, $0 < \alpha_i(x) \leq +\infty$, $\lim_{i \to +\infty} \alpha_i(x) = +\infty$, if $\alpha_i(x) < +\infty$ then $\alpha_i(x) < \alpha_{i+1}(x)$. It is implicitly understood that $\int_0^{+\infty} f_0(x, v)\, dv \leq 1$, $q_0(x) + \sum_{i \geq 1} 1_{\{\alpha_i(x) < +\infty\}} q_i(x) = 1$, $dF_x(\{+\infty\}) = 1 - \int_0^{+\infty} f_0(x, v)\, dv$.

When $j > k$, define $\sum_{i=j}^{i=k} = 0$ and $\prod_{i=j}^{i=k} = 1$.

When we write the interarrival distributions as in (B.1), the formulas such as the Kolmogorov equations are awful. Indeed, to obtain nice formulas, the hazard measures of the interarrival distributions must be nice (see, for instance, Sect. A.4). The following example, inspired by a multistage rocket that drops its stages, gives an idea of how to write such mixtures in a pleasant way. Consider an item whose intrinsic lifetime S has a probability distribution with density f. Moreover, this item suffers stress at times α_i $(1 \leq i \leq M)$, and the probability that the stress at time α_i causes the item's failure is p_i. Then the real lifetime probability distribution of the item is

$$\begin{aligned} dF(v) = \Bigg(& 1_{\{v < \alpha_1\}} + \sum_{i=2}^{M} (1 - p_1) \cdots (1 - p_{i-1})\, 1_{\{\alpha_{i-1} < v < \alpha_i\}} \\ & + (1 - p_1) \cdots (1 - p_M)\, 1_{\{v > \alpha_M\}} \Bigg) f(v)\, dv + p_1 \mathbb{P}(S > \alpha_1)\, \delta_{\alpha_1}(dv) \\ & + \sum_{i=2}^{M} (1 - p_1) \cdots (1 - p_{i-1})\, p_i\, \mathbb{P}(S > \alpha_i)\, \delta_{\alpha_i}(dv) \end{aligned}$$

C. Cocozza-Thivent, *Markov Renewal and Piecewise Deterministic Processes*, Probability Theory and Stochastic Modelling 100,
https://doi.org/10.1007/978-3-030-70447-6

$$+ (1 - p_1) \cdots (1 - p_M) \left(1 - \int_0^{+\infty} f(s)\, ds \right) \delta_{+\infty}(dv). \tag{B.2}$$

Denote by ℓ the hazard rate of S. Then the hazard measure of the real lifetime is

$$\frac{dF(v)}{dF([v, +\infty])} = \ell(v)\, dv + \sum_{i=1}^{M} 1_{\{\alpha_i < +\infty\}} p_i\, \delta_{\alpha_i}(dv).$$

As a matter of fact, the interarrival distributions (B.1) can be written as in (B.2).

Proposition B.1 *Consider:*

1. *a sequence* $(\alpha_i)_{i \geq 1}$ *with values in* $]0, +\infty]$ *such that if* $\alpha_i < +\infty$, *then* $\alpha_i < \alpha_{i+1}$ *and* $\lim_{i \to +\infty} \alpha_i = +\infty$,
2. *a sequence* $(q_i)_{i \geq 0}$ *of positive real numbers such that* $q_0 + \sum_{i \geq 1} 1_{\{\alpha_i < +\infty\}} q_i = 1$,
3. *a Borel function* $f_0 : \mathbb{R}_+ \to \mathbb{R}_+$ *such that* $\int_0^{+\infty} f_0(v)\, dv \leq 1$.

Define $\alpha_0 = 0$ *and*

$$m(dv) = q_0 \left(f_0(v)\, dv + (1 - \int_0^{+\infty} f_0(s)\, ds)\, \delta_{+\infty}(dv) \right) + \sum_{i \geq 1} 1_{\{\alpha_i < +\infty\}}\, q_i\, \delta_{\alpha_i}(dv).$$

Then the measure m *can be written as*

$$\begin{aligned} m(dv) = & \sum_{i \geq 0 : \alpha_i < +\infty} \prod_{j=1}^{i} (1 - p_j)\, 1_{\{\alpha_i < v < \alpha_{i+1}\}} f(v)\, dv \\ & + \sum_{i \geq 1} 1_{\{\alpha_i < +\infty\}}\, p_i \prod_{j=1}^{i-1} (1 - p_j)\, (1 - \int_0^{\alpha_i} f(s)\, ds)\, \delta_{\alpha_i}(dv) \\ & + \prod_{i \geq 1 : \alpha_i < +\infty} (1 - p_i) \left(1 - \int_0^{+\infty} f(s)\, ds \right) \delta_{+\infty}(dv), \end{aligned}$$

where $f : \mathbb{R}_+ \to \mathbb{R}_+$ *is a Borel function satisfying* $\int_0^{+\infty} f(v)\, dv \leq 1$, p_i *defined for* i *such that* $\alpha_i < +\infty$ *satisfies* $0 < p_i \leq 1$. *Moreover,* $p_i = 1$ *if and only if* $\alpha_{i+1} = +\infty$ *and* $\int_0^{\alpha_i} f_0(v)\, dv = 1$.

Moreover,

$$\frac{1_{\mathbb{R}_+}(v)\, m(dv)}{1 - m([0, v[)} = \ell(v)\, dv + \sum_{i \geq 1} 1_{\{\alpha_i < +\infty\}} p_i\, \delta_{\alpha_i}(dv),$$

where ℓ *is the hazard rate of* f, *i.e.,* $\ell(v) = f(v)/(1 - \int_0^v f(s)\, ds)$.

Proof We define f and the p_i by induction. We also prove by induction that $0 < p_i \leq 1$ and $p_i = 1$ iff $\alpha_{i+1} = +\infty$ and $\int_0^{\alpha_i} f_0(v)\,dv = 1$. Indeed, $f(v) = q_0\, f_0(v)$ for $v \leq \alpha_1$, $p_1 = q_1/(1 - q_0 \int_0^{\alpha_1} f_0(s)\,ds)$, $f(v) = q_0\, f_0(v)/(1 - p_1)\cdots(1 - p_i)$ for $\alpha_i < v \leq \alpha_{i+1} \leq +\infty$, $p_{i+1} = q_{i+1}/(1 - p_1)\cdots(1 - p_i)(1 - \int_0^{\alpha_{i+1}} f(v)\,dv)$ for $\alpha_{i+1} < +\infty$. ∎

Leading Role of Simple CSMPs

The following proposition shows that every CSMP taking values in E, with interarrival distributions that are mixtures of absolutely continuous probability distributions and countable numbers of Dirac measures, can be written as a function of a simple CSMP $(\bar{Z}_t, \bar{A}_t)_{t\geq 0}$ taking values in $E \times \mathbb{R}_+$ and such that $\bar{A}_0 = 0$.

Consider:

1. a Borel function $\ell : E \times \mathbb{R}_+ \to \mathbb{R}_+$ such that for all $x \in E$, there exists $\varepsilon(x)$ satisfying $\int_0^{\varepsilon(x)} \ell(x, s)\,ds < +\infty$,
2. a sequence $(\alpha_i)_{i\geq 1}$ of Borel functions defined on E with values in $]0, +\infty]$ such that if $\alpha_i(x) < +\infty$, then $\alpha_i(x) < \alpha_{i+1}(x)$ and $\lim_{i\to+\infty} \alpha_i(x) = +\infty$ for all $x \in E$,
3. a sequence $(p_i)_{i\geq 1}$ of Borel functions defined on E with values in $]0, 1]$ such that for all $x \in E$ satisfying $\alpha_j(x) < +\infty$, we have $\prod_{i=1}^{j-1}(1 - p_i(x)) \neq 0$,
4. a probability kernel β on $E \times \mathbb{R}_+$.

For $(x, u) \in E \times \mathbb{R}_+$, define $\alpha_0(x) = 0$ and $I(x, u) = \max\{j \geq 0 : \alpha_j(x) \leq u\}$, that is, $\alpha_{I(x,u)}(x) \leq u < \alpha_{I(x,u)+1}$.

Proposition B.2 *Let $(Z_t, A_t)_{t\geq 1}$ be a CSMP with kernel N given by*

$$N(x, dy, dv) = \Bigg(\prod_{i=1}^{I(x,v)}(1 - p_i(x))\,\ell(x, v)\,e^{-\int_0^v \ell(x,w)\,dw}\,dv$$
$$+ \sum_{j\geq 1} 1_{\{\alpha_j(x)<+\infty\}}\, p_j(x) \prod_{i=1}^{j-1}(1 - p_i(x))\,e^{-\int_0^{\alpha_j(x)} \ell(x,w)\,dw}\,\delta_{\alpha_j(x)}(dv)\Bigg)\,\beta(x, v; dy).$$

Define

$$\bar{\ell}((x, u), v) = \ell(x, u + v), \quad \bar{\alpha}(x, u) = \alpha_{I(x,u)+1}(x) - u,$$

$$\bar{\beta}((x, u), v; dy, dw) = \beta(x, u + v; dy)\,\delta_0(dw), \quad v < \bar{\alpha}(x, u),$$

$$\bar{\beta}((x, u), \bar{\alpha}(x, u); dy, dw) = p_{I(x,u)+1}(x)\,\beta(x, \alpha_{I(x,u)+1}(x); dy)\,\delta_0(dw)$$
$$+ (1 - p_{I(x,u)+1}(x))\,\delta_{x,\alpha_{I(x,u)+1}(x)}(dy, dw).$$

Then there exists a simple CSMP $((\bar{Z}_t^1, \bar{Z}_t^2), \bar{A}_t)_{t\geq 0}$ *taking values in* $E \times \mathbb{R}_+$*, with parameters* $\bar{\ell}, \bar{\alpha}, \bar{\beta}$*, such that* $\bar{A}_0 = 0$ *and* $Z_t = \bar{Z}_t^1$, $A_t = \bar{A}_t + \bar{Z}_t^2$ *for all* $t \geq 0$.

Proof Let $(Y_k, T_k)_{k\geq 1}$ be the Markov renewal process associated with the CSMP $(Z_t, A_t)_{t\geq 0}$. The idea is to add jumps, without changing the place where the process is, at times $T_k + \alpha_j(Y_k)$ for j such that $T_k + \alpha_j(Y_k) < T_{k+1}$, and to add a component that keeps the memory of the time elapsed since time T_k $(k \geq 1)$. More precisely, define the marked point process $((\xi_n, D_n), \tau_n)_{n\geq 1}$ with values in $E \times \mathbb{R}_+$ as follows: Assume $Z_0 = x$, $A_0 = u$. Let $(\tau_n)_{n\geq 1}$ be the sequence of the times $\alpha_j(x) - u$ for j such that $0 < \alpha_j(x) - u < T_1$, T_k, and $T_k + \alpha_j(Y_k)$ for j such that $T_k + \alpha_j(Y_k) < T_{k+1}$ $(k \geq 1)$, listed in increasing order. Define $\xi_n = Z_{\tau_n}$, $D_n = A_{\tau_n}$, and $\bar{N}$ as the kernel of a simple CSMP with parameters $\bar{\ell}, \bar{\alpha}, \bar{\beta}$. Tedious but easy calculus shows that $((\xi_n, D_n), \tau_n)_{n\geq 1}$ is a Markov renewal process with kernel $\bar{N}$ and that the probability distribution of (ξ_1, D_1, τ_1) is $\bar{N}((x, u); dy, dw, dv)$, that is, $\bar{A}_0 = 0$. Denote by $(\bar{Z}_t, \bar{A}_t)_{t\geq 0}$ the associated CSMP. The proof of $\bar{Z}_t = (Z_t, A_t - \bar{A}_t)$ is straightforward. ■

Appendix C
Proof of Convergence of the Scheme of Sect. 9.4

We give an outline of the convergence proofs for the scheme described in Sect. 9.4. We follow [41] while relaxing some assumptions. The notation and assumptions, not recalled here, are those of Sect. 9.4. For ease of notation, we assume that $E = \mathbb{R}^d$.

C.1 Uniqueness

We deduce from formula (6.10), Sect. 6.2 that the measures $\mu_i(dx, ds) = 1_{E_i}(x)\, \pi_s(dx)\, ds$ and $\bar{\sigma}_i = \sigma_i$ $(i \in I)$ are such that

$$
\begin{aligned}
0 = & \int_{\mathbb{R}^d} g(x,0)\, \pi_0(dx) + \sum_{i \in I} \int_{\mathbb{R}^d \times \mathbb{R}_+} \partial_{t,\phi} g(x,s)\, \mu_i(dx, ds) \\
& + \sum_{i \in I} \int_{\mathbb{R}^d \times \mathbb{R}_+} \lambda(x) \left(\sum_{j \in I} \int_{E_j} g_j(y,s)\, Q(x; dy) - g_i(x,s) \right) \mu_i(dx, ds) \\
& + \sum_{i \in I} \int_{\mathbb{R}^d \times \mathbb{R}_+} \left(\sum_{j \in I} \int_{E_j} g_j(y,s)\, \hat{q}_i(x; dy) - g_i(x,s) \right) \bar{\sigma}_i(dx, ds) \qquad \text{(C.1)}
\end{aligned}
$$

for all functions g and $(g_i)_{i \in I}$ satisfying the assumptions of Theorem 6.1, Sect. 6.2 for all $x \in E$, and such that for some $T > 0$ we have $g(x,t) = 0$ for all $(x,t) \in E \times [T, +\infty[$. We prove in this section that there are no other measures $(\mu_i, \bar{\sigma}_i)_{i \in I}$ for which (C.1) hold for all such functions g and $(g_i)_{i \in I}$.

For $T > 0$, define $C_b^T(\mathbb{R}^d \times \mathbb{R}_+)$ as the set of bounded continuous functions $g : \mathbb{R}^d \times \mathbb{R}_+ \to \mathbb{R}$ such that $g(x,t) = 0$ for all $(x,t) \in \mathbb{R}^d \times [T, +\infty[$ and

$$
H^{(T)} = \{(f_i)_{i \in I} : \forall i \in I,\ f_i \in C_b^T(\mathbb{R}^d \times \mathbb{R}_+),\ \sup_{i \in I} ||f_i||_\infty < +\infty\}.
$$

C. Cocozza-Thivent, *Markov Renewal and Piecewise Deterministic Processes*, Probability Theory and Stochastic Modelling 100,
https://doi.org/10.1007/978-3-030-70447-6

Define $\mathcal{T}$ as the set of functions $g : \mathbb{R}^d \times \mathbb{R}_+ \to \mathbb{R}$ such that there exist $T > 0$ and $(\xi_i)_{i\in I} \in H^{(T)}$, $(\zeta_i)_{i\in I} \in H^{(T)}$ with

1. $\forall i \in I, \forall x \in E_i, \forall t \in \mathbb{R}_+ : g(x,t) = g_i(x,t)$,
2. $\forall i \in I, \forall (x,t) \in \mathbb{R}^d \times \mathbb{R}_+$:

$$g_i(x,t) = 1_{\{\alpha_i(x)<+\infty\}}\, \zeta_i(\phi_i(x,\alpha_i(x)), t+\alpha_i(x)) - \int_0^{\alpha_i(x)} \xi_i(\phi_i(x,s), t+s)\, ds.$$

We define $g_i = \mathbb{T}_i(\xi_i, \zeta_i)$ and improperly $g = \mathbb{T}((\xi_i),(\zeta_i))$.

For $(\xi_i)_{i\in I}, (\zeta_i)_{i\in I} \in H^{(T)}$, we have

$$||\mathbb{T}((\xi_i),(\zeta_i))||_\infty \le \sup_{i\in I} ||g_i||_\infty \le \sup_{i\in I} ||\zeta_i||_\infty + T \sup_{i\in I} ||\xi_i||_\infty < +\infty. \tag{C.2}$$

For $x \in E_i$, $a \in \mathbb{R}_+$, $t < \alpha(x)$, we get

$$g_i(\phi(x,t), a+t) - g_i(x,a) = \int_0^t \xi_i(\phi(x,s), a+s)\, ds.$$

Therefore, for $(x,a) \in E_i \times \mathbb{R}_+$, $v \in [0,\alpha(x)] \to g_i(\phi(x,v), a+v)$ is absolutely continuous and

$$\forall (x,t) \in E_i \times \mathbb{R}_+, \quad \partial_{t,\phi} g(x,t) = \xi_i(x,t). \tag{C.3}$$

Moreover, if $z \in \Gamma_i$, then $\alpha_i(z) = 0$ and $\phi_i(z,0) = z$. Indeed, $z = \phi(x,\alpha(x))$ for some $x \in E_i$; letting $t_n \uparrow \alpha(x)$ and defining $y_n = \phi(x,t_n)$, we obtain $y_n \in E_i$, $(y_n, \alpha(y_n)) \to (z,0)$, and the continuities of α_i and ϕ_i give the results. Therefore,

$$\forall (x,t) \in \Gamma_i \times \mathbb{R}_+, \quad g_i(x,t) = \zeta_i(x,t). \tag{C.4}$$

Let $g = \mathbb{T}((\xi_i),(\zeta_i)) \in \mathcal{T}$, $g_i = \mathbb{T}_i(\xi_i,\zeta_i)$. From Eq. (C.1), we get

$$\begin{aligned} 0 = & \int_{\mathbb{R}^d} g(x,0)\, \pi_0(dx) + \sum_{i\in I} \int_{E_i\times\mathbb{R}_+} \xi_i(x,s)\, \pi_s(dx)\, ds \\ & + \sum_{i\in I} \int_{E_i\times\mathbb{R}_+} \lambda(x) \left(\sum_{j\in I} \int_{E_j} g_j(y,s)\, Q(x;dy) - g_i(x,s) \right) \pi_s(dx)\, ds \\ & + \sum_{i\in I} \int_{\mathbb{R}^d\times\mathbb{R}_+} \left(\sum_{j\in I} \int_{E_j} g_j(y,s)\, \hat{q}_i(x;dy) - \zeta_i(x,s) \right) \sigma_i(dx,ds). \end{aligned} \tag{C.5}$$

Lemma C.1 (Generalization of [41] Lemma 3.2) *Let $(\bar{\xi}_i, \bar{\zeta}_i)_{i\in I} \in H^{(T)} \times H^{(T)}$. Then there exists $(\xi_i, \zeta_i)_{i\in I} \in H^{(T)} \times H^{(T)}$ such that for all $i \in I$, setting $g_i = \mathbb{T}_i(\xi_i, \zeta_i)$,*

1. $\forall (x,t) \in \mathbb{R}^d \times \mathbb{R}_+$, $\bar{\xi}_i(x,t) =$
$\xi_i(x,t) + \lambda(x)\Big(\sum_{j\in I}\int_{E_j} g_j(y,t)\, Q(x;dy) - g_i(x,t)\Big)$,

2. $\forall (x,t) \in \mathbb{R}^d \times \mathbb{R}_+$, $\forall i \in I$, $\bar{\zeta}_i(x,t) = \sum_{j\in I}\int_{E_j} g_j(y,t)\,\hat{q}_i(x;dy) - \zeta_i(x,t)$.

Proof Define $\xi_i^{(n)}, \zeta_i^{(n)}$ $(i \in I, n \geq 0)$ by $\xi_i^{(0)} = \zeta_i^{(0)} = 0$ and the following recursive formulas: for all $i \in I, n \geq 0$, $(x,t) \in \mathbb{R}^d \times \mathbb{R}_+$,

$$g_i^{(n)}(x,t) = \mathbb{T}_i(\xi_i^{(n)}, \zeta_i^{(n)}), \tag{C.6}$$

$$\xi_i^{(n+1)}(x,t) = \bar{\xi}_i(x,t) - \lambda(x)\sum_{j\in I}\int_{E_j} g_j^{(n)}(y,t)\, Q_i(x;dy) + \lambda(x)\, g_i^{(n)}(x,t), \tag{C.7}$$

$$\zeta_i^{(n+1)}(x,t) = \sum_{j\in I}\int_{E_j} g_j^{(n)}(y,t)\,\hat{q}_i(x;dy) - \bar{\zeta}_i(x,t). \tag{C.8}$$

By induction, we have $(\xi_i^{(n)})_{i\in I}, (\zeta_i^{(n)})_{i\in I}, (g_i^{(n)})_{i\in I} \in H^{(T)}$.

We get $(g_i^{(n+1)})_{i\in I} = \Psi((g_i^{(n)})_{i\in I})$, where for $f = (f_i)_{i\in I} \in H^{(T)}$,

$$\begin{aligned}(\Psi f)_i(x,t) = (\widetilde{\Psi} f)_i(x,t) &- \int_0^{\alpha_i(x)} \bar{\xi}_i(\phi_i(x,s), t+s)\, ds \\ &- 1_{\{\alpha_i(x)<+\infty\}}\bar{\zeta}_i(\phi_i(x,\alpha_i(x)), t+\alpha_i(x)),\end{aligned}$$

$$\begin{aligned}(\widetilde{\Psi} f)_i(x,t) &= \int_0^{\alpha_i(x)} \lambda(\phi_i(x,s))\sum_{j\in I}\int_{E_j} f_j(y,t+s)\, Q_i(\phi_i(x,s);dy)\, ds \\ &- \int_0^{\alpha_i(x)} \lambda(\phi_i(x,s))\, f_i(\phi_i(x,s), t+s)\, ds \\ &+ 1_{\{\alpha_i(x)<+\infty\}}\sum_{j\in I}\int_{E_j} f_j(y, \alpha_i(x)+t)\,\hat{q}_i(\phi_i(x,\alpha_i(x));dy).\end{aligned}$$

Using Lebesgue's dominated convergence theorem, we deduce that Ψ maps $H^{(T)}$ to $H^{(T)}$.

<u>First Step</u>
Define $\hat{\alpha}_i(x) = \min(\alpha_i(x), T)$. Let $A > B$, where B is given in (9.14), Sect. 9.4. We define the following norm on $H^{(T)}$:

$$||f||_{A,B} = \sup_{i\in I}\sup_{(x,t)\in\mathbb{R}^d\times[0,T]} e^{At+B\hat{\alpha}_i(x)}\,|f_i(x,t)|.$$

The space $(H^{(T)}, ||\cdot||_{A,B})$ is a Banach space.

Let us prove that $\Psi^{\circ p} = \underbrace{\Psi \circ \cdots \circ \Psi}_{p\ times}$ is a contraction on this Banach space for some $A > 0$, $p \geq 1$. It is equivalent to prove that there exist $p \geq 1$ and $c \in]0, 1[$ such that

$$\forall f \in H^{(T)} : ||\widetilde{\Psi}^{\circ p}(f)||_{A,B} \leq c\, ||f||_{A,B}.$$

We have

$$\forall (x,t) \in \mathbb{R}^d \times \mathbb{R}_+ : |(\widetilde{\Psi} f)_i|(x,t) \leq \sum_{j \in I} \int_{\mathbb{R}^d \times \mathbb{R}_+} |f_j(y, t+s)|\, \mu_{i,j}(x, dy, ds),$$

where

$$\mu_{i,j}(x, dy, ds) = \nu_{i,j}(x, dy, ds) + \tilde{q}_{i,j}(x, dy, ds),$$

$$\nu_{i,j}(x, dy, ds) = \left(\Lambda 1_{E_j}(y)\, Q_i(\phi_i(x,s); dy) + 1_{\{i=j\}}\, \Lambda\, \delta_{\phi_i(x,s)}(dy)\right)\, 1_{\{s \leq T\}}\, ds,$$

$$\tilde{q}_{i,j}(x, dy, ds) = 1_{\{\alpha_i(x) < +\infty\}} 1_{E_j}(y)\, \hat{q}_i(\phi_i(x, \alpha_i(x)); dy)\, 1_{\{s \leq T\}}\, \delta_{\alpha_i(x)}(ds).$$

For all $i, j \in I$, $\mu_{i,j}$ and $\tilde{q}_{i,j}$ are transition measures from $\mathbb{R}^d$ to $\mathbb{R}^d \times \mathbb{R}_+$.

Given transition measures m_1 and m_2 from $\mathbb{R}^d$ to $\mathbb{R}^d \times \mathbb{R}_+$, define the transition measure $m_1 * m_2$ from $\mathbb{R}^d$ to $\mathbb{R}^d \times \mathbb{R}_+$ as

$$\int_{\mathbb{R}^d \times \mathbb{R}_+} \varphi(y,s)\, m_1 * m_2(x, dy, ds) = \int_{(\mathbb{R}^d \times \mathbb{R}_+)^2} \varphi(y_2, s_1 + s_2)\, m_1(x, dy_1, ds_1)\, m_2(y_1, dy_2, ds_2)$$

for every Borel function $\varphi : \mathbb{R}^d \times \mathbb{R}_+ \to \mathbb{R}_+$.

By induction, we get for $k \geq 2$ and $(x,t) \in \mathbb{R}^d \times \mathbb{R}_+$ that

$$|(\widetilde{\Psi}^{\circ k} f)_i|(x,t) \leq \sum_{j \in I} \int_{\mathbb{R}^d \times \mathbb{R}_+} |f_j(y, t+s)|\, \mu_{i,j}^{(k)}(x, dy, ds),$$

where $\mu_{i,j}^{(k)} = \sum_{\ell \in I} \mu_{i,\ell} * \mu_{\ell,j}^{(k-1)}$. We have $\mu_{i,j}^{(k)} = \nu_{i,j}^{(k)} + \tilde{q}_{i,j}^{(k)}$, $\nu_{i,j}^{(k)} = \sum_{\ell \in I} \mu_{i,\ell} * \nu_{\ell,j}^{(k-1)} + \sum_{\ell \in I} \nu_{i,\ell} * \tilde{q}_{\ell,j}^{(k-1)}$, $\tilde{q}_{i,j}^{(k)} = \sum_{\ell \in I} \tilde{q}_{i,\ell} * \tilde{q}_{\ell,j}^{(k-1)}$.

We get

$$\begin{aligned} e^{At} e^{B\hat{\alpha}_i(x)} |(\widetilde{\Psi}^{\circ k} f)_i|(x,t) &\leq e^{BT} \sum_{j \in I} \int_{\mathbb{R}^d \times \mathbb{R}_+} e^{A(t+s)} e^{B\hat{\alpha}_i(y))} |f_j(y, t+s)| \\ &\quad \times e^{-As} e^{-B\hat{\alpha}_i(y)}\, \mu_{i,j}^{(k)}(x, dy, ds) \\ &\leq e^{BT} ||f||_{A,B} \sum_{j \in I} \int_{\mathbb{R}^d \times \mathbb{R}_+} e^{-As}\, \mu_{i,j}^{(k)}(x, dy, ds). \end{aligned}$$

We have $\sum_{j\in I}\mu_{i,j}^{(k)}(\mathbb{R}^d\times\mathbb{R}_+)\le(2\wedge T+1)^k$, hence

$$\sum_{j\in I}\int_{\mathbb{R}^d\times\mathbb{R}_+} e^{-As}\,\nu_{i,j}^{(k)}(x,dy,ds)\le\frac{(2\wedge T+1)^k}{A}.$$

Furthermore,

$$\begin{aligned}&\textstyle\sum_{j\in I}\int_{\mathbb{R}^d\times\mathbb{R}_+} e^{-As}\,\tilde{q}_{i,j}^{(k)}(x,dy,ds)\le\sum_{j\in I}\int_{\mathbb{R}^d\times\mathbb{R}_+} e^{-Bs}\,\tilde{q}_{i,j}^{(k)}(x,dy,ds)\\ &\le\textstyle\sup_{\ell_1\in I}\sup_{y_1\in E_{\ell_1}}\sum_{\ell_2\in I}\cdots\sum_{\ell_{k-1}\in I}\int_{(\mathbb{R}^d\times\mathbb{R}_+)^{k-2}} e^{-B(\alpha(y_1)+\dots+\alpha(y_{k-1}))}\,1_{E_{\ell_2}}(y_2)\\ &\quad q_{\ell_1}(\phi(y_1,\alpha(y_1));dy_2)\cdots 1_{E_{\ell_{k-1}}}(y_{k-1})\,q_{\ell_{k-2}}(\phi(y_{k-2},\alpha(y_{k-2}));dy_{k-1}).\end{aligned}$$

Let a and m be as in (9.14), and let $c<1$, n_1 be such that $e^{BT}a^{n_1}\le c/2$ and $p=(m+1)n_1$. We get

$$e^{At}e^{B\hat{\alpha}_i(x)}|(\widetilde{\Psi}^{\circ p}f)_i|(x,t)\le||f_{A,B}||\left(e^{BT}\,\frac{(2\wedge T+1)^p}{A}+\frac{c}{2}\right),$$

and therefore

$$||\widetilde{\Psi}^{\circ p}f||_{A,B}\le||f||_{A,B}\left(\frac{e^{BT}(2\Lambda+1)^p}{A}+\frac{c}{2}\right).$$

Given A such that $e^{BT}(2\Lambda+1)^p/A\le c/2$, we get $||\widetilde{\Psi}^{\circ p}||_{A,B}\le c\,||f||_{A,B}$.

Second Step

Since $\Psi^{\circ p}$ is a contraction, the sequence $((g_i^{(np)})_{i\in I})_{n\ge1}$ converges to some $(\hat{g}_i^{(p)})_{i\in I}$, and for all $2\le k\le p$, the sequence $((g_i^{(np+k-1)})_{i\in I})_{n\ge1}$ converges to some $(\hat{g}_i^{(k-1)})_{i\in I}$. Using formulas (C.6), (C.7), (C.8), and Lebesgue's dominated convergence theorem, we conclude that for all $i\in I$:

- the sequences $(\xi_i^{(np+1)})_{n\ge1}$ and $(\zeta_i^{(np+1)})_{n\ge1}$ converge respectively to some $\hat{\xi}_i^{(1)}$ and $\hat{\zeta}_i^{(1)}$,
- for $2\le k<p$, the sequences $(\xi_i^{(np+k)})_{n\ge1}$ and $(\zeta_i^{(np+k)})_{n\ge1}$ converge respectively to some $\hat{\xi}_i^{(k)}$ and $\hat{\zeta}_i^{(k)}$,
- for $1\le k\le p$, $\hat{g}_i^{(k)}=\mathbb{T}_i(\hat{\xi}_i^{(k)},\hat{\zeta}_i^{(k)})$,
- for $(x,t)\in\mathbb{R}^d\times\mathbb{R}_+$

$$\hat{\xi}_i^{(1)}(x,t)=\bar{\xi}_i(x,t)-\lambda(x)\sum_{j\in I}\int_{E_j}\hat{g}_j^{(p)}(y,t)\,Q(x;dy)+\lambda(x)\,\hat{g}_i^{(p)}(x,t),$$

$$\hat{\zeta}_i^{(1)}(x,t)=\sum_{j\in I}\int_{E_j}g_j^{(p)}(y,t)\,\hat{q}_i(x;dy)-\bar{\zeta}_i(x,t),$$

- for $2 \leq k < p$ and $(x,t) \in \mathbb{R}^d \times \mathbb{R}_+$,

$$\hat{\xi}_i^{(k)}(x,t) = \bar{\xi}_i(x,t) - \lambda(x) \sum_{j \in I} \int_{E_j} \hat{g}_j^{(k-1)}(y,t)\, Q(x;dy) + \lambda(x)\, \hat{g}_i^{(k-1)}(x,t),$$

$$\hat{\zeta}_i^{(k)}(x,t) = \sum_{j \in I} \int_{E_j} \hat{g}_j^{(k-1)}(y,t)\, \hat{q}_i(x;dy) - \bar{\zeta}_i(x,t).$$

For all $i \in I$, define $\xi_i = \frac{1}{p}\sum_{k=1}^p \hat{\xi}_i^{(k)}$, $\zeta_i = \frac{1}{p}\sum_{k=1}^p \hat{\zeta}_i^{(k)}$, $g_i = \frac{1}{p}\sum_{k=1}^p \hat{g}_i^{(k)}$. This concludes the proof. ■

Theorem C.1 *If the family of nonnegative measures* $(\mu_i, \bar{\sigma}_i)_{i \in I}$ *satisfies, for all* $T \geq 0$, $\sum_{i \in I} \mu_i(\mathbb{R}^d \times [0,T]) < +\infty$, $\sum_{i \in I} \bar{\sigma}_i(\mathbb{R}^d \times [0,T]) < +\infty$, *and*

$$\begin{aligned} 0 = & \int_{\mathbb{R}^d} g(x,0)\, \pi_0(dx) + \sum_{i \in I} \int_{E_i \times \mathbb{R}_+} \xi_i(x,s)\, \mu_i(dx,ds) \\ & + \sum_{i \in I} \int_{\mathbb{R}^d \times \mathbb{R}_+} \lambda(x) \left(\sum_{j \in I} \int_{E_j} g_j(y,s)\, Q(x;dy) - g_i(x,s) \right) \mu_i(dx,ds) \\ & + \sum_{i \in I} \int_{\mathbb{R}^d \times \mathbb{R}_+} \left(\sum_{j \in I} \int_{E_j} g_j(y,s)\, \hat{q}_i(x;dy) - \zeta_i(x,s) \right) \bar{\sigma}_i(dx,ds). \end{aligned} \tag{C.9}$$

for all $g = \mathbb{T}((\xi_i),(\zeta_i)) \in \mathcal{T}$, $g_i = \mathbb{T}_i(\xi_i, \zeta_i)$, *then for all* $i \in I$, $\mu_i(dx,ds) = 1_{E_i}(x)\, \pi_s(dx)\, ds$ *and* $\bar{\sigma}_i = \sigma_i$.

Proof Let us assume that $(\mu_i^1, \bar{\sigma}_i^1)_{i \in I}$ and $(\mu_i^2, \bar{\sigma}_i^2)_{i \in I}$ satisfy Eq. (C.9) for all $t \geq 0$, $g = \mathbb{T}((\xi_i),(\zeta_i)) \in \mathcal{T}$, $g_i = \mathbb{T}_i(\xi_i, \zeta_i)$ $(i \in I)$. Then, using Lemma C.1, we get

$$\begin{aligned} & \sum_{i \in I} \int_{\mathbb{R}^d \times \mathbb{R}_+} \bar{\xi}_i(x,s)\, \mu_i^1(dx,ds) + \sum_{i \in I} \int_{\mathbb{R}^d \times \mathbb{R}_+} \bar{\zeta}_i(x,s)\, \bar{\sigma}_i^1(dx,ds) \\ & = \sum_{i \in I} \int_{\mathbb{R}^d \times \mathbb{R}_+} \bar{\xi}_i(x,s)\, \mu_i^2(dx,ds) + \sum_{i \in I} \int_{\mathbb{R}^d \times \mathbb{R}_+} \bar{\zeta}_i(x,s)\, \bar{\sigma}_i^2(dx,ds) \end{aligned}$$

for all $(\bar{\xi}_i)_{i \in I} \in H^{(T)}$, $(\bar{\zeta}_i)_{i \in I} \in H^{(T)}$. Therefore for all $i \in I$, we have $\bar{\sigma}_i^1 = \bar{\sigma}_i^2$ and $\mu_i^1 = \mu_i^2$.

Since $(1_{E_i}(x)\, \pi_s(dx)\, ds, \sigma_i)_{i \in I}$ satisfies (C.9) (see (C.5)), the proof of the theorem is complete. ■

C.2 Tightness

The existence and uniqueness of $(p_n^K)_{n\in\mathbb{N},K\in\mathcal{M}}$, a solution of (9.15) and (9.16), Sect. 9.4 satisfying $p_n^K \geq 0$, $\sum_{K\in\mathcal{M}} |K|\, p_n^{(K)} = 1$, are as in [41] Lemma 4.1 (Existence of solution).

Define $\mu_i^{\mathcal{D}}$ and $\sigma_i^{\mathcal{D}}$ $(i \in I)$ respectively as in (9.17) and (9.18), Sect. 9.4. In this section, we prove that the family of measures $\sum_{i\in I} \mu_i^{\mathcal{D}_k}$ and $\sum_{i\in I} \sigma_i^{\mathcal{D}_k}$ are tight when $\mathcal{D}_k = (\mathcal{M}_k, \delta t_k, \tau_k)$ $(k \geq 1)$ is a sequence of discretizations such that $\delta t_k \to 0$, $\tau_k \to 0$, $|\mathcal{M}_k|/\tau_k \to 0$.

Lemma C.2 (Finitness: generalization of [41] Lemma 4.2) *For $T > 0$, there exists $C_\sigma > 0$, depending only on T and the constants given in hypotheses (H3) such that for $|\mathcal{M}| < \tau$ and $\delta t \leq T$,*

$$\sum_{i\in I} \sigma_i^{\mathcal{D}}(\Gamma_i \times [0, T]) = \sum_{n\in\mathbb{N}:n\delta t\leq T} \delta t \sum_{K\in\mathcal{M}} q_K\, p_{n+1}^{(K)} \leq C_\sigma. \qquad \text{(C.10)}$$

Proof Let B, m and a be as in (9.14), Sect. 9.4. Let $\eta > 0$ be such that $e^{(m+1)B\eta}a < 1$. Let $N \in \mathbb{N}$ be such that $(N-1)\delta t < T \leq N\delta t$. We define $h = |\mathcal{M}| := \sup_{K\in\mathcal{M}} \sup_{x,y\in K} ||x-y||$.

If $\tau \geq \eta/(L_\alpha + 1)$, we have

$$\sum_{n=0}^{N-1} \delta t \sum_{K\in\mathcal{M}} p_{n+1}^{(K)} q_K \leq \sum_{n=0}^{N-1} \delta t \sum_{K\in\mathcal{M}} p_{n+1}^{(K)} \frac{|K|}{\tau} = \sum_{n=0}^{N-1} \delta t \frac{1}{\tau} \leq \frac{T+\delta t}{\tau} \leq \frac{2T(L_\alpha+1)}{\eta},$$

and (C.10) holds.

Let us now assume that $\tau \leq \eta/(L_\alpha + 1)$. Define

$$\theta_1(x) = e^{-B\alpha(x)}, \quad g_1(x) = 1, \qquad \text{(C.11)}$$

and by induction, for $\ell \geq 1$,

$$g_{\ell+1}(x) = \sum_{i\in I} 1_{E_i}(x)\, 1_{\{\alpha(x)<+\infty\}} \int e^{-B\,\alpha(y)}\, g_\ell(y)\, q_i(\phi(x,\alpha(x)); dy) \qquad \text{(C.12)}$$

$$= \sum_{i\in I} 1_{E_i}(x)\, 1_{\{\alpha(x)<+\infty\}} \sum_{i_1\in I} \cdots \sum_{i_\ell\in I} \int_{E_{i_1}\times\cdots\times E_{i_\ell}} e^{-B(\alpha(x_1)+\cdots+\alpha(x_\ell))}$$
$$q_i(\phi(x,\alpha(x)); dx_1)\, q_{i_1}(\phi(x_1,\alpha(x_1)); dx_2)\, \ldots\, q_{i_{\ell-1}}(\phi(x_{\ell-1},\alpha(x_{\ell-1})); dx_\ell),$$

$$\forall x \in E,\ \theta_{\ell+1}(x) = e^{-B\alpha(x)} g_{\ell+1}(x). \qquad \text{(C.13)}$$

The functions θ_ℓ and g_ℓ are nonnegative and bounded by 1. Thanks to hypothesis (H3) 2(a), θ_1 is Lipschitz continuous on each E_i $(i \in I)$ with Lipschitz constant not depending on i. By induction, thanks to hypotheses (H3) 1(b), 2(a), and 5(c), for all $\ell \geq 1$, the function g_ℓ is Lipschitz continuous on each E_i with Lipschitz constant not depending on i. Hence the same holds for θ_ℓ. Denote by γ_ℓ a common Lipschitz constant.

For each $K \in \mathcal{M}$, choose $x_K \in K$. Multiplying (9.16), Sect. 9.4 by $\delta t \theta_\ell(x_K)$ and summing over $n = 0, \ldots, N-1$ and $K \in \mathcal{M}$, we get

$$0 = \sum_{K \in \mathcal{M}} |K| \, (p_N^{(K)} - p_0^{(K)}) \theta_\ell(x_K)$$
$$+ \sum_{n=0}^{N-1} \sum_{K \in \mathcal{M}} p_{n+1}^{(K)} \delta t \sum_{L \in \mathcal{M}} (v_{KL} + \lambda_{KL} + q_{KL})(\theta_\ell(x_K) - \theta_\ell(x_L)).$$

Hence

$$\sum_{n=0}^{N-1} \delta t \sum_{K \in \mathcal{M}} p_{n+1}^{(K)} q_K \theta_\ell(x_K) \leq 1 + \sum_{n=0}^{N-1} \delta t \sum_{K \in \mathcal{M}} p_{n+1}^{(K)} \sum_{L \in \mathcal{M}} v_{KL} |\theta_\ell(x_L) - \theta_\ell(x_K)|$$
$$+ \Lambda T + \sum_{n=0}^{N-1} \delta t \sum_{K \in \mathcal{M}} p_{n+1}^{(K)} \sum_{L \in \mathcal{M}} q_{KL} \theta_\ell(x_L).$$

If $v_{KL} \neq 0$, then there exists $x_0 \in K$ such that $\alpha(x_0) > \tau$ and $\phi(x_0, \tau) \in L$. Hence thanks to hypothesis (H3) 1(a), we deduce that x_0 and $\phi(x_0, \tau)$ are in the same E_i and so are x_K and x_L. Thus thanks to hypothesis (H3) 1(b),

$$v_{KL} \neq 0 \Rightarrow ||x_K - x_L|| \leq 2h + L_\phi \tau \leq (2 + L_\phi)\, \tau. \tag{C.14}$$

We get

$$v_{KL} \neq 0 \Rightarrow |\theta_\ell(x_L) - \theta_\ell(x_K)| \leq \gamma_\ell \, (2 + L_\phi)\, \tau \tag{C.15}$$

and

$$\sum_{n=0}^{N-1} \delta t \sum_{K \in \mathcal{M}} p_{n+1}^{(K)} \sum_{L \in \mathcal{M}} v_{KL} \, |\theta_\ell(x_L) - \theta_\ell(x_k)| \leq 2\gamma_\ell \, (2 + L_\phi)\, T.$$

On the other hand, given $K \in \mathcal{M}_i$, we have

$$\sum_{L \in \mathcal{M}} q_{KL} \theta_\ell(x_L) = q_K \frac{1}{|\{x \in K : \alpha(x) \leq \tau\}|} \int_{\{x \in K : \alpha(x) \leq \tau\}} \sum_{L \in \mathcal{M}} \theta_\ell(x_L)\, q_i(\phi(x, \alpha(x)); L)\, dx,$$

and

$$
\begin{aligned}
&|\sum_{L\in\mathcal{M}} \theta_\ell(x_L)\, q_i(\phi(x,\alpha(x)); L) - \int_E \theta_\ell(y)\, q_i(\phi(x,\alpha(x)); dy)| \\
&\le \sum_{L\in\mathcal{M}} \int_L |\theta_\ell(x_L) - \theta_\ell(y)|\, q_i(\phi(x,\alpha(x)); dy)| \le \gamma_\ell\, h.
\end{aligned}
$$

Moreover, for $x \in K$, we have $\int_E \theta_\ell(y)\, q_i(\phi(x,\alpha(x)); dy) = g_{\ell+1}(x)$ and $|g_{\ell+1}(x) - g_{\ell+1}(x_K)| \le \gamma_{\ell+1} h$. Thus

$$
|\sum_{L\in\mathcal{M}} q_{KL}\, \theta_\ell(x_L) - q_K g_{\ell+1}(x_K)| \le q_K(\gamma_\ell + \gamma_{\ell+1})\, h, \tag{C.16}
$$

whence

$$
\sum_{n=0}^{N-1} \delta t \sum_{K\in\mathcal{M}} p_{n+1}^{(K)} \sum_{L\in\mathcal{M}} q_{KL}\, \theta_\ell(x_L) \le \sum_{n=0}^{N-1} \delta t \sum_{K\in\mathcal{M}} p_{n+1}^{(K)} q_K g_{\ell+1}(x_K) + 2(\gamma_\ell + \gamma_{\ell+1})T
$$

since $\sum_{K\in\mathcal{M}} p_{n+1}^{(K)} q_K \le 1/\tau$ and $h/\tau \le 1$.

Gathering the above results, we get

$$
\sum_{n=0}^{N-1} \delta t \sum_{K\in\mathcal{M}} p_{n+1}^{(K)} q_K \theta_\ell(x_K) \le 1 + \tilde{c}_\ell T + \sum_{n=0}^{N-1} \delta t \sum_{K\in\mathcal{M}} p_{n+1}^{(K)} q_K g_{\ell+1}(x_K)
$$

for some constant $\tilde{c}_\ell$.

If $q_K \ne 0$, then there exists $x_0 \in K$ such that $\alpha(x_0) \le \tau$. Hence $\alpha(x_K) \le L_\alpha h + \tau \le (L_\alpha + 1)\tau$. Since $\tau \le \eta/(L_\alpha + 1)$, we get $1 \le e^{B\eta} e^{-B\alpha(x_K)}$, and therefore

$$
\sum_{n=0}^{N-1} \delta t \sum_{K\in\mathcal{M}} p_{n+1}^{(K)} q_K \theta_\ell(x_K) \le 1 + \tilde{c}_\ell T + e^{B\eta} \sum_{n=0}^{N-1} \delta t \sum_{K\in\mathcal{M}} p_{n+1}^{(K)} q_K \theta_{\ell+1}(x_K).
$$

Thus, since

$$
\sum_{n=0}^{N-1} \delta t \sum_{K\in\mathcal{M}} p_{n+1}^{(K)} q_K \le e^{B\eta} \sum_{n=0}^{N-1} \delta t \sum_{K\in\mathcal{M}} p_{n+1}^{(K)} q_K \theta_1(x_K),
$$

we get by induction that

$$
\sum_{n=0}^{N-1} \delta t \sum_{K\in\mathcal{M}} p_{n+1}^{(K)} q_K \le \sum_{\ell=1}^{m} e^{\ell B\eta}(1 + \tilde{c}_\ell T) + e^{(m+1)B\eta} \sum_{n=0}^{N-1} \delta t \sum_{K\in\mathcal{M}} p_{n+1}^{(K)} q_K \theta_{m+1}(x_K).
$$

From hypothesis (H3) 5(e), we have $\theta_{m+1}(x) \le g_{m+1}(x) \le a$ for all $x \in E$, and hence

$$\sum_{n=0}^{N-1} \delta t \sum_{K\in\mathcal{M}} p_{n+1}^{(K)} q_K \le \frac{1}{1-ae^{(m+1)B\eta}} \sum_{\ell=1}^{m} e^{\ell B\eta}(1+\tilde{c}_\ell T).$$

This concludes the proof. ■

Let $\mathcal{A}$ be a family of nonnegative measures on a topological space F. This family is tight if for all $\varepsilon > 0$, there exists a compact K_ε of F such that $m(K_\varepsilon^c) < \varepsilon$ for all $m \in \mathcal{A}$.

Lemma C.3 (Tightness of $\sum_{i\in I} \mu_i^{\mathcal{D}}$: *[41] Lemma 4.3) Let $T > 0$ and let $\widetilde{\mathcal{D}}$ denote the set of discretizations $\mathcal{D}$ such that $\delta t < T$, $\tau < 1$, and $|\mathcal{M}|/\tau < 1$.*

Then for all $\varepsilon > 0$, there exists $R > 0$ such that

$$\forall \mathcal{D} \in \widetilde{\mathcal{D}} : \int_{(\mathbb{R}^d\setminus B(0,R))\times[0,T]} \sum_{i\in I} \mu_i^{\mathcal{D}}(dx, dt) \le \varepsilon, \tag{C.17}$$

which implies that the family of uniformly bounded measures $(\sum_{i\in I} \mu_i^{\mathcal{D}})_{\mathcal{D}\in\widetilde{\mathcal{D}}}$ on $E \times [0, T]$ is tight.

Once (C.10), Sect. C.2 has been obtained, the proof of Lemma C.3 is as in [41].

Lemma C.4 (Tightness of $\sum_{i\in I} \sigma_i^{\mathcal{D}}$: generalization of [41] Lemma 4.4) *Let $T > 0$ and let $(\mathcal{D}_k)_{k\ge1} = ((\mathcal{M}_k, \delta t_k, \tau_k))_{k\ge1}$ be a sequence of discretizations such that $\delta t_k \to 0$, $\tau_k \to 0$, $|\mathcal{M}_k|/\tau_k \to 0$ as $k \to +\infty$.*

Then the family of uniformly bounded measures $(\sum_{i\in I} \sigma_i^{\mathcal{D}_k})_{k\ge1}$ on $E \times [0, T]$ is tight, i.e., for all $\varepsilon > 0$, there exists $R > 0$ such that

$$\forall k \ge 1 : \sum_{i\in I} \int_{(\mathbb{R}^d\setminus B(0,R))\times[0,T]} \sum_{i\in I} \sigma_i^{\mathcal{D}_k}(dx, dt) \le \varepsilon.$$

Proof We come back to the notation of the proof of Lemma C.2: $\eta > 0$ and $N \in \mathbb{N}$ are such that $a\, e^{mB\eta} < 1$, $(N-1)\,\delta t < T \le N\,\delta t$, $h = |\mathcal{M}| := \sup_{x\in K} \sup_{x,y\in K} ||x - y||$, while g_ℓ and θ_ℓ are defined by (C.11), (C.12), and (C.13).

Let $\varepsilon > 0$ be given. Let ε_1 (which will be defined later in (C.27)) depend only on ε and the constants given in the hypotheses.

Since $\delta t_k \to 0$, $\tau_k \to 0$, $|\mathcal{M}_k|/\tau_k \to 0$ as $k \to +\infty$, there exists k_0 such that for all $k \ge k_0$, $\delta t_k < T$, $|\mathcal{M}_k| < \tau_k < 1$, $|\mathcal{M}_k| \le \varepsilon_1$, $\tau_k \le \eta/(L_\alpha + 1)$. The family of uniformly bounded measures $(\sum_{i\in I} \sigma_i^{\mathcal{D}_k})_{k<k_0}$ on $E \times [0, T]$ is finite, and therefore there exists R_0 such that

$$\forall k < k_0 : \sum_{i\in I} \int_{(\mathbb{R}^d\setminus B(0,R_0))\times[0,T]} \sigma_i^{\mathcal{D}_k}(dx, dt) \le \varepsilon.$$

It remains to prove that $\sum_{i\in I} \int_{(\mathbb{R}^d\setminus B(0,R_1))\times[0,T]} \sigma_i^{\mathcal{D}_k}(dx, dt) \le \varepsilon$ for some R_1 and all $k \ge k_0$.

To facilitate the notation, we omit the index k, and from now on, we may assume that

$$\delta t < T, \quad h < \tau < 1, \quad h \le \varepsilon_1, \quad \tau \le \eta/(L_\alpha + 1). \tag{C.18}$$

Define $R_K = \inf\{||x|| : x \in K\}$.

For $x \in K$ and $\alpha(x) \le \tau$ we have $||\phi(x, \alpha(x))|| \le ||x|| + L_\phi \tau \le R_K + 1 + L_\phi$. Hence we get $\sum_{i \in I} \int_{(\mathbb{R}^d \setminus B(0,R_1)) \times [0,T]} \sigma_i^{\mathcal{D}}(dx, dt) \le \sum_{n=0}^{N-1} \delta t \sum_{K \in \mathcal{M}} 1_{\{R_K \ge R_1 - 1 - L_\phi\}} p_{n+1}^{(K)} q_K$, and we are going to prove that there exists R such that

$$\sum_{n=0}^{N-1} \delta t \sum_{\substack{K \in \mathcal{M}, \\ R_K \ge R}} p_{n+1}^{(K)} q_K \le \varepsilon. \tag{C.19}$$

Thanks to hypotheses (H3) 4(c), 5(d) and Lemma C.3, we can find $r_{\varepsilon_1} \ge L_\phi + 3$ such that

$$\forall x \in E, \ \forall r \ge r_{\varepsilon_1} - 1 : \quad \int_{\{y \in E : ||y|| \ge ||x|| + r\}} Q(x; dy) \le \varepsilon_1, \tag{C.20}$$

$$\forall i \in I, \ \forall x \in \Gamma_i, \ \forall r \ge r_{\varepsilon_1} - 1 - L_\phi : \quad \int_{\{y \in E : ||y|| \ge ||x|| + r\}} q_i(x; dy) \le \varepsilon_1, \tag{C.21}$$

$$\forall r \ge r_{\varepsilon_1} : \quad \sum_{n=0}^{N-1} \delta t \sum_{\substack{K \in \mathcal{M}, \\ K \cap B(0,r) = \emptyset}} |K| \, p_{n+1}^{(K)} \le \varepsilon_1, \tag{C.22}$$

$$\forall r \ge r_{\varepsilon_1} : \quad \int_{E \setminus B(0,r)} \pi_0(dx) \le \varepsilon_1. \tag{C.23}$$

Let $\ell \ge 1$. For $K \in \mathcal{M}$, let $x_K \in K$ and $\widehat{\theta}_\ell^{(K)} = \theta_\ell(x_K)\,(1 - e^{-(R_K - r'_\ell)_+})$, where $r'_\ell > 0$ will be chosen below. Let us assume for the time being that the following statement has been proved:

$$r'_\ell \ge 3\, r_{\varepsilon_1} \Rightarrow \sum_{n=0}^{N-1} \delta t \sum_{K \in \mathcal{M}} p_{n+1}^{(K)} q_K \widehat{\theta}_\ell^{(K)} \le c_\ell\, \varepsilon_1 + \sum_{n=0}^{N-1} \delta t \sum_{\substack{K \in \mathcal{M}, \\ R_K \ge r'_\ell - r_{\varepsilon_1}}} p_{n+1}^{(K)} q_K\, g_{\ell+1}(x_K), \tag{C.24}$$

where c_ℓ does not depend on the discretization.

Let $r > 0$. If $q_K \ne 0$, then there exists $x_0 \in K$ such that $\alpha(x_0) \le \tau$. Hence $\alpha(x_K) \le \tau + L_\alpha\, h \le \eta$ (see (C.18)), and therefore $\widehat{\theta}_\ell^{(K)} \ge e^{-B\eta}(1 - e^{-r})\, 1_{\{R_K \ge r'_\ell + r\}} g_\ell(x_K)$.

Define $a_r = e^{-B\eta}(1 - e^{-r})$ and r'_ℓ $(\ell \ge 2)$ from r'_1 as $r'_\ell = r'_1 - (\ell - 1)(r + r_{\varepsilon_1})$. From (C.24) we get

$$a_r \sum_{n=0}^{N-1} \delta t \sum_{\substack{K\in\mathcal{M},\\ R_K\geq r'_\ell+r}} p_{n+1}^{(K)} q_K g_\ell(x_K) \leq c_\ell\,\varepsilon_1 + \sum_{n=0}^{N-1} \delta t \sum_{\substack{K\in\mathcal{M},\\ R_K\geq r'_{\ell+1}+r}} p_{n+1}^{(K)} q_K g_{\ell+1}(x_K).$$

Since the sequence $(r'_\ell)_{\ell\geq 1}$ is decreasing and $g_1 = 1$, we get recursively for all $p \geq 1$ that if $r'_p \geq 3r_{\varepsilon_1}$, then

$$\sum_{n=0}^{N-1} \delta t \sum_{\substack{K\in\mathcal{M}:\\ R_K\geq r'_1+r}} p_{n+1}^{(K)}\, q_K \leq \sum_{\ell=1}^{p} \frac{c_\ell}{a_r^\ell}\,\varepsilon_1 + \frac{1}{a_r^p} \sum_{n=0}^{N-1} \delta t \sum_{\substack{K\in\mathcal{M}:\\ R_K\geq r'_{p+1}+r}} p_{n+1}^{(K)} q_K g_{p+1}(x_K). \tag{C.25}$$

Hypothesis (H3) 5(e) gives $g_{mn_1+1}(x) \leq a^{n_1}$ for all $x \in E$ and $n_1 \geq 1$. Thus we deduce from Lemma C.2 and (C.25) with $p = nm_1$ that

$$r'_1 \geq (nm_1 - 1)r + (nm_1 + 2)r_{\varepsilon_1} \Rightarrow \sum_{n=0}^{N-1} \delta t \sum_{\substack{K\in\mathcal{M}:\\ R_K\geq r'_1+r}} p_{n+1}^{(K)}\, q_K \leq \sum_{\ell=1}^{mn_1} \frac{c_\ell}{a_r^\ell}\,\varepsilon_1 + C_\sigma \left(\frac{a}{a_r^m}\right)^{n_1}. \tag{C.26}$$

Recall that η is such that $a\, e^{mB\eta} < 1$. Choose r such that $a\, e^{mB\eta} < (1 - e^{-r})^m$; hence $a/a_r^m < 1$. Take n_1, ε_1, and R satisfying

$$\left(\frac{a}{a_r^m}\right)^{n_1} \leq \frac{\varepsilon}{2C_\sigma}, \quad \varepsilon_1 = \frac{\varepsilon}{2\sum_{\ell=1}^{nm_1} c_\ell/a_r^\ell}, \qquad R = nm_1 r + (nm_1+2)r_{\varepsilon_1}. \tag{C.27}$$

We then get (C.19) from (C.26).

It remains to prove (C.24). We multiply (9.16), Sect. 9.4 by $\delta t\, \widehat{\theta}_\ell^{(K)}$ and sum over $n = 0, \ldots, N-1$ and $K \in \mathcal{M}$. We get

$$\sum_{K\in\mathcal{M}} |K|\,(p_N^{(K)} - p_0^{(K)})\,\widehat{\theta}_\ell^{(K)} + \sum_{n=0}^{N-1} \sum_{K\in\mathcal{M}} p_{n+1}^{(K)} \delta t \sum_{L\in\mathcal{M}} (v_{KL} + \lambda_{KL} + q_{KL})(\widehat{\theta}_\ell^{(K)} - \widehat{\theta}_\ell^{(L)}) = 0.$$

We deduce

$$\sum_{n=0}^{N-1} \delta t \sum_{K\in\mathcal{M}} p_{n+1}^{(K)} q_K \widehat{\theta}_\ell^{(K)} \leq$$

$$\sum_{K\in\mathcal{M}} |K|\, p_0^{(K)} \hat{\theta}_\ell^{(K)} + \sum_{n=0}^{N-1} \delta t \sum_{K\in\mathcal{M}} p_{n+1}^{(K)} \sum_{L\in\mathcal{M}} v_{KL} |\widehat{\theta}_\ell^{(L)} - \widehat{\theta}_\ell^{(K)}| \tag{C.28}$$

$$+ \sum_{n=0}^{N-1} \delta t \sum_{K\in\mathcal{M}} p_{n+1}^{(K)} \sum_{L\in\mathcal{M}} \lambda_{KL} \widehat{\theta}_\ell^{(L)} + \sum_{n=0}^{N-1} \delta t \sum_{K\in\mathcal{M}} p_{n+1}^{(K)} \sum_{L\in\mathcal{M}} q_{KL} \widehat{\theta}_\ell^{(L)}. \tag{C.29}$$

Thanks to (C.23) and $r'_\ell \geq 3\, r_{\varepsilon_1}$, we have

$$\sum_{K \in \mathcal{M}} |K|\, p_0^{(K)} \widehat{\theta}_\ell^{(K)} \leq \sum_{\substack{K \in \mathcal{M} \\ R_K > r'_\ell}} |K|\, p_0^{(K)} \leq \varepsilon_1. \tag{C.30}$$

Let us consider the last term of (C.28). From (C.14), (C.15), and (C.18), we get

$$v_{KL} \neq 0 \Rightarrow |\widehat{\theta}_\ell^{(L)} - \widehat{\theta}_\ell^{(K)}| \leq (\gamma_\ell + 1)(3 + L_\phi)\, \tau\, 1_{\{R_K > r'_\ell - 3 - L_\phi\}},$$

since $R_K \leq r'_\ell - 3 - L_\phi$ gives $R_K \leq r'_\ell$ and $R_L \leq r'_\ell$, whence $\widehat{\theta}_\ell^{(L)} = \widehat{\theta}_\ell^{(K)} = 0$. Thus, thanks to (C.22) and $r'_\ell \geq 3\, r_{\varepsilon_1}$, we have

$$\sum_{n=0}^{N-1} \delta t \sum_{K \in \mathcal{M}} p_{n+1}^{(K)} \sum_{L \in \mathcal{M}} v_{KL} |\widehat{\theta}_\ell^{(L)} - \widehat{\theta}_\ell^{(K)}| \leq (\gamma_\ell + 1)(3 + L_\phi)\tau \sum_{n=0}^{N-1} \delta t \sum_{\substack{K \in \mathcal{M} \\ R_K > r'_\ell - 3 - L_\phi}} p_{n+1}^{(K)} \frac{1}{\tau} |K|$$
$$\leq (\gamma_\ell + 1)(3 + L_\phi)\, \varepsilon_1. \tag{C.31}$$

Let us now consider the first term of (C.29). We have

$$\sum_{L \in \mathcal{M}} \lambda_{KL} \widehat{\theta}_\ell^{(L)} \leq \sum_{\substack{L \in \mathcal{M}, \\ R_L > R_K + r_{\varepsilon_1}}} \lambda_{KL} + \sum_{\substack{L \in \mathcal{M}, \\ R_L \leq R_K + r_{\varepsilon_1}}} \lambda_{KL} (1 - e^{-(R_L - r'_\ell)_+})$$
$$\leq \Lambda \int_K Q(x; \{y : ||y|| > R_K + r_{\varepsilon_1}\})\, dx + \Lambda\, |K|\, (1 - e^{-(R_K + r_{\varepsilon_1} - r'_\ell)_+})$$
$$\leq \Lambda \int_K Q(x; \{y : ||y|| > ||x|| + r_{\varepsilon_1} - 1\})\, dx + \Lambda\, |K|\, (1 - e^{-(R_K + r_{\varepsilon_1} - r'_\ell)_+}).$$

Thus we deduce from (C.20) that $\sum_{L \in \mathcal{M}} \lambda_{KL} \widehat{\theta}_\ell^{(L)} \leq \varepsilon_1 \Lambda |K| + \Lambda |K| 1_{\{R_K + r_{\varepsilon_1} > r'_\ell\}}$, and thanks to (C.22), we get for $r'_\ell \geq 3 r_{\varepsilon_1}$,

$$\sum_{n=0}^{N-1} \delta t \sum_{K \in \mathcal{M}} p_{n+1}^{(K)} \sum_{L \in \mathcal{M}} \lambda_{KL} \widehat{\theta}_\ell^{(L)} \leq \Lambda\, \varepsilon_1 \sum_{n=0}^{N-1} \delta t \sum_{K \in \mathcal{M}} |K|\, p_{n+1}^{(K)}$$
$$+ \Lambda \sum_{n=0}^{N-1} \delta t \sum_{K \in \mathcal{M}, R_K > r'_\ell - r_{\varepsilon_1}} |K|\, p_{n+1}^{(K)} \leq 2\Lambda T \varepsilon_1 + \Lambda\, \varepsilon_1. \tag{C.32}$$

Let us consider the second term of (C.29). We have

$$\sum_{L \in \mathcal{M}} q_{KL} \widehat{\theta}_\ell^{(L)} \leq \sum_{\substack{L \in \mathcal{M}, \\ r'_\ell < R_L \leq R_K + r_{\varepsilon_1}}} q_{KL} \theta_\ell(x_L) + \sum_{\substack{L \in \mathcal{M}, \\ R_L > R_K + r_{\varepsilon_1}}} q_{KL}. \tag{C.33}$$

First, thanks to (C.16), (C.18), and Lemma C.2, we get

$$\begin{aligned}
&\sum_{n=0}^{N-1} \delta t \sum_{K\in\mathcal{M}} p_{n+1}^{(K)} \sum_{L\in\mathcal{M}, r'_\ell < R_L \le R_K + r_{\varepsilon_1}} q_{KL}\theta_\ell(x_L) \\
&\le \sum_{n=0}^{N-1} \delta t \sum_{K\in\mathcal{M}, R_K > r'_\ell - r_{\varepsilon_1}} p_{n+1}^{(K)} q_K g_{\ell+1}(x_K) \\
&\quad + \sum_{n=0}^{N-1} \delta t \sum_{K\in M} p_{n+1}^{(K)} q_K (\gamma_\ell + \gamma_{\ell+1}) h \\
&\le \sum_{n=0}^{N-1} \delta t \sum_{K\in\mathcal{M}, R_K > r'_\ell - r_{\varepsilon_1}} p_{n+1}^{(K)} q_K g_{\ell+1}(x_K) + C_\sigma(\gamma_\ell + \gamma_{\ell+1})\,\varepsilon_1 .
\end{aligned} \tag{C.34}$$

Second, for $K \in \mathcal{M}_i$ and $x \in K$ such that $\alpha(x) \le \tau$, we have $z = \phi(x, \alpha(x)) \in \Gamma_i$ and $||z - x|| \le L_\phi\, \alpha(x) \le L_\phi \tau \le L_\phi$. Therefore $||z|| \le ||x|| + L_\phi \le R_K + 1 + L_\phi$. Let L be such that $R_L > R_K + r_{\varepsilon_1}$ and $y \in L$. We have $||y|| \ge R_L > R_K + r_{\varepsilon_1} \ge ||z|| - 1 - L_\phi + r_{\varepsilon_1}$. Thus, thanks to (C.21), we obtain $\sum_{L\in\mathcal{M}, R_L > R_K + r_{\varepsilon_1}} q_{KL} \le f_q(r_{\varepsilon_1} - 1 - L_\phi)\, q_K \le \varepsilon_1\, q_K$, and hence Lemma C.2 yields

$$\sum_{n=0}^{N-1} \delta t \sum_{K\in\mathcal{M}} p_{n+1}^{(K)} \sum_{\substack{L\in\mathcal{M},\\ R_L > R_K + r_{\varepsilon_1}}} q_{KL} \le C_\sigma\, \varepsilon_1 . \tag{C.35}$$

Gathering (C.33), (C.34), and (C.35), we get

$$\sum_{n=0}^{N-1} \delta t \sum_{K\in\mathcal{M}} p_{n+1}^{(K)} \sum_{L\in\mathcal{M}} q_{KL} \widehat{\theta}_\ell^{(L)} \le \sum_{n=0}^{N-1} \delta t \sum_{\substack{K\in\mathcal{M},\\ R_K > r'_\ell - r_{\varepsilon_1}}} p_{n+1}^{(K)} q_K g_{\ell+1}(x_K) + C_\sigma(\gamma_\ell + \gamma_{\ell+1} + 1)\,\varepsilon_1 . \tag{C.36}$$

Now gathering (C.30), (C.31), (C.32), and (C.36), we get (C.24) with

$$c_\ell = 1 + (\gamma_\ell + 1)(3 + L_\phi) + \Lambda(2T + 1) + C_\sigma(\gamma_\ell + \gamma_{\ell+1} + 1).$$

This ends the proof of Lemma C.4. ■

C.3 Convergence

Let $T > 0$ be given. We are going to prove the weak convergence on $\mathbb{R}^d \times [0, T]$ of the measures $\mu_i^{\mathcal{D}}(dx, dt)$ and $\sigma_i^{\mathcal{D}}(dx, dt)$ respectively to $1_{E_i}(x)\,\pi_t(dx)\,dt$ and $\sigma_i(dx, dt)$ under hypothesis $\delta t \to 0$, $\tau \to 0$, $|\mathcal{M}|/\tau \to 0$.

Let $\mathcal{D}_m = (\mathcal{M}_m, \delta t_m, \tau_m)$ be such that $\delta t_m \to 0$, $\tau_m \to 0$, $|\mathcal{M}_m|/\tau_m \to 0$. Owing to Prokhorov's theorem, to Theorem C.1, Lemmas C.2, C.3 and C.4, we have to prove that if $\mu_i^{\mathcal{D}_m}$ and $\sigma_i^{\mathcal{D}_m}$ $(i \in I)$ are sequences that converge weakly on $\mathbb{R}^d \times [0, T]$ respectively to μ_i and $\bar{\sigma}_i$, then $(\mu_i, \bar{\sigma}_i)_{i \in I}$ satisfies Eq. (C.9), Sect. C.2 for all $g = \mathbb{T}((\xi_i), (\zeta_i)) \in \mathcal{T}$, $g_i = \mathbb{T}_i(\xi_i, \zeta_i)$.

The assumption that I is finite is required only in this section, and more precisely in Lemma C.7. We think that it is a technical assumption. It can be removed if, for instance, we are able to prove that for all T, $\mu_i^{\mathcal{D}}(E \times [0, T]) + \sigma_i^{\mathcal{D}}(\Gamma_i \times [0, T]) \le c_i^T$, c_i^T not depending on $\mathcal{D}$ and $\sum_{i \in I} c_i^T < +\infty$. That is why the definition of $\mathcal{T}_r$ given below and Lemma C.6 are written in a way that allows I to be countable and not only finite.

Define H_r as the set of $(f_i)_{i \in I} \in \cup_{T>0} H^{(T)}$ such that for all $i \in I$, f_i is infinitely differentiable with derivatives bounded by constants that do not depend on i.

Define $\mathcal{T}_r$ as the set of functions $g = \mathbb{T}((\xi_i), (\zeta_i))$ such that $(\xi_i)_{i \in I}, (\zeta_i)_{i \in I} \in H_r$.

Lemma C.5 *Let $(\xi_i)_{i \in I}, (\zeta_i)_{i \in I} \in H^{(T)}$ for some $T > 0$.*
There exist $(\xi_i^{(n)})_{i \in I}, (\zeta_i^{(n)})_{i \in I} \in H_r$ such that:

1. $\sup_{n \in \mathbb{N}, i \in I} ||\xi_i^{(n)}||_\infty < +\infty$, $\sup_{n \in \mathbb{N}, i \in I} ||\zeta_i^{(n)}||_\infty < +\infty$,
2. *for all* $(x, t) \in \mathbb{R}^d \times \mathbb{R}_+$, $i \in I$,

$$\xi_i^{(n)}(x, t) \underset{n \to \infty}{\longrightarrow} \xi_i(x, t), \quad \zeta_i^{(n)}(x, t) \underset{n \to \infty}{\longrightarrow} \zeta_i(x, t),$$

3. *for all* $(x, t) \in \mathbb{R}^d \times \mathbb{R}_+$, $i \in I$,

$$\mathbb{T}_i(\xi_i^{(n)}, \zeta_i^{(n)})(x, t) \underset{n \to \infty}{\longrightarrow} \mathbb{T}_i(\xi_i, \zeta_i)(x, t),$$

$$\mathbb{T}((\xi_i^{(n)}), (\zeta_i^{(n)}))(x, t) \underset{n \to \infty}{\longrightarrow} \mathbb{T}((\xi_i^{(n)}), (\zeta_i^{(n)}))(x, t)).$$

Proof Let κ be a nonnegative infinitely differentiable function on $\mathbb{R}^d \times \mathbb{R}_-$ with compact support and such that

$$\int_{\mathbb{R}^d \times \mathbb{R}_-} \kappa(x, t)\, dx\, dt = 1.$$

Now define

$$\xi_i^{(n)}(x, t) = n^{d+1} \int_{\mathbb{R}^d \times \mathbb{R}_+} \xi_i(y, s)\, \kappa(n(x - y), n(t - s))\, dy\, ds,$$

$$\zeta_i^{(n)}(x,t) = n^{d+1} \int_{\mathbb{R}^d\times\mathbb{R}_+} \zeta_i(y,s)\,\kappa(n(x-y), n(t-s))\,dy\,ds,$$

$$g_i^{(n)} = \mathbb{T}_i(\xi_i^{(n)}, \zeta_i^{(n)}), \quad g^{(n)} = \mathbb{T}((\xi_i^{(n)}), (\zeta_i^{(n)})).$$

We have $||\xi_i^{(n)}||_\infty \le ||\xi_i||_\infty$ and $||\zeta_i^{(n)}||_\infty \le ||\zeta_i||_\infty$, and therefore assertion 1 is proved.

If (ξ_i), $(\zeta_i) \in H^{(T)}$, and $\kappa(x,u) = 0$ for all $x \in \mathbb{R}^d$ and $|u| > A$, then $(\xi_i^{(n)})$, $(\zeta_i^{(n)}) \in H^{(T+A)}$ for all $n \in \mathbb{N}$. Moreover, each $\xi_i^{(n)}$ (respectively $\zeta_i^{(n)}$) is infinitely differentiable, and the uniform norm of each derivative is bounded above by $\sup_{i\in I} ||\xi_i||$ multiplied by a constant that depends only on n and κ. Hence $(\xi_i^{(n)})_{i\in I}, (\zeta_i^{(n)})_{i\in I} \in H_r$ for all $n \in \mathbb{N}$.

Assertion 2 comes from usual results on regularization by convolution, and assertion 3 is a consequence of Lebesgue's dominated convergence theorem. ■

Thanks to Lemma C.5 and Lebesgue's dominated convergence theorem, we have only to prove that $(\mu_i, \bar{\sigma}_i)_{i\in I}$ satisfies (C.9) for all $g \in \mathcal{T}_r$.

Lemma C.6 *Let $(\xi_i)_{i\in I}, (\zeta_i)_{i\in I} \in H_r$, and $g_i = \mathbb{T}_i(\xi_i, \zeta_i)$. Then for all $i \in I$:*

1. *g_i is Lipschitz continuous with a Lipschitz constant, namely L_g, that can be chosen independently of i,*
2. *$\partial_t g_i$ is Lipschitz continuous with a Lipschitz constant, namely $L_{t,g}$, that can be chosen independent of i,*
3. *the function $\partial_\phi g_i$ defined as $\partial_\phi g_i = \partial_{t,\phi} g_i - \partial_t g_i$ is uniformly bounded in i and Lipschitz continuous with a Lipschitz constant that can be chosen independent of i, and there exists $L_{\phi,g}$ that can be chosen independent of i such that for all $(x,t) \in E \times \mathbb{R}_+$, $\tau \in]0, \alpha(x)[$:*

$$\left| \frac{1}{\tau}(g_i(\phi(x,\tau), n\delta t) - g_i(x, n\delta t)) - \partial_\phi g_i(x, n\delta t) \right| \le \tau\, L_{\phi,g}.$$

The proof of Lemma C.6 follows the proofs of [41] Lemma 4.6(2) and Lemma 4.7.

Lemma C.7 *Assume that $\mu_i^{\mathcal{D}_m}$ and $\sigma_i^{\mathcal{D}_m}$ $(i \in I)$ are sequences that converge weakly on $\mathbb{R}^d \times [0,T]$ respectively to μ_i and $\bar{\sigma}_i$. Then $(\mu_i, \bar{\sigma}_i)_{i\in I}$ satisfies (C.9), Sect. C.1 for all $g = \mathbb{T}((\xi_i), (\zeta_i)) \in \mathcal{T}_r$, $g_i = \mathbb{T}_i(\xi_i, \zeta_i)$.*

Proof Let $g = \mathcal{T}((\xi_i), (\zeta_i)) \in \mathcal{T}_r$ be such that $(\xi)_{i\in I}, (\zeta_i)_{i\in I} \in H^{(T)}$ and $g_i = \mathbb{T}_i(\xi_i, \zeta_i)$. Let L_g be a Lipschitz constant for all g_i $(i \in I)$ and a bound for all $\partial_\phi g_i$ $(i \in I)$. Let $N \in \mathbb{N}$ be such that $(N-1)\,\delta t \le T < N\delta t$. Define

$$\forall K \in \mathcal{M},\ \forall n \in \mathbb{N}\quad g_n^{(K)} = \int_K g(x, n\delta t)\,dx.$$

We multiply (9.16), Sect. 9.4 by $\delta t\, g_n^{(K)}$ and sum over $K \in \mathcal{M}, n \in \mathbb{N}$. For readability, we sometimes omit the subscript m, among others on the quantities $(p_n^K)_{n\in\mathbb{N}}$, $h = |\mathcal{M}|$, δt, τ, and so on. We get

$$T_1^m + T_2^m + T_3^m + T_4^m + T_5^m = 0,$$

with

$$\begin{aligned}
T_1^m &= -\sum_{K\in\mathcal{M}} |K| p_0^{(K)} g_0^{(K)},\\
T_2^m &= -\sum_{n\in\mathbb{N}}\sum_{K\in\mathcal{M}} |K| p_{n+1}^{(K)} (g_{n+1}^{(K)} - g_n^{(K)})\\
&= -\sum_{i\in I}\sum_{n\in\mathbb{N}}\sum_{K\in\mathcal{M}_i} p_{n+1}^{(K)} \int_K \int_0^{\delta t} \partial_t g_i(x, n\delta t + s)\, ds\, dx,\\
T_3^m &= \sum_{n\in\mathbb{N}} \delta t \sum_{K\in\mathcal{M}}\sum_{L\in\mathcal{M}} (v_{KL} p_{n+1}^{(K)} - v_{LK} p_{n+1}^{(L)})\, g_n^{(K)},\\
T_4^m &= \sum_{n\in\mathbb{N}} \delta t \sum_{K\in\mathcal{M}}\sum_{L\in\mathcal{M}} \lambda_{KL}\, p_{n+1}^{(K)} (g_n^{(K)} - g_n^{(L)}),\\
T_5^m &= \sum_{n\in\mathbb{N}} \delta t \sum_{K\in\mathcal{M}}\sum_{L\in\mathcal{M}} q_{KL}\, p_{n+1}^{(K)} (g_n^{(K)} - g_n^{(L)}).
\end{aligned}$$

Study of the Limit of T_1^m

Since each g_i is Lipschitz continuous with the same Lipschitz constant L_g, we get

$$\Big|\sum_{K\in\mathcal{M}} \int_K g(x,0)\, \pi_0(dx) + T_1^m\Big| = \sum_{i\in I}\sum_{K\in\mathcal{M}_i} \int_K |g_i(x,0) - g_0^{(K)}|\, \pi_0(dx) \le L_g h,$$

which implies

$$\lim_{m\to\infty} T_1^m = -\int_E g(x,0)\, \pi_0(dx).$$

Study of the limit of T_2^m

Define $\tilde{T}_2^m$ as

$$\begin{aligned}
\tilde{T}_2^m &= -\sum_{i\in I}\sum_{n\in\mathbb{N}} \delta t \sum_{K\in\mathcal{M}_i} p_{n+1}^{(K)} \int_K \partial_t g_i(x, n\delta t)\, dx\\
&= -\sum_{i\in I} \int_{\mathbb{R}^d\times[0,T]} \partial_t g_i(x,t)\, \mu_i^{\mathcal{D}_m}(dx, dt).
\end{aligned}$$

Since $\mu_i^{\mathcal{D}_m}$ converges weakly to μ_i, $\partial_t g_i$ is continuous and bounded, and I is finite, we have

$$\lim_{m\to\infty} \tilde{T}_2^m = -\sum_{i\in I} \int_{\mathbb{R}^d\times[0,T]} \partial_t g_i(x,t)\,\mu_i(dx,dt) = -\sum_{i\in I} \int_{\mathbb{R}^d\times\mathbb{R}_+} \partial_t g_i(x,t)\,\mu_i(dx,dt).$$

On the other hand, thanks to Lemma C.6, we get

$$\begin{aligned} |T_2^m - \tilde{T}_2^m| &\leq \sum_{i\in I}\sum_{n=0}^{N-1}\sum_{K\in\mathcal{M}_i} p_{n+1}^{(K)} \int_K \int_0^{\delta t} |\partial_t g_i(x; n\delta t + s) - \partial_t g_i(x, n\delta t)|\, ds\, dx \\ &\leq (T+\delta t) L_{t,g}\,\delta t, \end{aligned}$$

and thus

$$\lim_{m\to\infty} T_2^m = -\sum_{i\in I} \int_{\mathbb{R}^d\times\mathbb{R}_+} \partial_t g_i(x,t)\,\mu_i(dx,dt).$$

Study of the Limit of T_3^m

We have $T_3^m = \sum_{n=0}^{N-1} \delta t \sum_{K\in\mathcal{M}} p_{n+1}^{(K)} T_{30}^{(K)}$ with $T_{30}^{(K)} = \sum_{L\in\mathcal{M}} v_{KL}(g_n^{(K)} - g_n^{(L)}) = T_{31}^{(K)} + T_{32}^{(K)} + T_{33}^{(K)}$, where

$$\begin{aligned} T_{31}^{(K)} &= \sum_{L\in\mathcal{M}} \frac{1}{\tau} \int_{\{x\in K, \alpha(x)>\tau, \phi(x,\tau)\in L\}} \frac{1}{|K|} \int_{\{y\in K\}} (g(y, n\delta t) - g(x, n\delta t))\, dy\, dx, \\ T_{32}^{(K)} &= \int_{\{x\in K, \alpha(x)>\tau\}} \frac{1}{\tau}(g(x, n\delta t) - g(\phi(x,\tau), n\delta t))\, dx, \\ T_{33}^{(K)} &= \sum_{L\in\mathcal{M}} \frac{1}{\tau} \int_{\{x\in K, \alpha(x)>\tau, \phi(x,\tau)\in L\}} \frac{1}{|L|} \int_{\{y\in L\}} (g(\phi(x,\tau), n\delta t) - g(y, n\delta t))\, dy\, dx. \end{aligned}$$

If $K \in \mathcal{M}_i$, we have

$$\begin{aligned} |T_{31}^{(K)}| &\leq \sum_{L\in\mathcal{M}} \frac{1}{\tau} \int_{\{x\in K, \alpha(x)>\tau, \phi(x,\tau)\in L\}} \frac{1}{|K|} \int_{\{y\in K\}} |g_i(y, n\delta t) - g_i(x, n\delta t)|\, dy\, dx \\ &\leq \frac{h}{\tau} L_g |K|. \end{aligned}$$

In the same way, $|T_{33}^{(K)}| \leq h\, L_g |K|/\tau$. Moreover, define

$$\forall i \in I, \forall K \in \mathcal{M}_i, \quad \tilde{T}_{32}^{(K)} = -\int_{\{x\in K, \alpha(x)>\tau\}} \partial_\phi g_i(x, n\delta t)\, dx.$$

Then thanks to Lemma C.6, we get for $K \in \mathcal{M}_i$,

$$|\tilde{T}_{32}^{(K)} - T_{32}^{(K)}| \leq \int_{\{x\in K, \alpha(x)>\tau\}} \left| \frac{1}{\tau}(g_i(\phi(x,\tau), n\delta t) - g_i(x, n\delta t)) - \partial_\phi g_i(x, n\delta t) \right| dx$$
$$\leq \tau\, L_{\phi,g}\, |K|.$$

Thus defining $\bar{T}_3^m = -\sum_{n=0}^{N-1} \delta t \sum_{K\in\mathcal{M}} p_{n+1}^{(K)} \int_{\{x\in K:\alpha(x)>\tau\}} \partial_\phi g_i(x, n\delta t)\, dx$, we have

$$\lim_{m\to\infty} |T_3^m - \bar{T}_3^m| = 0.$$

On the other hand, define

$$\tilde{T}_3^m = -\sum_{n=0}^{N-1} \delta t \sum_{i\in I} \sum_{K\in\mathcal{M}_i} p_{n+1}^{(K)} \int_K \partial_\phi g_i(x, n\delta t)\, dx$$
$$= -\sum_{i\in I} \int_{\mathbb{R}^d\times[0,T]} (\xi_i(x,t) - \partial_t g_i(x,t))\, \mu_i^{\mathcal{D}_m}(dx, dt).$$

We then have (as for $\tilde{T}_2^m$)

$$\lim_{m\to\infty} \tilde{T}_3^m = -\sum_{i\in I} \int_{\mathbb{R}^d\times\mathbb{R}_+} (\xi_i(x,t) - \partial_t g_i(x,t))\, \mu_i(dx, dt).$$

Finally, Lemma C.2, Sect. C.2 gives $|\bar{T}_3^m - \tilde{T}_3^m| \leq L_g C_\sigma \tau$, and hence

$$\lim_{m\to\infty} T_3^m = -\sum_{i\in I} \int_{\mathbb{R}^d\times\mathbb{R}_+} (\xi_i(x,t) - \partial_t g_i(x,t))\, \mu_i(dx, dt).$$

Study of the Limit of T_4^m

Define

$$\tilde{T}_4^m = \sum_{n=0}^{N-1} \delta t \sum_{K\in\mathcal{M}} p_{n+1}^{(K)} \int_K \lambda(x) \left(g(x, n\delta t) - \int_E g(y, n\delta t)\, Q(x; dy) \right) dx$$
$$= \sum_{i\in I} \int_{\mathbb{R}^d\times[0,T]} \lambda(x) \left(g_i(x,t) - \sum_{j\in I} \int_{E_j} g_j(y,t)\, Q_i(x; dy) \right) \mu_i^{\mathcal{D}_m}(dx, dt).$$

Hypothesis (H3) 4 and I finite yield

$$\lim_{m\to\infty} \tilde{T}_4^m = \sum_{i\in I} \int_{\mathbb{R}^d\times[0,T]} \lambda(x) \left(g_i(x,t) - \sum_{j\in I} \int_{E_j} g_j(y,t)\, Q_i(x;dy) \right) \mu_i(dx,dt)$$

$$= \sum_{i\in I} \int_{\mathbb{R}^d\times\mathbb{R}_+} \lambda(x) \left(g_i(x,t) - \sum_{j\in I} \int_{E_j} g_j(y,t)\, Q_i(x;dy) \right) \mu_i(dx,dt).$$

Moreover, for $z \in M \subset \mathcal{M}_i$, we have $|g(z,n\delta t) - g_n^{(M)}| \le h\, L_g$. Therefore $|\tilde{T}_4^m - T_4^m| \le 2\Lambda h L_g\,(T+\delta t) \xrightarrow[m\to\infty]{} 0$, and thus

$$\lim_{m\to\infty} T_4^m = \sum_{i\in I} \int_{\mathbb{R}^d\times\mathbb{R}_+} \lambda(x) \left(g_i(x,t) - \sum_{j\in I} \int_{E_j} g_j(y,t)\, Q_i(x;dy) \right) \mu_i(dx,dt).$$

Study of the Limit of T_5^m

In the same way as for T_4^m, define

$$\tilde{T}_5^m = \sum_{i\in I} \sum_{n=0}^{N-1} \delta t \sum_{K\in\mathcal{M}_i} p_{n+1}^{(K)} \frac{1}{\tau} \int_{\{x\in K:\alpha(x)\le\tau\}} \Bigg(g_i(\phi(x,\alpha(x)), n\delta t)$$
$$- \sum_{j\in I} \int_{E_j} g_j(y,n\delta t)\, q_i(\phi(x,\alpha(x));dy) \Bigg) dx$$
$$= \sum_{i\in I} \int_{\mathbb{R}^d\times[0,T]} \left(g_i(x,t) - \sum_{j\in I} \int_{E_j} g_j(y,t)\, \hat{q}_i(x;dy) \right) \sigma_i^{\mathcal{D}_m}(dx,dt).$$

Thanks to hypothesis (H3) 5 and I finite, we get

$$\lim_{m\to\infty} \tilde{T}_5^m = \sum_{i\in I} \int_{\mathbb{R}^d\times\mathbb{R}^d} \left(g_i(x,t) - \sum_{j\in I} \int_{E_j} g_j(y,t)\, q_i(x;dy) \right) \sigma_i(dx,dt).$$

Moreover,

$$|\tilde{T}_5^m - T_5^m| \le \sum_{i\in I} \sum_{n=0}^{N-1} \sum_{K\in\mathcal{M}_i} p_{n+1}^{(K)} \frac{1}{\tau} \int_{\{x\in K:\alpha(x)\le\tau\}} \Bigg(|g_i(\phi(x,\alpha(x)), n\delta t) - g_n^{(K)}|$$
$$+ \sum_{j\in I} \sum_{L\in\mathcal{M}_j} \int_L |g_n^{(L)} - g_j(y,n\delta t)|\, q_i(\phi(x,\alpha(x));dy) \Bigg).$$

For $x \in K$ and $\alpha(x) \le \tau$ we have $|g_i(\phi(x,\alpha(x)), n\delta t) - g_n^{(K)}| \le L_g(h + L_\phi \tau)$. Thus we deduce from Lemma C.2 that

$$|\tilde{T}_5^m - T_5^m| \leq (2h + L_\phi \tau)\, L_g \sum_{n=0}^{N-1} \delta t \sum_{K \in \mathcal{M}} p_{n+1}^{(K)}\, q_K \leq (2h + L_\phi \tau)\, L_g\, C_\sigma \underset{m\to\infty}{\longrightarrow} 0,$$

and therefore

$$\lim_{m\to\infty} T_5^m = \sum_{i \in I} \int_{\mathbb{R}^d \times \mathbb{R}_+} \left(g_i(x,t) - \sum_{j \in I} \int_{E_j} g_j(y,t)\, q_i(x;dy) \right) \sigma_i(dx,dt).$$

This concludes the proof. ■

References

1. Air liquide.: Exercise submitted to ESRA technical committee on dependability modelling (2003)
2. Albrecher, H., Thonhauser, S.: Optimality results for dividend problems in insurance. Rev. R. Acad. Cien. Serie A. Math **103**(2), 295–320 (2009)
3. Alsmeyer, G.: On the markov renewal theorem. Stoch. Processes Appl. **50**, 37–56 (1994). Corrected version:http://wwwmath.uni-muenster.de/statistik/alsmeyer/Publikationen/
4. Alsmeyer, G.: The Markov renewal theorem and related results. Markov Process. Relat. Fields **3**, 103–127 (1997)
5. Asmussen, S.: Applied Probability and Queues. Wiley, New York (1992)
6. Azéma, J., Duflo, M., Revuz, D.: Mesure invariante des processus de Markov récurrents, Séminaire de Probabilités III, Université de Strasbourg, Springer. Lect. Notes Math. **88**, 24–33 (1968)
7. Azaïs, R.M., Bardet, J.B., Génadot, A., Krell, N., Zitt, P.A.: Piecewise deterministic Markov process - recent results. ESAIM: Proc. EDP Sci., J. MAS **2012**(44), 276–290 (2014)
8. Bect J., *Processus de Markov diffusifs par morceaux : outils analytiques et numériques*, Thèse de doctorat, Ecole Doctorale "Sciences et Technologies de l'Information, des Télécommunications et des Systèmes", Université Paris-Sud (2007)
9. Benaïm, M., Le Borgne, S., Malrieu, F., Zitt, P.A.: On the stability of planar randomly switched systems. Ann. Appl. Probab. **24**(1), 292–311 (2014)
10. Benaïm, M., Le Borgne, S., Malrieu, F., Zitt, P.A.: Qualitative properties of certain piecewise deterministic Markov processes. Annales de l'Institut Henri Poincaré **51**, 1040–1075 (2015)
11. Billingsley, P.: Convergence of Probability Measures, Wiley Series in Probability and Mathematical Statistics. Wiley, New York (1968)
12. Billingsley, P.: Probability and Measure. Wiley, New York (1979)
13. Bloch-Mercier, S.: Stationary availability of a semi-Markov system with random maintenance. Appl. Stoch. Models Busin. Ind. **16**, 219–234 (2000)
14. Boel, R., Varaiya, P., Wong, E.: Martingales on jump processes i: representation results. SIAM J. Control. Optim. **13**(5), 999–1021 (1975)
15. Boel, R., Varaiya, P., Wong, E.: Martinagales on jump processes ii: applications. SIAM J. Control. Optim. **13**(5), 1022–1060 (1975)
16. Borovkov, K., Vere-Jones, D.: Explicit formulae for stationary distributions of stress release processes. J. Appl. Probab. **37**(2), 315–321 (2000)

C. Cocozza-Thivent, *Markov Renewal and Piecewise Deterministic Processes*,
Probability Theory and Stochastic Modelling 100,
https://doi.org/10.1007/978-3-030-70447-6

17. Boxma, O., Kaspi, H., Kella, O., Perry, D.: On/off storage systems with state-dependent input, output, and switching rates. Probab. Eng. Inf. Sci. **19**(1), 1–14 (2005)
18. Boxma, O., Perry, D., Stadje, W., Zacks, S.: A Markovian growth-collapse model. Adv. Appl. Probab. **38**(1), 221–243 (2006)
19. Brémaud P.M.: A martingale approach to point processes. Berkeley EECS Department Memorandum, No. UCB/ERL M345 (1972)
20. Brémaud, P.: Point Processes and Queues, Martingale Dynamics. Springer, Berlin (1981)
21. Brémaud, P., Foss, F.: Ergodicity of a stress release point process seismic model with aftershocks. Markov Process. Relat. Fields **16**, 389–408 (2010)
22. Browne, S., Sigman, K.: Work-modulated queues with applications to storage processes. J. Appl. Probab. **29**(3), 699–712 (1992)
23. Buckwar, E., Riedler, M.G.: An exact stochastic hybrid model of excitable membranes including spatio-temporal evolution. J. Math. Biol. **63**(6), 1051–1093 (2011)
24. Bujorianu, M.L., Lygeros, J.: General stochastic hybrid systems: modeling and optimal control. In: Proceedings of the 43rd IEEE Conference on Decision and Control, vol. 2, pp. 1872–1877. Atlantis, Bahamas, Décembre (2004)
25. Chafaï, D., Malrieu, F., Paroux, K.: On the long time behavior of the TCP window size process. Stoch. Process. Appl. **120**(8), 1518–1534 (2010)
26. Choquet, G.: Cours d'analyse. Tome II, Topologie, Masson (1964)
27. Çinlar, E.: Introduction to Stochastic Processes. Prentice-Hall, New Jersey (1975)
28. Çinlar, E., Pinsky, M.: A stochastic integral in storage theory. Z. Wahrs. verw. Geb. **17**, 227–240 (1971)
29. Cocozza-Thivent, C.: Processus stochastiques et fiabilité des systèmes, Collection Mathématiques and Applications, vol. 28. Springer, Berlin (1997)
30. Cocozza-Thivent, C.: A model for a dynamic preventive maintenance policy. J. Appl. Math. Stoch. Anal. **13**, 321–346 (2000)
31. Cocozza-Thivent, C., Eymard, R.: Marginal distributions of a semi-Markov process and their computations. In: Pham et, H., Yamada, S., (Eds.), Ninth ISSAT International Conference on Reliability and Quality in Design, International Society of Science and Applied Technology (2003). (ISBN : 0-9639998-8-5)
32. Cocozza-Thivent C., Eymard R.: Approximation of the marginal distributions of a semi-Markov process using a finite volume scheme. ESAIM: M2AN **38**(5), 853–875 (2004)
33. Cocozza-Thivent, C., Eymard, R.: Algorithmes de fiabilité dynamique. In: Proceedings of $\lambda\mu$ 15, Lille (2006)
34. Cocozza-Thivent, C., Kalashnikov, V.: The failure rate in reliability: approximations and bounds. J. Appl. Math. Stoch. Anal. **9**(4), 497–530 (1996)
35. Cocozza-Thivent, C., Roussignol, M.: Semi-Markov processes for reliability studies. ESAIM Probab. Stat. **1**, 207–223 (1997)
36. Cocozza-Thivent, C., Roussignol, M.: A general framework for some asymptotic reliability formulas. Adv. Appl. Probab. **32**, 446–467 (2000)
37. Cocozza-Thivent, C., Eymard, R., Mercier, S.: Numerical scheme to solve integro-diffrential equations in the dynamic reliability field. In: Proceedings of PSAM7-ESREL'04, Berlin (2004)
38. Cocozza-Thivent, C., Eymard, R., Mercier, S.: Méthodologie et algorithmes pour la quantification de petits systèmes redondants. In: Proceedings of $\lambda\mu$ 14, Bourges (2004)
39. Cocozza-Thivent, C., Eymard, R., Mercier, S., Roussignol, M.: Characterization of the marginal distributions of Markov processes used in dynamic reliability. J. Appl. Math. Stoch. Anal. Article ID 92156, 18 pages (2006)
40. Cocozza-Thivent, C., Eymard, R., Mercier, S.: A finite volume scheme for dynamic reliability models. IMA J. Numer. Anal. **26**(3), 446–471 (2006)
41. Cocozza-Thivent, C., Eymard, R., Goudenège, L., Roussignol, M.: Numerical methods for piecewise deterministic Markov processes with boundary. IMA J. Numer. Anal. **37**(1), 170–208 (2017)
42. Costa, O.L.V.: Stationary distribution for piecewise-deterministic Markov processes. J. Appl. Probab. **27**, 60–73 (1990)

43. Costa, O.L.V.: Impulse control of piecewise-deterministic processes via linear programming. IEEE Trans. Autom. Control **AC-36**, 371–375 (1991)
44. Costa, O.L.V., Davis, M.H.A.: Impluse control of piecewise-deterministic processes. Math. Control Signals Syst. **2**, 187–206 (1989)
45. Costa, O.L.V., Dufour, F.: Stability and ergodicity of piecewise deterministic Markov processes. SIAM J. Control. Optim. **47**(2), 1053–1077 (2008)
46. Crudu, A., Debussche, A., Radulescu, O.: Hybrid stochastic simplifications for multiscale gene networks. BMC Syst. Biol. 3–89 (2009)
47. Crudu, A.Debussche, A., Muller A.: Radulescu, O.: Convergence of stochastic gene networks to hybrid piecewise deterministic processes. Ann. Appl. Prob. **22**(5), 1822–1859 (2012)
48. Dacunha-Castelle, D., Duflo, M., Probabilités et statistiques, 2. : problèmes à temps mobile, Masson, 1993 (2ème édition)
49. Dassios, A., Embrechts, P.: Martingales and insurance risk. Commun. Stat. Stoch. Models **5**(2), 181–217 (1989)
50. Davis, M.H.A.: Piecewise-deterministic Markov processes: a general class of non-diffusion stochastic models. J. R. Stat. Soc. Ser. B **46**, 353–388 (1984)
51. Davis, M.H.A.: Markov Models and Optimization, Monographs on Statistics and Applied Probability 49. Chapman & Hall, London (1993)
52. De Saporta, B., Dufour, F., Zhang, H., Elegbede, C.: Optimal stopping for the predictive maintenance of a structure subject to corrosion. J. Risk Reliab. **226**(2), 169–181 (2012)
53. De Saporta, B., Dufour, F., Zhang, H.: Numerical Methods for Simulation and Optimization of Piecewise Deterministic Markov Processes. Mathematics and Statistics. Wiley, New York (2015)
54. Dieudonné, J.: Fondements de l'analyse moderne. Gauthier-Villars, Paris (1965)
55. Dufour, F., Dutuit, Y., Zhang, H., Innal, F.: Fiabilité dynamique et processus déterministes par morceaux. In: Proceedings of $\lambda\mu$ 15, Lille (2006)
56. Embrechts, P., Schmidli, H.: Ruin estimation for a general insurance risk model. Adv. Appl. Probab. **26**(2), 404–422 (1994)
57. Ethier, S.N., Kurtz, T.G.: Markov Processes: Characterization and Convergence. Wiley, New York (1986)
58. Eymard, R., Mercier, S.: Comparison of numerical methods for the assessment of production availability of a hybrid system. Reliab. Eng. Syst. Safety **93**(1), 169–178 (2008). Jan
59. Eymard, R., Mercier, S., Prignet, A.: An implicit finite volume scheme for a scalar hyperbolic problem with measure data related to piecewise deterministic Markov processes. J. Comput. Appl. Math. **222**(2), 293–323 (2008)
60. Fontbona, J., Guérin, H., Malrieu, F.: Long time behavior of telegraph processes under convex potentials. Stoch. Process. Appl. **126**(10), 3077–3101 (2016)
61. Goreac, D., Rotenstein, E.: Infection time in multistable gene networks. A backward stochastic variational inequality with nonconvex switch-dependent reflection approach. Set-Valued Variat. Anal. (2016)
62. Graham, C., Robert, P.: Interacting multi-class transmissions in large stochastic networks. Ann. Appl. Probab. **19**(6), 2334–2361 (2009)
63. Harrison, J.M., Resnick, S.I.: The stationary distribution and first exit probabilities of a storage process with general release rule. Math. Oper. Res. **1**, 347–358 (1976)
64. Harrison, J.M., Resnick, S.I.: The recurrence classification of risk and storage processes. Math. Oper. Res. **3**(1), 57–66 (1978)
65. Holley, R.A., Stroock, D.W.: A martingale approach to infinite systems of interacting processes. Ann. Probab. **4**, 195–228 (1976)
66. Jacobsen, M.: Point Process Theory and Applications. Marked Point and Piecewise Deterministic Processes, Probability and Its Applications. Birkhäuser (2006)
67. Jacod, J.: Théorèmes de renouvellement et classification pour les chaines semi-markoviennes. Annales de l'Institut Henri Poincaré, section B **7**(2), 83–129 (1971)
68. Jacod, J.: Corrections et compléments à l'article "Théorèmes de renouvellement et classification pour les chaines semi-markoviennes". Annales de l'Institut Henri Poincaré, section B **10**(2), 201–209 (1974)

69. Jacod, J., Processus de Markov, application à la dynamique des populations. cours de Master de Mathématiques, Spécialité : Probabilités et Applications, Université Pierre et Marie Curie, 2004–2005. http://www.proba.jussieu.fr/cours/dea/M2-2004.jj.pdf
70. Jacod, J., Shiryaev, A.N.: Limit Theorems for Stochastic Processes. Grandlehren der Mathematischen Wissenshaften, 2nd edn. Springer, Berlin (2003)
71. Jacod, J., Skorokhod, A.V.: Jumping Markov Processes. Annales de l'Institut Henri Poincaré, section B **32**(1), 11–67 (1996)
72. Jamison, B., Orey, S.: Markov chains recurrent in the sense of Harris. Z. Wahr. verw. Geb. **8**, 41–48 (1967)
73. Kalashnikov, V.: Quantitative estimates in queueing. In: Dshalalow, J., (ed.), Advances in Queueing. Theory, Methods and Open Problems, pp. 407-428. CRC Press, Boca Raton (1995)
74. Koureta, P., Koutroumpas, K., Lygeros, J., Lygerou, Z.: Stochastic Hybrid Modeling of Biochemical Processes, Stochastic Hybrid Systems, Chapter 9. CRC Press, Boca Raton (2007)
75. Labeau, P.E., Dutuit, Y., Fiabilité dynamique et disponibilité de production : un cas illustratif. In: Proceedings of $\lambda\mu$ 14, **2**, Bourges (2004)
76. Lair, W., Mercier, S., Roussignol, M., Ziani R.: Piecewise deterministic markov processes and maintenance modelling: application to maintenance of a train air conditioning system. Proc. Inst. Mech. Eng. O J. Risk Reliab. **225**(2), 199–209 (2011)
77. Last, G.: Ergodicity properties of stress release, repairable system and workload models. Adv. Appl. Probab. **36**(2), 471–498 (2004)
78. Last, G., Szekli, R.: Stochastic comparison of repairable systems. J. Appl. Probab. **35**, 348–370 (1998)
79. Lawley, S.D., Mattingly, J.C., Reed, M.C.: Sensitivity to switching rates in stochastically switched ODEs. Commun. Math. Sci. **12**(7), 1343–1352 (2014)
80. Löpker, A., Palmowski, Z.: On time reversal of piecewise deterministic markov processes. Electron. J. Probab. **18**(13), 1–29 (2013)
81. Lonchampt, J.: Evaluation de stocks par un modèle de file d'attente : indicateurs du risque. In: Proceedings of $\lambda\mu$ 16, Avignon (2008)
82. Lorton, A.: Contribution aux approches hybrides pour le pronostic à l'aide de processus de Markov déterministes par morceaux. Université de Technologie de Troyes, France (2012). PHD thesis
83. Malrieu, F.: Some simple but challenging Markov processes. Ann. Fac. Sci Toulouse Math. (6) **24**(4), 857–883 (2015)
84. Ogata, Y., Vere-Jones, D.: Inference for earthquake models: a self-correcting model. Stoch. Process. Appl. **17**, 337–347 (1984)
85. Orey, S.: Lecture Notes on Limit Theorems for Markov Chain Transition Probabilities. Von Nostrand Reinhold Co. (1971)
86. Pascucci, A.: PDE and Martingale Methods in Option Pricing. Springer, Berlin (2011)
87. Protter, P.: Stochastic Integration and Differential Equations, 2nd edn. Springer, Heidelberg (2005)
88. Rolski, T., Schmidli, H., Schmidt, V., Teugels, J.: Stochastic Processes for Insurance and Finance. Wiley Series in Probability and Statistics. Wiley, New York (1999)
89. Rudin, W.: Real and Complex Analysis, 3rd edn. Mc-Graw-Hill, New York (1987)
90. Rudnicki, R., Tyran-Kaminńska, M.: Piecewise Deterministic Markov Processes in Biological Models. Semigroups of Operators - Theory and Applications. Poland (2013)
91. Schäl, M.: Sprungprozesse, Lecture Notes. University of Bonn (1977)
92. Schäl, M.: On piecewise deterministic Markov control processes: Control of jumps and of risk processes in insurance. Insur. Math. Econ. **22**(1), 75–91 (1998)
93. Shurenkov, V.M.: On the theory of Markov renewal. Theory Probab. Appl. **29**, 247–265 (1985)
94. Vere-Jones, D.: Earthquake prediction - a statistician's view. J. Phys. Earth **26**, 129–146 (1978)
95. Vere-Jones, D.: On the variance properties of stress release models. Austral. J. Stat. **30A**, 123–135 (1988)
96. Waters, H., et al.: Permanent Health Insurance, CMI Report 12, Institute of Actuaries, London (1991)

97. Wobst, R.: On Jump Processes with Drift, Diss. Math. Polish Scientific Publishers, Warszawa (1983)
98. Zheng, X.: Ergodic theorems for stress release processes. Stoch. Process. Appl. **37**, 239–258 (1991)

Index

C. Cocozza-Thivent, *Markov Renewal and Piecewise Deterministic Processes*,
Probability Theory and Stochastic Modelling 100,
https://doi.org/10.1007/978-3-030-70447-6

Lightning Source UK Ltd.
Milton Keynes UK
UKHW020813060721
386706UK00001B/48

9 783030 704469